Metal-Semiconductor Schottky Barrier Junctions and Their Applications

Metal-Semiconductor Schottky Barrier Junctions and Their Applications

Edited by
B. L. Sharma
Solid State Physics Laboratory
Delhi, India

Plenum Press · New York and London

Library of Congress Cataloging in Publication Data

Main entry under title:

Metal–semiconductor Schottky barrier junctions and their applications.

Bibliography: p.
Includes index.
1. Diodes, Schottky-barrier—Addresses, essays, lectures. I. Sharma, B. L.
TK7871.89.S35M48 1984 621.3815′2 84-1723

ISBN-13: 978-1-4684-4657-9 e-ISBN-13: 978-1-4684-4655-5
DOI: 10.1007/978-1-4684-4655-5

Softcover reprint of the hardcover 1st edition 1984
A Division of Plenum Publishing Corporation
233 Spring Street, New York, N.Y. 10013

Contributors

Y. Anand M/A-COM Gallium Arsenide Products, Inc., Burlington, Massachusetts

R. Z. Bachrach Xerox Palo Alto Research Center, Palo Alto, California

Stephen J. Fonash Engineering Science Program, The Pennsylvania State University, University Park, Pennsylvania

S.C. Gupta Solid State Physics Laboratory, Delhi, India

R.J. Nemanich Xerox Palo Alto Research Center, Palo Alto, California

H. Preier Fraunhofer-Institut für Physikalische Messtechnik, Freiburg, Federal Republic of Germany

Dieter K. Schroder Department of Electrical and Computer Engineering, Arizona State University, Tempe, Arizona

B.L. Sharma Solid State Physics Laboratory, Delhi, India

M.J. Thompson Xerox Palo Alto Research Center, Palo Alto, California

James A. Turner Plessey Research (Caswell) Ltd., Allen Clark Research Centre, Caswell, Towcester, Northants, England

M.S. Tyagi Department of Electrical Engineering, Indian Institute of Technology, Kanpur, India

Preface

The present-day semiconductor technology would be inconceivable without extensive use of Schottky barrier junctions. In spite of an excellent book by Professor E.H. Rhoderick (1978) dealing with the basic principles of metal–semiconductor contacts and a few recent review articles, the need for a monograph on "Metal–Semiconductor Schottky Barrier Junctions and Their Applications" has long been felt by students, researchers, and technologists. It was in this context that the idea of publishing such a monograph by Mr. Ellis H. Rosenberg, Senior Editor, Plenum Publishing Corporation, was considered very timely. Due to the numerous and varied applications of Schottky barrier junctions, the task of bringing it out, however, looked difficult in the beginning. After discussions at various levels, it was deemed appropriate to include only those typical applications which were extremely rich in R & D and still posed many challenges so that it could be brought out in the stipulated time frame. Keeping in view the larger interest, it was also considered necessary to have the different topics of Schottky barrier junctions written by experts.

This monograph is divided into eight chapters. The first chapter reviews the physics of Schottky barrier junctions, while the second deals with the interface chemistry and structure of Schottky barrier formation. Chapter 3 emphasizes the design considerations, fabrication processes, and characterization aspects of metal–silicon and metal–gallium arsenide junctions in a general way. Chapters 4–8 are concerned with specific applications. Amongst these, Chapter 4 encompasses a number of optoelectronic structures which employ not only Schottky barrier junctions but also Schottky barrier-type junctions. Chapters 5, 6, 7, and 8 deal with various theoretical and practical aspects of Schottky photodiodes, microwave Schottky diodes, MESFETs, and Schottky barrier gate CCDs, respectively. Considering the timely importance of metal–amorphous silicon junctions, a chapter on such junctions and their applications is also included at the end. In a monograph of this type it is difficult to maintain the interwoven nature of the text and ensure a uniformity of presentation and notations, especially when the contributors are from different parts of the world. For this, I regret any inconvenience to those readers who prefer a textbook-type approach.

Finally, I am indebted to the contributors for their self-contained contributions and to Mr. Ellis H. Rosenberg and his staff, without whose cooperation it would have been difficult to bring out this monograph in the stipulated time.

B.L. SHARMA

Contents

1. PHYSICS OF SCHOTTKY BARRIER JUNCTIONS

M.S. Tyagi

2. INTERFACE CHEMISTRY AND STRUCTURE OF SCHOTTKY BARRIER FORMATION

R.Z. Bachrach

3. FABRICATION AND CHARACTERIZATION OF METAL–SEMICONDUCTOR SCHOTTKY BARRIER JUNCTIONS

B.L. Sharma

4. SCHOTTKY-BARRIER-TYPE OPTOELECTRONIC STRUCTURES

Stephen J. Fonash

5. SCHOTTKY BARRIER PHOTODIODES

S.C. Gupta and H. Preier

6. MICROWAVE SCHOTTKY BARRIER DIODES

Y. Anand

7. METAL–SEMICONDUCTOR FIELD EFFECT TRANSISTORS

James A. Turner

8. SCHOTTKY BARRIER GATE CHARGE-COUPLED DEVICES

Dieter K. Schroder

9. SCHOTTKY BARRIERS ON AMORPHOUS Si AND THEIR APPLICATIONS

R.J. Nemanich and M.J. Thompson

1

Physics of Schottky Barrier Junctions

M.S. Tyagi

1. INTRODUCTION

A rectifying metal–semiconductor contact is known as a Schottky barrier after W. Schottky, who first proposed a model for barrier formation. Our knowledge of metal–semiconductor diodes is more than a century old. F. Braun,[1] in 1874, reported the rectifying nature of metallic contacts on copper, iron, and lead sulfide crystals. Although numerous experimental and theoretical studies have been carried out since then, our understanding of the metal–semiconductor junctions is still far from complete. This is perhaps due to the fact that their performance is highly process dependent.

Point contact diodes which employed a sharpened metallic wire in contact with an exposed semiconductor surface were used as radio wave detectors in the early days of wireless telegraphy. They were, however, subsequently replaced by vacuum diodes developed in the early 1920s. During the Second World War the point contact diode again became important because of its use as frequency converter and as low-level microwave detector diode. A complete account of these developments is given in the classic book of Torrey and Whitmer.[2] Point contact rectifiers proved highly unreliable in their characteristics and were subsequently replaced by rectifiers obtained by deposition of a thin metallic film on a properly prepared surface of semiconductor. These contacts have shown much superior characteristics and our present understanding of the behavior of metal–semiconductor contacts is obtained from studies on such devices.

The first significant step towards understanding the rectifying action of

M.S. Tyagi • Department of Electrical Engineering, Indian Institute of Technology, Kanpur, India.

metal–semiconductor contact was the realization by Schottky *et al.*[3] of a potential barrier at the interface between the metal and the semiconductor. Schottky[4] and Mott[5] subsequently explained the mechanism of barrier formation and also proposed models for calculating the barrier height and the shape of the barrier. Another significant advance in our understanding of Schottky barrier contacts was made during the Second World War when Bethe[6] proposed thermionic emission as the means of current transport over the barrier.

The 1960s have seen a great revival of research and development work on Schottky barrier diodes. This activity was inspired to a considerable extent by the importance of metallic contacts in semiconductor technology. As a result of this development, further areas of applications of Schottky barriers emerged. Details of these applications are given in other chapters of this monograph. The work during the 1970s has been mainly in two directions. Firstly, the knowledge gained from the research and development work in the previous decade has been utilized in industrial production of devices using Schottky barriers. Secondly, intensive efforts have been made to gain a more complete understanding of the metal–semiconductor interface.

This introductory chapter is devoted to the basic physics and electrical characteristics of rectifying metal–semiconductor contacts. In Section 2, the status of our present knowledge about the origin of the barrier height is discussed. This area encompasses some of the very poorly understood problems in solid-state physics. Section 3 provides a brief description of the methods of measuring barrier heights. Results of measurements are discussed in Section 4. Capacitance–voltage characteristics of Schottky barriers is the subject of Section 5. The various mechanisms of current transport and the resulting current–voltage characteristics are described in Section 6 and transient behavior is discussed in Section 7. The chapter closes with a brief discussion of low-resistance Schottky contacts in Section 8.

An extensive account of earlier work on metal–semiconductor contacts is given by Henisch.[7] A number of recent reviews on the subject are also available. Some of these are concerned mainly with the physics of Schottky barriers[8–10] while others have also discussed the technology and applications of these devices.[11,12]

2. ORIGINS OF BARRIER HEIGHT

2.1. Schottky–Mott Theory of Ideal Metal–Semiconductor Contact

The potential barrier, which forms when a metal is contacted with a semiconductor, arises from the separation of charges at the metal–

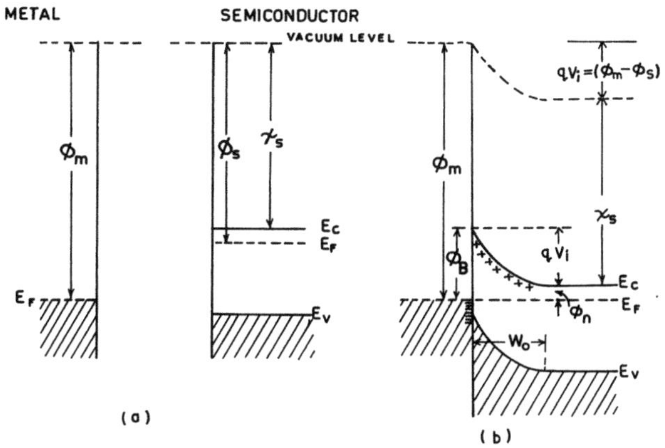

Figure 1. Electron energy band diagrams of metal contact to *n*-type semiconductor with $\phi_m > \phi_s$. (a) Neutral materials separated from each other and (b) thermal equilibrium situation after the contact has been made.

semiconductor interface such that a high-resistance region devoid of mobile carriers is created in the semiconductor. The earliest model put forward to explain the barrier height is that of Schottky and Mott. According to this model the barrier results from the difference in the work functions of the two substances. The energy band diagrams in Fig. 1 illustrate the process of barrier formation. Figure 1a shows the electron energy band diagram of a metal of work function ϕ_m and an *n*-type semiconductor of work function ϕ_s which is smaller than ϕ_m. The work function of a metal is defined as the amount of energy required to raise an electron from the Fermi level to the vacuum level. The vacuum level is the energy level of an electron just outside the metal with zero kinetic energy and is the reference level in Fig. 1a. The work function ϕ_m has a volume contribution due to the periodic potential of the crystal lattice and a surface contribution due to the possible existence of a dipole layer at the surface. The work function ϕ_s of the semiconductor is defined similarly and is a variable quantity because the Fermi level in the semiconductor varies with the doping. An important surface parameter which does not depend on doping is the electron affinity χ_s defined as the energy difference of an electron between the vacuum level and the lower edge of the conduction band. The work functions ϕ_m and ϕ_s and the electron affinity χ_s are usually expressed in electron volts (eV). Note that the semiconductor shown in Fig. 1a does not contain any charges at the surface so that the band structure of the surface is the same as that of the bulk and there is no band bending.

Figure 1b shows the energy band diagram after the contact is made and equilibrium has been reached. When the two substances are brought into intimate contact electrons from the conduction band of the semiconductor, which have higher energy than the metal electrons, flow into the metal till the Fermi level on the two sides is brought into coincidence. As the electrons move out of the semiconductor into the metal, the free electron concentration in the semiconductor region near the boundary decreases. Since the separation between the conduction band edge E_c and the Fermi level E_F increases with decreasing electron concentration and in thermal equilibrium E_F remains constant throughout, the conduction band edge E_c bends up as shown in Fig. 1b. The conduction band electrons which cross over into the metal leave a positive charge of ionized donors behind, so the semiconductor region near the metal gets depleted of mobile electrons. Thus a positive charge is established on the semiconductor side of the interface and the electrons which cross over into the metal form a thin sheet of negative charge contained within the Thomas–Fermi screening distance from the interface (≈ 0.5 Å). Consequently an electric field is established from the semiconductor to metal in Fig. 1b. Note that the width of space charge layer in the semiconductor is appreciable because the donor concentration in the semiconductor is several orders of magnitude smaller than the electron concentration in the metal.

Let us now investigate how much the energy bands in the semiconductor will bend up. It should be evident that since the band gap of the semiconductor is not changed by making contact with the metal, the valence band edge E_v will move up parallel to the conduction band edge E_c. Also the vacuum level in the semiconductor will follow the same variations as E_c. This is because the electron affinity of the semiconductor is assumed to remain unchanged even after the metal contact is made. Thus, for a metal–semiconductor system in thermal equilibrium the important point which determines the barrier height is that the vacuum level must remain continuous across the transition region. Hence, the vacuum level from the semiconductor side must approach the vacuum level on the metal side gradually to preserve the continuity. The amount of band bending, then, is just equal to the difference between the two vacuum levels, which is equal to the difference between the two work functions. This difference is given by $qV_i = (\phi_m - \phi_s)$, where V_i is expressed in volts and is known as contact potential difference or the built-in potential of the junction. qV_i obviously is the potential barrier which an electron moving from the semiconductor into the metal has to surmount. However, the barrier looking from the metal towards the semiconductor is different and is given by

$$\phi_B = (\phi_m - \chi_s) \tag{1}$$

Since

$$\phi_s = \chi_s + \phi_n$$

We have

$$\phi_B = (qV_i + \phi_n) \tag{1a}$$

where $\phi_n = (E_c - E_F)$ represents the penetration of the Fermi level in the band gap of the semiconductor and q is the electronic charge. Equation (1) was stated by Schottky[4] and independently by Mott.[5] In subsequent discussion it will be referred to as the Schottky limit. In obtaining equation (1), it has been assumed that the surface dipole contributions to ϕ_m and ϕ_s remain unchanged after the metal makes contact with the semiconductor. Before discussing it further, let us consider the nature of the contact obtained in Fig. 1b.

The exact shape of the potential barrier can be calculated from the charge distribution within the space charge layer. In most cases, the height ϕ_B of the barrier is orders of magnitude larger than the thermal voltage kT/q, and the space charge region in the semiconductor becomes a high-resistivity depletion region devoid of mobile carriers. The shape of the barrier is then determined from the donor distribution in the semiconductor. Schottky[4] assumed the semiconductor to be uniformly doped up to the metal interface, which gives rise to a uniform charge density in the depletion region. The electric field strength for this constant space charge rises linearly with distance from the edge of the space charge layer and the resulting parabolic barrier is known as a Schottky barrier (see Section 5 for details). Mott[5] assumed a thin layer of semiconductor, devoid of any charge, sandwiched between a uniformly doped semiconductor and the metal. The electric field strength in the thin region is constant and the potential increases linearly across this region. This type of barrier is known as the Mott barrier. The Mott barrier is encountered in situations where a thin layer of low doped, nearly intrinsic semiconductor is interposed between a metal and a heavily doped semiconductor.

It will now be shown that the contact of Fig. 1b is a rectifying contact.

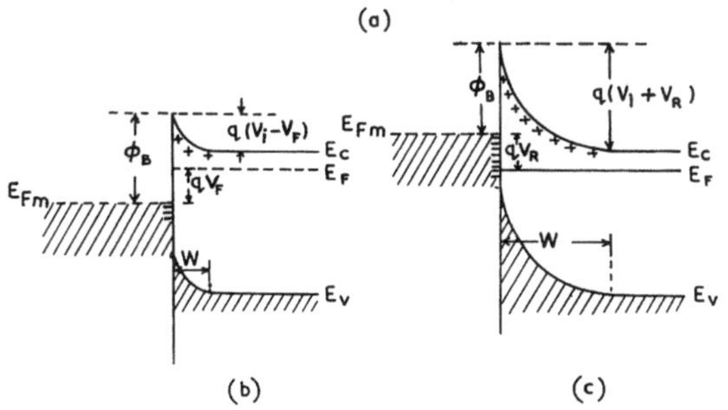

Figure 2. Electron energy band diagrams of rectifying metal contact on n-type semiconductor. (a) Thermal equilibrium situation; (b) forward bias; and (c) reverse bias.

Figure 2a represents the thermal equilibrium energy band diagram of the contact. At equilibrium, the rate at which the electrons cross over the barrier from the semiconductor into the metal is balanced by the rate at which the electrons cross the barrier in the opposite direction and no net current flows. Since the depletion region in the semiconductor has few mobile carriers, its resistance is very high in comparison to the resistances of the metal and the neutral semiconductor and practically all the externally applied voltage appears across this region. The applied voltage would alter the equilibrium band diagram by changing the total curvature of the bands and by modifying the potential drop across the depletion region. When the semiconductor is made negative with respect to the metal by a voltage $V = V_F$, the depletion region width is reduced and the voltage across this region decreases from V_i to $(V_i - V_F)$ as shown in Fig. 2b. The electrons on the semiconductor side now see a reduced barrier and as a result the electron flux from the semiconductor towards the metal is increased above its value under thermal equilibrium. The electron flow from the metal towards the semiconductor, however, is not changed from its equilibrium value. This is because practically no voltage drop occurs across the metal so that ϕ_B remains unaffected by the bias voltage. Thus, for a negative bias on the semiconductor there is a net flow of electrons from the semiconductor towards the metal causing a current flow from the metal to semiconductor. For this polarity, the junction is said to be forward biased. The forward current increases exponentially with the voltage V_F.

The energy band diagram for a reverse biased contact is shown in Fig. 2c. Here the semiconductor is biased positive with respect to the metal by a voltage $V = -V_R$ and the potential drop across the depletion region is increased to $(V_i + V_R)$. The electron flow from the semiconductor towards metal is reduced below its equilibrium value while the flow from the metal side remains practically unchanged. This leads to a current flowing in the opposite direction (i.e., from semiconductor to the metal) which is small compared to the forward current. Thus the contact under discussion is a rectifying metal–semiconductor contact. Note that the energy band diagrams of Figs. 2b and 2c correspond to nonequilibrium conditions and do not have a single Fermi level. The Fermi energy in the region from which the electrons flow is higher than the Fermi energy of the region into which the electrons enter.

The above description applies only to n-type semiconductor whose work function is less than the metal work function ϕ_m. The electron energy band diagrams for an n-type semiconductor with $\phi_m < \phi_s$ are shown in Fig. 3. Figure 3a shows the energy bands for separated materials. After the contact is made electrons flow from the metal into the conduction band of the semiconductor, leaving behind a positive charge on the metal and causing an accumulation of electrons on the semiconductor side of the boundary. When equilibrium is reached the Fermi level in the semiconductor is raised by an

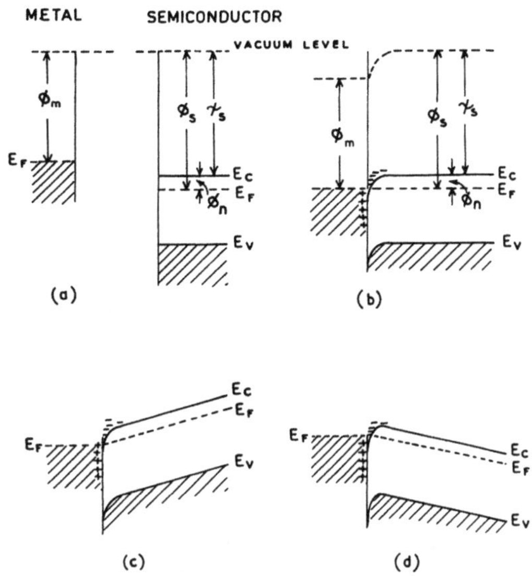

Figure 3. Electron energy band diagrams of metal contact on *n*-type semiconductor with $\phi_m < \phi_s$. (a) Neutral materials separated from each other; (b) contact under thermal equilibrium; (c) negative bias on the semiconductor; and (d) positive bias on the semiconductor.

amount $(\phi_s - \phi_m)$ as shown in Fig. 3b. The accumulation layer charge in the semiconductor is confined to a thickness of the order of Debye length and is essentially a surface charge. Since the concentration of electrons in the metal is very large, the positive charge on the metal side is also a surface charge contained within a distance of about 0.5Å from the metal–semiconductor interface. It is clear that no depletion region is formed in the semiconductor and there is no potential barrier for the electron flow either from the semiconductor towards the metal or in the opposite direction. The electron concentration is increased in the region near the interface and the highest resistivity region in the system is the bulk semiconductor region. Practically all the externally applied voltage appears across this bulk region as shown in Figs. 3c and 3d for the two directions of current flow. It is thus obvious that the current is determined by the resistance of the bulk region and is independent of the direction of the applied bias. Such a non rectifying contact is often referred to as an ohmic contact.

The foregoing discussion has shown that in case of *n*-type semiconductor, a metal–semiconductor contact is rectifying if $\phi_m > \phi_s$ and is nonrectifying if $\phi_m < \phi_s$. The opposite is true for a metal *p*-type semiconductor contact. As an example consider the case when $\phi_m < \phi_s$. The energy band diagrams for separated materials are shown in Fig. 4a. When the two are brought into intimate contact electrons flow from the metal into the semiconductor till the Fermi level on the two sides are aligned. These electrons are minority carriers in the *p*-type semiconductor. After reaching the semiconductor they recombine with holes giving rise to a space charge layer of ionized acceptors as

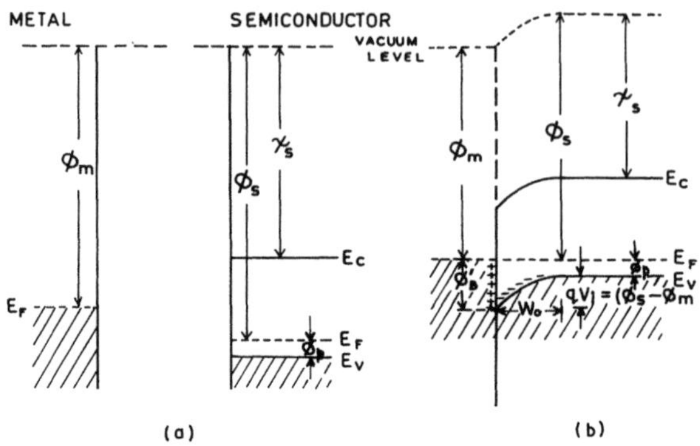

Figure 4. Electron energy band diagrams of metal contact on *p*-type semiconductor with $\phi_m < \phi_s$. (a) Neutral and separated materials and (b) thermal equilibrium situation after the contact has been made.

shown in Fig. 4b. The concentration of holes in the space charge region is negligibly small compared to the acceptor concentration. It follows, therefore, that on the semiconductor side of the contact the space charge region consists of a depletion layer whose thickness W_0 depends on the concentration of ionized acceptor atoms. This situation is similar to that shown in Fig. 1b, except that the role of ionized donors is now taken over by the ionized acceptors. Because the current in a *p*-type semiconductor is carried by holes we have to look for barrier for holes in the band diagram of Fig. 46. As seen from this figure the barrier height ϕ'_B for holes is given by the relation

$$\phi'_B = \chi_s + E_g - \phi_m \tag{2}$$

where E_g represents the band gap of the semiconductor. From Eqs. (1) and (2) one obtains $\phi_B + \phi'_B = E_g$. Using considerations similar to those in Fig. 4 it can be seen that a metal–*p*-type semiconductor contact is nonrectifying if $\phi_m > \phi_s$.

Experimental results have shown that a large majority of metal–semiconductor combinations form rectifying contacts with potential barriers. Moreover, Schottky barrier contacts on *p*-type semiconductors in general have smaller barrier heights and are rarely used in practical devices. In the rest of this chapter we shall be mainly concerned with Schottky barrier contacts on *n*-type semiconductors.

2.2. Modifications to Schottky Theory

Practical metal–semiconductor contacts do not appear to obey the above guidelines. In particular Eq. (1) shows that the barrier height ϕ_B increases linearly with the metal work function ϕ_m. Strong dependence of barrier height on ϕ_m is observed only in predominantly ionic semiconductors. In many covalent semiconductors the barrier height is a less sensitive function of ϕ_m than given by Eq. (1) and in some cases it is almost independent of ϕ_m.

The insensitivity of barrier height to the metal work function in covalently bonded semiconductors was first explained by Bardeen,[13] who pointed out the importance of localized surface states in determining the barrier height. At the surface of a semiconductor the periodicity of the crystal lattice is terminated. In a covalent crystal the surface atoms have neighbors only on the semiconductor side; on the vacuum side there are no neighbors with whom the surface atoms can make covalent bonds. Thus, each of the surface atoms has one broken covalent bond in which only one electron is present and the other is missing. The broken covalent bonds are known as dangling bonds. Dangling bonds give rise to localized energy states at the surface of the semiconductor with energy levels lying in the forbidden gap. These surface states are usually continuously distributed in the band gap and are characterized by a neutral level ϕ_0. The position of this neutral level is such that when there is no band bending in the semiconductor the states are occupied by electrons up to ϕ_0 making the surface electrically neutral. The states below ϕ_0 are donorlike because they are neutral when occupied and are positive when empty. Obviously the states above ϕ_0 behave as acceptorlike. On clean surfaces of covalent semiconductors the density of surface states equals the density of surface atoms. Adsorbed layers of foreign atoms may considerably reduce this density by completing the broken covalent bonds.

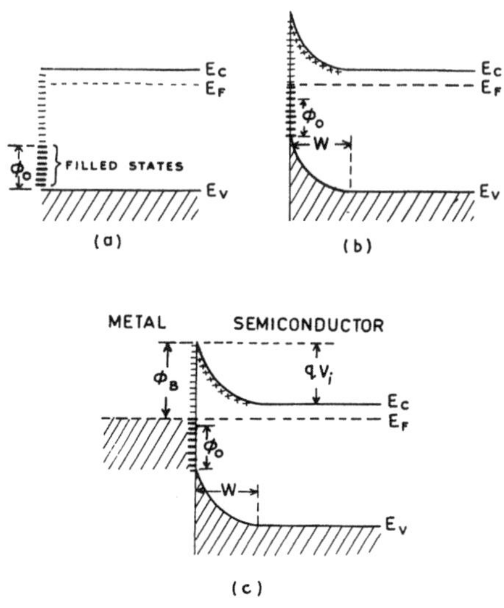

Figure 5. Electron energy band diagrams of *n*-type semiconductor with surface states. The diagrams show (a) flat band at the surface, (b) surface in thermal equilibrium with the bulk, and (c) semiconductor in contact with a metal.

The surface states modify the charge in the depletion region and thus affect the barrier height. Figure 5a shows the electron energy band diagram of an n-type semiconductor under flat band condition. This situation is one of nonequilibrium, and equilibrium is reached when electrons from the semiconductor adjacent to the surface occupy states above ϕ_0 and the Fermi level at the surface aligns with that in the bulk. The surface then becomes negatively charged and a depletion layer consisting of ionized donors is created in the semiconductor region near the surface. Because of this dipole formation a potential barrier looking from the surface towards the semiconductor is obtained even in the absence of a metal contact as shown in Fig. 5b. When a metal is now brought in contact with the semiconductor and equilibrium is reached, the Fermi level in the semiconductor must change by an amount equal to the contact potential by exchanging charge with the metal. If the density of surface states at the semiconductor surface is very large then the charge exchange takes place largely between the metal and the surface states, and the space charge in the semiconductor remains almost unaffected.

As a result the barrier height in Fig. 5c becomes independent of the metal work function and is given by

$$\phi_B = (E_g - \phi_0) \tag{3}$$

In this case the barrier height is said to be "pinned" by surface states. Equation (3) will be referred to as the Bardeen limit.

Mead[14] has suggested that semiconductors can be divided into two categories. Covalently bonded semiconductors like Si, Ge, and GaAs have a large density of surface states in the band gap, and the barrier height in these semiconductors is pinned by the surface states. For ionically bonded semiconductors there are few surface states in the band gap and the barrier height is primarily determined by the difference $(\phi_m - \chi_s)$. Because of a wide scatter in the reported values of the work function,[8] the exact choice of ϕ_m for a given metal becomes uncertain. To come to terms with this difficulty Aven and Mead[15] replaced ϕ_m with the electronegativity χ_m of the metal, which is related to ϕ_m by the empirical relation[16]. $\phi_m = A\chi_m + B$ with $A = 2.27$ and $B = 0.34$. From the analysis of empirical data on a large number of clean crystalline semiconductors, the Schottky barrier height ϕ_B can be expressed as[8,14]

$$\phi_B = S^*(s)\chi_m + \phi_0(s) \tag{4}$$

where m and s refer to the metal and the semiconductor, respectively, $\phi_0(s)$ represents the contribution of the surface states and the interface index $S^*(s) = d\phi_B/d\chi_m$ gives the dependence of barrier height on the metal elec-

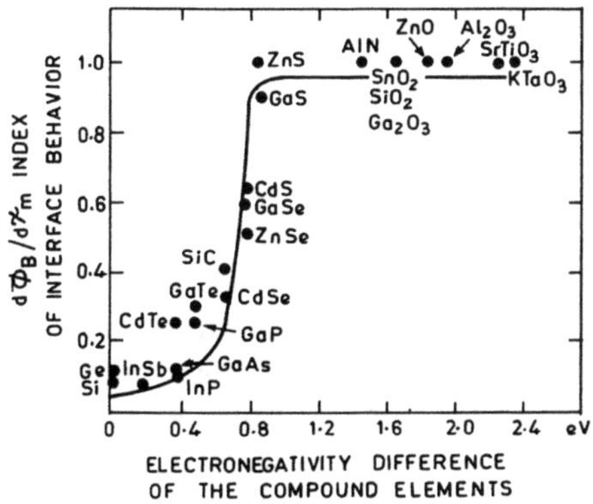

Figure 6. Plot showing the interface index $S^* = d\phi_B/d\chi_m$ as a function of electronegativity difference between cation and anion of compound semiconductors (after Kurtin *et al.*[17]).

tronegativity. For covalently bonded materials S^* is nearly zero and ϕ_B is substantially independent of χ_m. Ionic semiconductors on the other hand have a large value of S^* and ϕ_B increases linearly with χ_m. Kurtin *et al.*[17] have suggested that S^* is a function of the electronegativity difference $\Delta\chi = (\chi_A - \chi_B)$ in the semiconductor, which is a measure of the ionicity of the material. A plot of S^* as a function of $\Delta\chi$ obtained by Kurtin *et al.*[17] is shown in Fig. 6. This curve has been the subject of many studies but neither the sharp transition of S^* at $\Delta\chi \approx 0.7$ nor the saturation of S^* for $\Delta\chi$ in excess of unity have been fully understood. In recent years it is being realized that Mead's empirical relation Eq. (4), even if true in some cases, does not provide a general description of all types of metal–semiconductor contacts. A more precise knowledge of the microscopic structure of the metal–semiconductor interface and of interfacial reactions of the metal–semiconductor atoms is necessary to characterize the behavior of meatal contacts on semiconductors. Let us, therefore, have a fresh look at different types of interfaces that can occur between metals and semiconductors.

2.3. Classification of Metal–Semiconductor Interfaces

The interfaces between metals and nonmetals have been classified into four broad types according to the resulting interfacial atomic configuration.[18] These types are the following: (1) The nonmetal is an insulator (or a semiconductor) and the metal is physisorbed on its surface. (2) The nonmetal is a highly polarizable (dielectric constant $\epsilon_r > 7$) semiconductor such as silicon and the metal makes a weak chemical bond but does not react with it to form a bulk compound. (3) The highly polarizable semiconductor reacts with the metal and forms one or more chemical compounds. (4) A thin film of native oxide is left during the surface preparation of a highly polarizable semiconductor which prevents an intimate contact between the metal and the conductor. The film is referred to as interfacial layer.

Type (1) interface is an ideal Schottky barrier contact in which the barrier height varies directly with the metal work function in accordance to Eq. (1). Type (2) approximates to a "Bardeen barrier" provided that the surface states are assumed to be distributed in space inside the semiconductor to allow a potential drop across this region. In the clean contacts of this type one would expect the barrier height to show a weak dependence on ϕ_m. Type (3) interface represents a case of strong chemical bonding between the metal and the semiconductor and hence we would expect the barrier height to depend on some quantity related to chemical or metallurgical reactions at the interface. Type (4) contact is the one which is most frequently encountered in actual metal–semiconductor devices. We will first discuss the contacts involving type (3) and type (4) interfaces in the next two subsections. Clean contacts having interfaces of type (1) and type (2) will be reexamined in Section 2.6.

2.4. Contacts on Reactive Interfaces

Reactive interfaces that have been extensively studied are the transition metal silicide–silicon interfaces,[9,18–20] Pd–Si interface,[21] and metal–CdS and metal–CdSe systems.[22] In the transition metal silicide–silicon system the interfacial reaction at the silicide and silicon interface causes the interface to move into the interior of the silicon lattice away from the surface imperfections and contaminations. As a result the barrier height does not depend on the surface properties of the semiconductor and the metal work function.

Andrews and Phillips[18] have analyzed the barrier heights for silicides on the basis of the nature of chemical bond and found that they vary linearly with the heat of formation of the silicide (normalized per transition metal atom per formula unit). The explanation provided for this is that the transition metal–silicon bonds are long and interactions are weak; it is, therefore, very likely that the degree of hybridized bonding between transition metal and silicon atoms varies linearly with the heat of formation. Recently, Ottaviani *et al.*[19] have questioned the correlation of barrier height with the heat of formation because the former is an interfacial property whereas the latter is a bulk property of the silicide and it is difficult to see how the two could be correlated. They have proposed that the barrier height in metal silicide system is dominated by an interfacial composition which is a reacted layer of two solids in contact. This reactive layer controls not only the barrier height but also the interaction energy and the eutectic temperature of the interfacial compound. Hence, it should be possible to correlate the barrier height with the eutectic temperature of silicon and the metal–silicide. Figure 7 shows their plot of barrier height as a function of the eutectic temperature for the transition

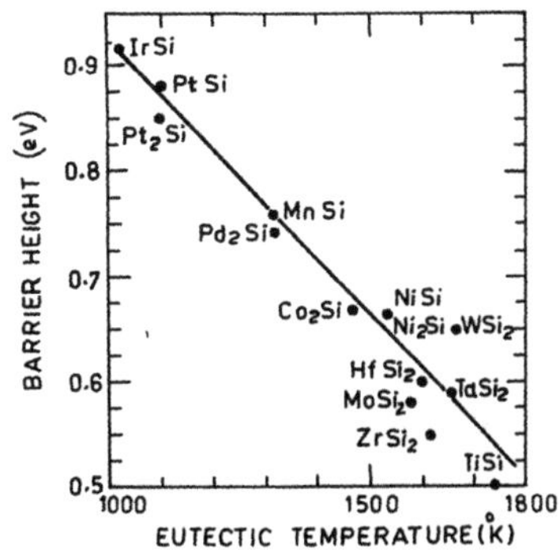

Figure 7. A plot of Schottky barrier height against eutectic temperature for transition-metal–silicide–Si systems (after Ottaviani *et al.*[19]).

metal–silicide–silicon systems for a number of metals. It is seen that the barrier height decreases linearly with the increasing eutectic temperature. The physical meaning of this behavior can be understood by noting that as the reacted interface layer becomes metallic, its melting point decreases while the barrier height is increased.

2.5. Contacts with Surface States and an Insulating Interfacial Layer

Contacts of this type frequently occur in semiconductor devices and an excellent account of such contacts is given by Northrop and Rhoderick[23] and more recently by Rhoderick.[10]

In most metal–semiconductor contacts the semiconductor surface before metal deposition is prepared by chemical cleaning and a thin insulating oxide layer is invariably left on the surface of the semiconductor. The thickness of this interfacial layer depends upon the method of surface preparation and for a good Schottky contact must be less than about 20 Å. The energy band diagram of a contact with interfacial oxide layer is shown in Fig. 8. This diagram can be

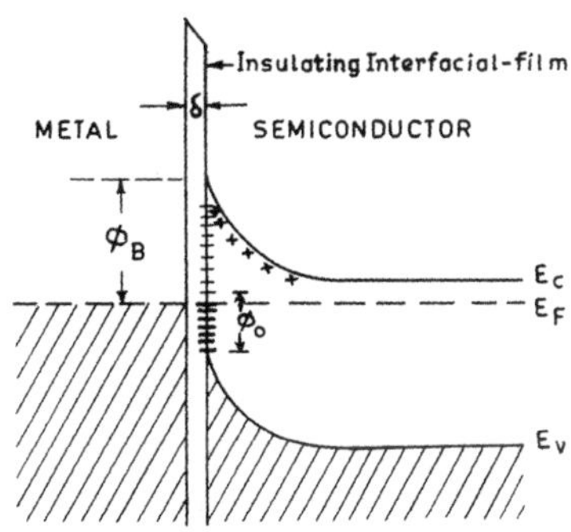

Figure 8. Electron energy band diagram of a metal–semiconductor contact with surface states and interfacial layer.

obtained using the same principles as used earlier in Section 2.1, namely, that the Fermi level is the same throughout the system and the vacuum level is continuous across the interface. In Fig. 8 the potential drops linearly across the interfacial oxide layer because this layer is assumed to be an ideal insulator devoid of any charge. It has also been assumed that the lower edge of the conduction band of the insulator lies below the vacuum level. When the interfacial layer is thin enough, (i.e., $< 20\,\text{Å}$), the potential drop across it is negligibly small compared to that in the semiconductor depletion region. Such a thin layer is transparent to the electrons as the electrons can tunnel through it in either direction. Because of these reasons, the barrier height ϕ_B and the contact potential difference V_i remain almost unaffected by the presence of a thin interfacial layer.

Contacts having a thin insulating layer between metal and semiconductor are easier to understand and analyze theoretically than the clean intimate contacts. This is because the insulating layer decouples the metal from the semiconductor so that each of them can be treated as a separate system. One can then regard the interface states as a property of the particular semiconductor–insulator combination and ignore any modification in the surface dipole contributions to the work functions of the metal and the semiconductor. These simplifications obviously are not possible in the case of clean contacts.

A generalized analysis of metal–semiconductor contact in the presence of surface states and interfacial layer was first made by Cowley and Sze.(24) Rhoderick(9) has reexamined their work and has shown that the expression for ϕ_B derived by Cowley and Sze gives the flat band barrier height ϕ_B^0 which is obtained when enough forward voltage V_F is applied to the Schottky barrier junction to reduce the voltage drop $(V_i - V_F)$ in the semiconductor to zero. Under this condition the charge in the depletion region vanishes and the charge on the metal side is balanced by the charge in the interface states on the semiconductor side. This flat band barrier height is given by(24)

$$\phi_B^0 = C_1(\phi_m - \chi_s) + (1 - C_1)(E_g - \phi_0) \tag{5a}$$

$$= C_1\phi_m + C_2 \tag{5b}$$

where

$$C_1 = \frac{\epsilon_i}{\epsilon_i + q^2\delta D_s}$$

Here $\epsilon_i = \epsilon_r\epsilon_0$ is the permittivity of the insulating layer, δ is its thickness, q is the electronic charge, and D_s is the density of interface states per unit area per eV. The position of the neutral level ϕ_0 is measured from the top of the valence band. From Eq. (5a) it is seen that as D_s tends to zero, C_1 becomes unity and ϕ_B^0 tends to the Schottky limit of Eq. (1). On the other hand as D_s approaches a large value, ϕ_B^0 tends to the Bardeen limit given by Eq. (2). A similar analysis for a rectifying contact on a *p*-type semiconductor leads to the expression

$$\phi_{B'}^0 = C_1(E_g - \phi_m + \chi_s) + (1 - C_1)\phi_0 \tag{6}$$

It is seen that for a given metal–semiconductor system the sum of the flat band barrier heights $\phi_B^0 + \phi_{B'}^0 = E_g$, assuming that ϵ_i, δ and D_s are the same in both the cases. The relation has been verified in experimental studies on a large number of semiconductors.

The expression for the flat band barrier height is modified when there is a charge present in the interfacial oxide layer. Considering the case of the *n*-type semiconductor, if there is a fixed charge Q_{ox} per unit area within the oxide layer then Eq. (5a) gets replaced by[25]

$$\phi_B^0 = C_1(\phi_m - \chi_s) + (1 - C_1)(E_g - \phi_0) - \frac{C_1 \delta Q_{ox}}{\epsilon_i} \tag{7}$$

The effect of the oxide charge is thus to change the barrier height from its ideal value given by Eq. (5a). When Q_{ox} is positive Eq. (7) gives a lower value of ϕ_B^0 than Eq. (5a) while for negative Q_{ox} the barrier height is increased. The presence of oxide charge Q_{ox} changes the intercept C_2 in Eq. (5b) and a method of determining ϕ_0 and Q_{ox} based on this observation has been suggested recently.[26] For Schottky barriers on semiconductors like Si and GaAs which have a large density of surface states the contribution of the last term in Eq. (7) to ϕ_B^0 remains relatively small for all practical values of δ if the oxide layer has less than 5×10^{11} charges/cm^2.

The flat band barrier height ϕ_B^0 in Eq. (5) is obtained when there is no electric field in the semiconductor. In general, there is a band bending in the semiconductor which gives rise to an electric field and this affects the barrier height making it a function of the voltage across the depletion region. Since there exists an electric field in the semiconductor at zero bias, the zero bias barrier height is different from the flat band barrier height.

There may be a number of causes which make the barrier height depend upon the electric field in the depletion region, and we shall consider only two of them. Firstly confining our attention to the potential barrier of Fig. 8, we note that the presence of electric field in the semiconductor alters the potential drop in the interfacial layer and thus modifies the barrier height. It has been shown[9] that the electric field decreases the barrier height in accordance with the relation

$$\phi_B = \phi_B^0 - \alpha|\mathscr{E}_m| \tag{8}$$

where $\mathscr{E}_m$ is the electric field intensity immediately inside the semiconductor surface and α is given by

$$\alpha = \frac{\delta \epsilon_s}{\epsilon_i + q^2 \delta D_s}$$

with ϵ_s representing the permittivity of the semiconductor. For good Schottky barriers the interfacial layer thickness δ is less than 20 Å and the difference between ϕ_B and ϕ_B^0 is insignificant at all forward biases and also at small reverse biases.

The other effect which makes the barrier height depend upon the electric field in the depletion region is the image force barrier lowering. This effect does not depend upon the presence of the interfacial oxide layer and occurs even when such a layer is not present. The image force barrier lowering can be understood by referring to Fig. 9. When an electron is at a distance x from the metal there exists an electric field perpendicular to the metal surface. This field may be calculated by assuming a hypothetical positive image charge q located at a distance $(-x)$ inside the metal. The force of attraction F between the electron and its image charge is $q^2/4\pi\epsilon_d(2x)^2$ and the electron has a negative potential energy $Fx = -q^2/16\pi\epsilon_d x$ relative to that of an electron at infinity, as shown by the dotted curve in Fig. 9. This potential energy must be added to the barrier energy $-q\mathscr{E}x$ to obtain the total energy of the electron. It is seen from Fig. 9 that the maximum in energy occurs at a distance x_m from the metal surface and it can be shown that the magnitude $\Delta\phi_B$ of the barrier lowering is given by[27]

$$\Delta\phi_B = \left[\frac{q^3 N_d}{8\pi^2 \epsilon_d^2 \epsilon_s}(V_i - V)\right]^{1/4} \tag{9}$$

Here N_d is the donor concentration in the semiconductor, and V is the applied

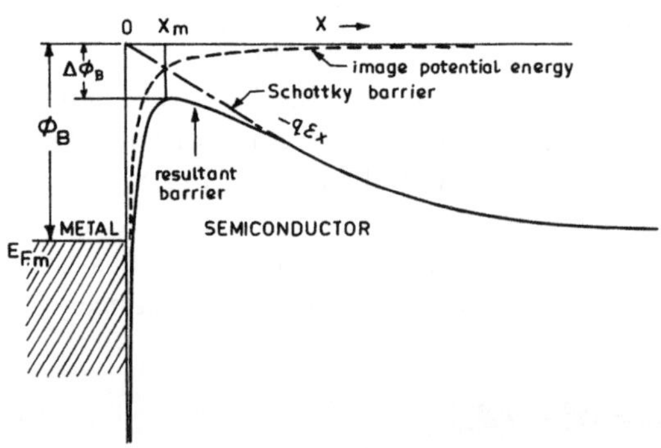

Figure 9. Electron energy diagram showing the image force lowering of the barrier.

voltage. The image force permittivity ϵ_d may be different from the static permittivity ϵ_s of the semiconductor. This is because the electron approaches the barrier with the thermal velocity ($\approx 10^7$ cm/sec) and if its transit time through the barrier region is small compared to the dielectric relaxation time, then the semiconductor does not get fully polarized. However, it is found that in practical situations the electron transit time through the barrier region is sufficiently large to justify $\epsilon_d = \epsilon_s$.

The image force lowering of the barrier results from the field produced by an electron and will be absent when there is no electron present in the semiconductor conduction band near the top of the barrier. Hence, when the barrier height is measured by a method which does not require movement of the electron over the barrier, the obtained value of ϕ_B is not lowered by the image force.

2.6. Contacts on Vacuum Cleaved Surfaces

As mentioned earlier, Schottky barrier junctions are usually fabricated by depositing metal layers onto chemically cleaned semiconductor surfaces. However, metallic contacts on vacuum cleaved surfaces have been investigated extensively mainly to understand the mechanism of barrier formation. These clean contacts are prepared by cleaving the semiconductor crystals in ultrahigh vacuum and evaporating metal layers onto the freshly created surface immediately after cleaving. The clean contacts on vacuum cleaved surfaces are more difficult to analyze theoretically because there is no insulating interfacial layer between the metal and the semiconductor which isolates the semiconductor interface states from the electron states in the metal. Thus the metal–semiconductor combination in such contacts has to be treated as a single system. One would like to know what factors determine the barrier height of these clean contacts and if the Bardeen model is applicable to them.

The Bardeen model[13] assumes that there are surface states intrinsic to the semiconductor which pin the Fermi level at the surface at about one third of the band gap above the valence band in covalent semiconductors. These states remain unchanged after the semiconductor is contacted with the metal and thus determine the barrier height. Heine[28] questioned the pinning of Fermi level by intrinsic surface states in a semiconductor covered with metal and suggested that in clean contacts the conduction electrons in the metal can tunnel into the forbidden gap of the semiconductor. The wave functions of these electrons will decay exponentially into the semiconductor with a decay length of the order of 10Å. Unlike Bardeen states,the Heine tail states are not localized but extend into the semiconductor. They can be thought of as replacing the surface states associated with the termination of the periodic potential[29] in the sense that the former will change into the latter if the metal is separated from the semiconductor.

The most comprehensive first treatment of clean metal–semiconductor contacts is that of Benett and Duke,[30] who considered the perturbations in the electron concentration at the metal–semiconductor boundary due to abrupt changes in the crystal potential. However, no calculations of barrier height for actual metal–semiconductor combinations was made. A detailed study of the effect of Heine tails on the barrier height in metal–semiconductor contacts has been carried out by Pellegrini.[31] A number of other models for the pinning of Fermi level in intimate metal–semiconductor contacts have been proposed. According to Inkson[32] the band gap of the semiconductor is reduced near the metal–semiconductor boundary and at the interface the top of the valence band and the bottom of the conduction band coalesce and E_F is pinned at this point. Other pinning models are those of Philips[33] and of Harrison.[34] However, none of these models have received experimental confirmation.

The effect of metal deposition on the density of semiconductor surface states has been investigated in recent years both theoretically and experimentally. Louie and Cohen[35] have made detailed calculations for Si–Al Schottky barriers. These calculations have shown that the intrinsic surface states on Si were quenched by a several monolayers thick metal and new metal-induced gap states (MIGS) appeared which were responsible for the pinning of the Fermi level at the surface. These MIGS were similar to Heine states except that their decay length is not about 10 Å but is only of the order of a bond length. Similar calculations have also been extended to III–V and II–VI compounds to study the effect of ionicity on the barrier height.[36]

Experimental evidence of metal-induced surface states is rather extensive.[37–39] It is now believed that in case of metallic contacts on clean surfaces of Ge, Si, and III–V compound semiconductors the intrinsic states are unimportant and the barrier height is primarily determined by metal-induced interface states. However, whether these states arise from the penetration of metal electron wave functions into the semiconductor forbidden gap[28] or are created by chemical defects formed near the interface by metal deposition is still an unsettled issue. There appears to be considerable recent evidence which points to the latter possibility. This is brought out by the following pieces of experimental findings:

(a) Studies on cleaved (110) surfaces of III–V compound semiconductors have revealed that there are no surface states in the band gap of these semiconductors[40,41] and the Fermi level is pinned by the metal-induced surface states. Moreover, it is observed that as little as one tenth of monolayer of an evaporated metal (or of oxygen) was enough to pin the Fermi level.[42] Following these observations, Spicer *et al.*[42,43] have proposed a model of barrier formation on III–V compound semiconductors based on the formation of defect levels due to deposition of foreign atoms. When a metal atom is adsorbed on the surface of the semiconductor the heat of condensation of atoms creates chemical defects. Such defects may be either missing group III or group V ions or antisites in which a pair of atoms have exchanged sites. These defects give rise to donor and acceptor type interface states in the band gap of the semiconductor which are responsible for pinning the Fermi level. Recently, confirmation of the Spicer *et al.* model[42] has been obtained by chemisorption studies of Ge on GaAs (110) surfaces.[44]

(b) It has been observed[45] that when metal films thicker than about 450 Å are evaporated onto cleaved surfaces of covalent semiconductors in ultrahigh vacuum, an intermixing of metal and semiconductor atoms takes place even at room temperature. Thus, ideal sharp boundaries at the metal–semiconductor interface never exist in practical contacts and there may be a gradual transition form metal to semiconductor over a few interatomic spacings due to interdiffusion effects.[46] The chemical bonding in this interfacial region is different from that in the bulk semiconductor and this may give rise to interfacial states in the band gap of the semiconductor and the subsequent pinning of the Fermi level.[20]

(c) A recent study of silver contacts on GaAs surfaces cleaned by ion bombardment and annealing has shown[47] that the surface Fermi level is pinned by donor and acceptor type surface states located near the center of the gap. The barrier height on these surfaces is determined by the surface defects produced by ion bombardment and is not modified by metal deposition. Moreover, the surface states produced by ion bombardment correspond to those produced by metal deposition.[42] It thus appears that the defects which create the interface states are characteristic of the semiconductor and the structure of the metal is not at the origin of the barrier height. Nevertheless, the metal can play a role in influencing the junction properties by alteration of the first few atomic layers on the semiconductor, i.e., by creating deep traps or by modification of the doping near the surface, etc. The presence of an oxide layer or of a layer perturbed by ion bombardment acts as a buffer to interdiffusion and prevents the degradation of the interface by the metal.

3. MEASUREMENT OF BARRIER HEIGHT

3.1. Capacitance–Voltage Measurement

The capacitance–voltage characteristics of a Schottky barrier are discussed in Section 5.2. In this method the diode capacitance is measured as a function of applied reverse bias. When a small ac voltage of a few millivolts is applied to a reverse biased diode the depletion region capacitance C is given by the relation

$$C = S\left[\frac{\epsilon_s q N_d}{2(V_i + V_R - kT/q)}\right]^{1/2} \tag{10}$$

where S is the diode cross-sectional area, ϵ_s is the permittivity of the semiconductor, V_R is the applied reverse voltage, and all other symbols have their usual meanings. In this equation it is assumed that the diode does not have an appreciable interfacial oxide layer and that the n-type semiconductor has a uniform donor concentration N_d. It is seen from the above equation that a plot of $1/C^2$ versus V_R gives a straight line with slope $2/S^2\epsilon_s q N_d$ and an intercept on the voltage axis $V_0 = (V_i - kT/q)$. The slope of the straight line can be used to determine the dopant concentration N_d and since $qV_i = (\phi_B - \phi_n)$ the barrier height ϕ_B is obtained as

$$\phi_B = (qV_0 + \phi_n + kT) \tag{11}$$

The kT factor comes from the contribution of majority carriers to the space charge and Eq. (11) does not include the image force barrier lowering. A comprehensive discussion of determination of barrier height from C–V measurements has been given by Goodman,[48] who has also pointed out the limitations of the method.

3.2. Current–Voltage Measurement

As will be seen in Section 6, in Schottky barrier diodes made on high-mobility semiconductors such as Si and GaAs the current is due to thermionic emission of electrons over the barrier and the current as a function of applied bias V is given by the relation

$$I = I_0\left[\exp\left(\frac{qV}{nkT}\right) - 1\right] \tag{12}$$

where $I_0 = SA^*T^2\exp(-\phi_B/kT)$. Here n is the diode ideality factor and A^* is the modified Richardson constant for the semiconductor. For forward values of V in excess of $3kT/q$, a plot of $\ln I$ against V gives a straight line. The value of I_0 can be obtained by extrapolating the straight line to $V = 0$. Knowing I_0, A^*, the diode cross-sectional area S, and the temperature T, the barrier height ϕ_B can be determined. The value obtained in this way is the zero bias barrier height and includes the image force barrier lowering $\Delta\phi_B$.

In case A^* is not known it is still possible to deduce the barrier height. The diode forward $I-V$ characteristics are obtained at a number of temperatures and I_0 determined at each temperature. A plot of $\ln(I_0/T^2)$ against $1/T$ then gives a straight line whose slope gives the barrier height ϕ_B and the intercept on the I_0/T^2 axis gives the value of A^*. This method gives the barrier height at 0 K which is slightly larger than the room-temperature barrier height.

3.3. Photoelectric Measurement

The photoelectric method is the most accurate and direct method of determining ϕ_B. When a monochromatic light is incident on a metal in contact with the semiconductor and the photon energy $h\nu$ is larger than the barrier height but smaller than the band gap of the semiconductor, the incident photons will excite some electrons from the metal over the barrier. The resulting photocurrent I_{ph} for $(h\nu - \phi_B) \gg 3kT$ is given by the Fowler theory[49]

$$I_{ph} = B(h\nu - \phi_B)^2 \tag{13}$$

where B is a constant of proportionality. If $\sqrt{I_{ph}}$ is plotted as a function of $h\nu$, a straight line is obtained whose intercept on the $h\nu$ axis directly gives the barrier height ϕ_B.

4. RESULTS OF BARRIER HEIGHT MEASUREMENTS

4.1. Chemically Prepared Surfaces

Silicon is the most widely used material for making Schottky barrier diodes and more information is available on etched surfaces of silicon than on any other semiconductor. Next to Si comes GaAs. Measurements are also available on other III–V compounds and to a lesser extent on some II–VI compound semiconductors. Results of various measurements have shown that the barrier height depends upon the method of surface preparation and there is considerable variation in the barrier height measured by different workers for the same metal–semiconductor combination.

There does not appear to be a systematic dependence of barrier height on the measurement method in silicon and germanium. However, in compound semiconductors GaAs,[50,51] GaP,[52] and ZnSe[53,54] it has been observed that the barrier heights obtained by the $C-V$ method are consistently larger than those obtained from the $I-V$ and the photoelectric measurements. This difference is partly caused by the fact that the $C-V$ method gives the flat band barrier height while the other two methods give the zero bias barrier height which is lower than the flat band value because of the image force barrier lowering. However, the difference is too large to be accounted for by the image force alone. It was observed in many cases[50,51,54] that the barrier heights obtained from I–V measurements could be brought into agreement with those deduced from C–V measurement when the former is multiplied by n, the diode ideality factor. Thus the ideality factor n is also to be included in the expression for the saturation current I_0. This shows that the discrepancy is probably caused by the presence of a relatively thick interfacial layer.[52]

A close correlation of ϕ_B with the metal work function is not possible especially in case of covalent semiconductors Si, Ge, GaAs and other III–V compounds. The difficulty in correlation arises partly because of lack of our knowledge about the microscopic nature of the metal–semiconductor interface and partly because of uncertainty in the exact values of the metal work function. There is a wide spread in the reported experimental values of metal work function.[8] Moreover, the work function is sensitive to surface contamination and it is doubtful if the evaporated thin metallic film has the same work function as the metal in the bulk. In spite of these difficulties, conclusions can be drawn that for a given semiconductor, metals with high values of ϕ_m tend to produce larger barrier height than the metals with low ϕ_m values. If results from a single source using one specific method of surface preparation are considered, a straight line could be fitted to the ϕ_B versus ϕ_m data by the method of least squares giving the relation of Eq. (5b). The parameters C_1 and C_2 have been used to estimate the interface state density and the position of the neutral level.[8,51,54] The limiting values of C_1 and C_2 for a number of semiconductors are given[12] in Table 1. The observed wide scatter in these values is obviously caused by the fact that different workers have used different methods of surface preparation.

TABLE 1. Limiting Values of C_1 and C_2 Determined from the Available Room Temperature Barrier Height Data[a]

Semiconductor	Slope parameter C_1	Intercept parameter C_2
Ge	$0.18 \geqslant C_1 \geqslant 0.05$	$-0.23 \leqslant C_2 \leqslant 0.11$
Si	$0.19 \geqslant C_1 \geqslant 0.11$	$-0.34 \leqslant C_2 \leqslant 0.35$
GaAs	$0.22 \geqslant C_1 \geqslant 0.06$	$-0.4 \leqslant C_3 \leqslant 0.075$
GaP	$0.21 \geqslant C_1 \geqslant 0.07$	$0.45 \leqslant C_2 \leqslant 0.8$
InP(78 K)	$0.37 \geqslant C_1 \geqslant 0.24$	$-1 \leqslant C_2 \leqslant 0.81$
ZnS	$0.66 \geqslant C_1 \geqslant 0.21$	$-1.63 \leqslant C_2 \leqslant 0.91$
ZnSe	$0.24 \geqslant C_1 \geqslant 0.17$	$-0.25 \leqslant C_2 \leqslant 0.57$

[a] After Sharma and Gupta.[12]

In recent years there has been a growing realization[12] that characterization of a semiconductor in terms of C_1 and C_2 cannot lead to much useful information because the basic physical processes responsible for barrier formation have not been completely understood. Thus, it appears that not much purpose is served by attempting to fit straight lines to the experimental ϕ_B versus ϕ_m data by the method of least squares.

4.2. Vacuum Cleaved Surfaces

Results of barrier height measurement on vacuum cleaved surfaces are available for Si,[55–57] GaAs,[58] and to some extent on II–VI semiconductors CdS[58] and ZnSe.[59] Available information on barrier heights of vacuum cleaved and chemically etched surfaces of various metal–semiconductor combinations is collected by Milnes and Feucht,[8] and recently by Rhoderick[10] and Sze.[27]

Barrier heights of Schottky contacts on vacuum cleaved surfaces are much less sensitive to the choice of metal than the barrier heights on etched surfaces. A traditional explanation for this difference has been in terms of Bardeen model which assumes that the cleaved surfaces of covalently bonded semiconductors have a large density of intrinsic surface states.[13] The native oxide on the surface of the semiconductor satisfies most of the dangling bonds and as a result the density of intrinsic surface states is reduced considerably. It now appears that for most of the semiconductors the intrinsic surface states do not play a dominant role in determining the Schottky barrier height so that the role of the interfacial oxide has to be reexamined.

An extensive and careful study of Schottky contacts on vacuum cleaved surfaces of *n*-type silicon has been made by Thanailakis[56] and by Thanailakis and Rasul.[57] Their results have shown that the Schottky barrier height is not a monotonically increasing function of ϕ_m but depends upon the details of the band structure of the metal. In fact for some metals ϕ_B was found to decrease with increasing value of ϕ_m. They observed that the barrier height deduced from the C–V measurements are consistently higher than those obtained from I–V measurements and that the image force barrier lowering alone cannot explain this difference. These workers tried to explain their results in terms of barrier models proposed by Heine[28] and by Inkson.[32] Figure 10 shows the charge distribution and the resulting energy band diagram of a clean contact based on the Heine model. Because of the tunneling of the conduction band electrons from the metal into the forbidden gap of the semiconductor the net charge is positive in the metal and is negative in the semiconductor. The effect of the exponentially decaying tails of the metal electron wave functions into the semiconductor is to distort the form of the barrier from its parabolic form as

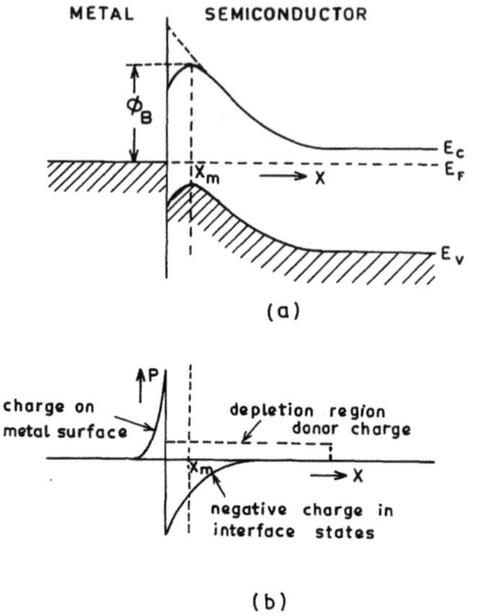

Figure 10. Electron energy band diagram of a clean metal contact on *n*-type semiconductor. The contact is nonideal and the semiconductor region near the interface contains energy states which are storing negative charge in this region. (a) Energy band diagram and (b) charge distribution.

shown in Fig. 10a. The $I-V$ and the photoelectric methods measure the maximum height ϕ_B while the barrier height deduced from the $C-V$ measurements is the extrapolated barrier height that would arise when the barrier remains parabolic right up to the metal as shown by the dotted curve.

The above qualitative explanation of the energy band diagram of Fig. 10a in terms of penetration of metal electron wave functions into the semiconductor is tentative. A plausible explanation for this diagram is also possible by considering the room-temperature interdiffusion effects at the metal–semiconductor boundary.[60] The interfacial region created by intermixing of metal semiconductor atoms across the boundary contains chemical defects which give rise to extrinsic interface states in the band gap of the semiconductor. If these states are such that they contain a net negative charge then the energy band diagram of Fig. 10a will result.

4.3. Concluding Remarks

Our understanding of the mechanism of barrier formation at the metal–semiconductor contact is still far from complete although considerable progress in understanding the physical processes occurring at the interface has been made during the last few years. Ideal conditions as postulated by Schottky and Bardeen hardly ever exist at the interface, and the dipole layer depends upon the spatial arrangement of constituent atoms at the interface and whether or not they form chemical bonds with each other. Thus, the type of interface appears to play an important role in the formation of the barrier.

For reactive interfaces where the metal forms compounds with the semiconductor the barrier height may be correlated with parameters related to the metallurgical reaction such as heat of formation or the eutectic temperature of the interfacial compound.

Other type of contacts most frequently encountered in devices are those in which there is a thin oxide layer between the metal and the semiconductor. In these contacts the barrier height depends upon the method of preparation of semiconductor surface and tends to increase with increasing value of the metal work function although the relation between ϕ_B and ϕ_m is not strictly linear. The presence of oxide produces interface states which depends only on the oxide semiconductor combination and the modification of these states by the presence of the metal is prevented to a large extent. The perturbing action of the metal comes through the penetration of the metal electron wave functions into the semiconductor forbidden gap and by interdiffusion of atoms across the boundary. Both these effects are eliminated by the presence of the interfacial oxide, though which of these mechanisms dominates the barrier formation is not very clear at present. Recent experimental studies on III–V compound semiconductors have suggested that the barrier is established by chemical defects produced by deposition of metal atoms on the surface of the semiconductor.[42] There are reasons to believe that the same is also true for group IV semiconductors.[98]

Metallic contacts on vacuum cleaved nonreactive interfaces are more difficult to understand than contacts on chemically cleaned surfaces because of the perturbing effects of the metal on the interface states. This is especially so in case of covalently bonded semiconductors where the perturbing effects of the metal as discussed above are more serious than in case of ionic semiconductors.

Experimental results on barrier height for various semiconductors do not suggest a close correlation of barrier height ϕ_B with the metal work function. This is true for contacts with interfacial oxide layer and also for clean contacts. Difficulties in establishing this correlation arise partly because the barrier height depends upon the method of surface preparation and partly because of uncertainty in the exact value of the metal work function. To obviate the second difficulty Mead *et al.*[14,15] proposed the replacement of the work function by the electronegativity of the metal, which has a unique value for each metal.[16] However, as Rhoderick[9] has correctly pointed out, the electronegativity is a property of a single atom and not of a solid. Moreover, it has also been observed that metals having the same electronegativity give rise to different barrier height when evaporated onto silicon[9] and also onto ZnSe.[54] In view of these observations one does not know how far the empirical relation[16] $\phi_m = A\chi_m + B$ is justified.

Mead's suggestion[14] that the barrier height in covalently bonded semiconductors is pinned by surface states and that surface states are unimportant in ionic semiconductors, presents too simplistic a view. In this classification no distinction has been made between the contacts on oxide-covered surfaces and the clean contacts on vacuum-cleaved surfaces. As pointed out by Rhoderick,[10] some semiconductors may belong to either of the two categories depending upon the method of surface preparation. Even for clean contacts such a viewpoint is not acceptable because as we have already mentioned there are no intrinsic surface states on vacuum-cleaved (110) surfaces of III–V compounds though they are covalently bonded.

Since the barrier formation in a large majority of semiconductors is not dominated by the density of intrinsic surface states the role played by the ionicity in determining the barrier height has to be reexamined. That the strong pinning of barrier height is not observed in semiconductors with predominantly ionic bonding is an established fact and the plot of Fig. 6 broadly indicates this. It has also been observed[61] that barrier height for holes on common III–V and II–VI compound semiconductors contacted with gold increases with increasing electronegativity of anion. Brillson[22] has recently observed that it is the chemical activity of the constituents at the metal–semiconductor interface which determines the barrier height. According to him the Kurtin *et al.*[17] transition of Fig. 6 can be described equally well in terms of chemical reactivity of the semiconductor. He finds as good a correlation of barrier heights of Fig. 6 with the heat of reaction of the metal–semiconductor system as with the electronegativity difference $\Delta\chi$. Covalent semiconductors (for which $\Delta\chi$ is small) have a lower heat of reaction than the ionic semiconductors and thus are more likely to react with metals than the ionic semiconductors. The sharp transition in the interface index S^* could also be explained as occurring at a critical value of heat of reaction. These observations strongly suggest that the local charge distribution associated with interfacial chemical reaction rather than the density of intrinsic surface states at the semiconductor surface determines the metal–semiconductor barrier height. The interface index S^* is small for covalent semiconductors because they readily react with metals and S^* is large for ionic semiconductors because of their nonreactivity.

The matter does not appear to be settled as yet. For example, it is not clear whether the interface index S^* is a function of ionicity or band gap. Most materials with large ionicity also have a large band gap, and as Cohen[62] has pointed out, correlation of ionicity and band gap are hard to determine separately. A recent study[45] of room temperature interfacial reactions in a number of semiconductors has shown that metal films exhibit interfacial intermixing with semiconductor of band gap $E_g < 2.5\,\text{eV}$ or a dielectric constant higher than 8. No distinction has been made between a ionic or a covalent semiconductor in this respect. Diamond has a large band gap but zero value of $\Delta\chi$. Hence, it would be of considerable interest to measure the interface index for diamond and SiC crystals to decide if the interface parameter S^* depends upon the ionicity or on the band gap. As for the plot of Fig. 6, it is interesting to note that Schlütter,[63] who has reexamined the available data on $\Delta\chi$ and S^*, finds neither the sharp transition of S^* at $\Delta\chi = 0.7$ not its saturation at $S^* = 1$.

5. CAPACITANCE–VOLTAGE CHARACTERISTICS

5.1. Electric Field and Potential Distribution in the Depletion Region

The electric field and potential distribution in the depletion region of a Schottky barrier junction depend upon the barrier height, the applied voltage, and the impurity concentration. These dependences are frequently needed and can be obtained by the solution of a one-dimensional Poisson equation. Figure 11a shows the energy band diagram of a reverse biased Schottky barrier junction made on an n-type semiconductor. We assume the semiconductor to be nondegenerate and uniformly doped and divide it into a space charge region and a neutral region devoid of any space charge. At any point in the semiconductor the Poisson equation can be written as

$$\frac{d^2\phi}{dx^2} = -\frac{q}{\epsilon_s}[N_d + p(x) - n(x)] \tag{14}$$

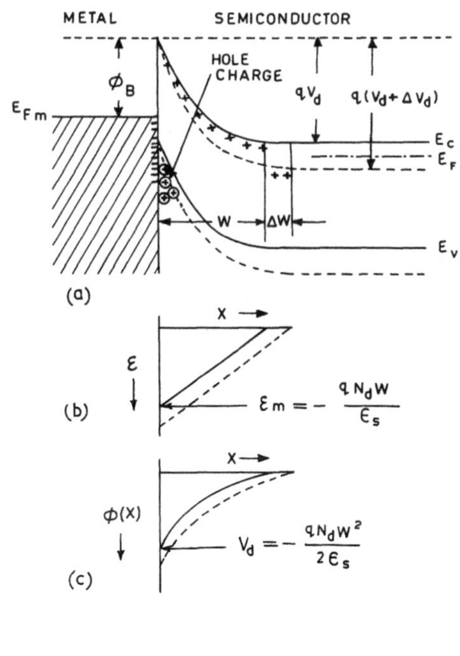

Figure 11. Electric field and potential distributions in the depletion region of a Schottky barrier. (a) Energy band diagram; (b) electric field distribution; and (c) potential distribution. The solid lines correspond to a reverse bias V_R and the dashed lines to $V_R + \Delta V_R$.

where ϵ_s is the semiconductor permittivity, N_d is the donor concentration, and $n(x)$ and $p(x)$ are the electron and the hole concentrations at any point x in the semiconductor, respectively. It is assumed that all the donors are ionized. Taking the potential ϕ to be zero in the neutral bulk region of the semiconductor at the edge of the space charge layer one can write

$$\begin{aligned} n(x) &= n_0 \exp[q\phi(x)/kT] \\ p &= p_0 \exp[-q\phi(x)/kT] \end{aligned} \tag{15}$$

where n_0 and p_0 represent the equilibrium electron hole concentrations in the neutral semiconductor. Substituting the values of $n(x)$ and $p(x)$ in Eq. (14) one obtains

$$\frac{d^2\phi}{dx^2} = -\frac{q}{\epsilon_s}\left\{N_d - n_0 \exp\left[\frac{q\phi(x)}{kT}\right] + p_0 \exp\left[-\frac{q\phi(x)}{kT}\right]\right\} \tag{16}$$

A closed form solution of this equation is not possible. An additional simplifying assumption made in the analysis is the so-called depletion approximation. In this approximation the free carrier concentrations are assumed to fall abruptly from their equilibrium values n_0 and p_0 in the bulk neutral region to a negligibly small value in the barrier space charge region. In reality this transition occurs smoothly over a distance in which the bands bend by about $3kT$ but the calculations made using the depletion approximation are sufficiently accurate for most purposes. Thus, using the depletion approximation Eq. (16) can be written as

$$\begin{aligned} \frac{d^2\phi}{dx^2} &= -\frac{q}{\epsilon_s}N_d \qquad 0 < x < W \\ &= 0 \qquad x > W \end{aligned} \tag{17}$$

where W represents the width of the depletion region. Integrating Eq. (17) with respcet to x and using the condition that $d\phi/dx = 0$ at $x = W$ we obtain the electric field $\mathscr{E}(x)$ in the depletion region

$$\mathscr{E}(x) = -\frac{d\phi}{dx} = \mathscr{E}_m\left(1 - \frac{x}{W}\right) \tag{18a}$$

where

$$\mathscr{E}_m = -\frac{qN_d}{\epsilon_s}W \tag{18b}$$

is the maximum electric field which occurs at $x=0$. A second integration with the boundary condition $\phi=0$ at $x=W$ leads to the following relation:

$$\phi(x)=-\frac{qN_d}{2\epsilon_s}W^2\left(1-\frac{x}{W}\right)^2 \tag{19}$$

Thus the potential varies parabolically with the distance in the depletion region and has its maximum value $\phi(0)=V_d$ given by

$$V_d=(V_i-V)=-\frac{qN_d}{2\epsilon_s}W^2 \tag{20}$$

where V is the externally applied voltage. For a forward bias $V=V_F$ and for a reverse bias $V=-V_R$. The negative sign in the above equation shows that the potential at $x=0$ is negative with respect to that at $x=W$.

The depletion region width W is obtained from the above relation:

$$W=\left(\frac{2\epsilon_s}{qN_d}|V_i-V|\right)^{1/2} \tag{21}$$

The width of depletion region at zero bias is obtained by putting $V=0$. From Eq. (21) it is seen that W decreases below its value W_0 in case of a forward bias and increases above W_0 in case of a reverse bias. Figure 11b and 11c show the electric field and the potential distributions for the reverse biased Schottky barrier junction.

In the depletion approximation we have neglected the electron and hole concentrations in comparison to the donor concentration N_d. In a strongly n-type semiconductor, the hole concentration is negligible, but at the edge of the depletion region $x=W$ the electron concentration $n(W)=n_0=N_d$ and decreases exponentially with the decreasing potential $\phi(x)$. It should be clear that the concentration $n(x)$ becomes negligibly small compared to N_d when $q\phi(x)$ in Eq. (15) is $-4kT$ or less. Thus, the depletion approximation is valid only when the potential drop V_d across the depletion region is large compared to about $4kT/q$.

5.2. Depletion Region Capacitance

5.2.1. Ideal Schottky Barrier

A change in the voltage across the Schottky barrier junction causes a change in the width of the depletion region, and this change is accomplished by the movement of charge carriers into the space charge layer or out of this region. This change in the depletion region charge gives rise to a capacitance. Referring to Fig. 11a and neglecting the charge in the surface states, there are three sources of charge in the barrier region. Firstly, there is the charge Q_d in the depletion region which results from the movement of electrons out of the semiconductor into the metal. Secondly, there is a charge Q_m on the metal surface which is caused by the electrons that have crossed from the semiconductor into the metal. Finally, if the band bending is sufficiently large, there will occur a charge Q_h due to holes which exist in the semiconductor region just adjacent to the metal contact. Electrical neutrality in the junction region requires that $Q_d + Q_m + Q_h = 0$, where each of these charges represents charge per unit area of the junction. Suppose now that the bias across the junction is increased by a small amount ΔV_d. This increase in the reverse bias causes a movement of electrons out of the semiconductor near the depletion layer edge causing an increase in the depletion region width from W to $W + dW$. The positive voltage on the semiconductor also causes a small decrease in the hole charge Q_h which results from the movement of holes from the semiconductor to the metal side. Finally to maintain the space charge neutrally the negative charge Q_m on the metal also increases. Since the charges Q_m and Q_h have no dielectric layer between them they can be treated as forming the charge on one side of the depletion region while Q_d is the opposite charge required to balance these charges. Similarly, a small decrease in the reverse bias causes a decrease in the width of the depletion region and a consequent decrease in the magnitudes of Q_d and $Q_m + Q_h$. The space charge layer capacitance C' per unit area is defined by the relation

$$C' = \frac{dQ_d}{dV_d} = -\frac{d}{dV_d}(Q_m + Q_h) \tag{22}$$

Let us neglect the effect of minority carriers and take $Q_h = 0$ so that $Q_d = -Q_m$. Applying Gauss law at the metal–semiconductor boundary we obtain

$$\epsilon_s \mathscr{E}_m = Q_d \tag{23}$$

The maximum electric field strength is assumed to occur at $x = 0$. The field $\mathscr{E}_m$ has been calculated in Eq. (18b) assuming the depletion approximation. A more accurate expression for $\mathscr{E}_m$ is obtained by integrating Eq. (16) assuming that band bending is small so that $p(x)$ is negligible everywhere and

$$\frac{d^2\phi}{dx^2} = -\frac{q}{\epsilon_s}\left\{N_d - n_0 \exp\left[\frac{q\phi(x)}{kT}\right]\right\} \tag{24}$$

Multiplying both sides of this relation by $2d\phi/dx$ and integrating from $x = 0$ to

$x = W$, with $\phi(0) = -V_d$ and $\phi(W) = 0$ and assuming $N_d = n_0$ we obtain

$$\left(\frac{d\phi}{dx}\right)^2_{x=0} = \mathscr{E}_m^2 = \frac{2q}{\epsilon_s} N_d \left(V_d - \frac{kT}{q}\right) \tag{25}$$

where $V_d = (V_i - V)$ is the voltage drop across the depletion region and the built-in voltage V_i is taken to be positive. The depletion region charge Q_d per unit area is then given by

$$Q_d = \epsilon_s \mathscr{E}_m = \left[2q\epsilon_s N_d \left(V_d - \frac{kT}{q}\right)\right]^{1/2} \tag{26}$$

and the depletion region capacitance C can be written as

$$C = S\frac{dQ_d}{dV_d} = S\left[\frac{q\epsilon_s N_d}{2[V_i - (kT/q) - V]}\right]^{1/2} \tag{27}$$

where S represents the area of the Schottky barrier contact. Since charge Q_d varies with the voltage in a nonlinear manner the capacitance is a nonlinear function of voltage and can be defined only as a differential capacitance presented to a small change ΔV_d in the depletion region voltage. The capacitance is measured by superimposing a small ac voltage on the steady-state dc bias. Note that the term kT/q in Eq. (26) is the contribution of majority carriers (in this case electrons) to the space charge. When this term is omitted the result is equivalent to the depletion approximation and the capacitance C then can be expressed as

$$C = \frac{S\epsilon_s}{W} \tag{28}$$

This relation shows that the Schottky barrier junction capacitance can be looked upon as the capacitance of a parallel plate capacitor whose separation between the two plates is equal to the depletion region width W. This is an important result and is valid even when the dopant concentration N_d is not constant but varies with the distance into the semiconductor.

Measurements of the depletion region capacitance under forward bias is difficult because the diode is conducting and the capacitance is shunted by a large conductance. However, the capacitance can be easily measured as a function of the reverse bias $V = -V_R$. Equation (27) predicts that a plot of $1/C^2$ as a function of applied voltage V is a straight line whose slope and intercept can be used to determine the dopant concentration N_d and the built-in voltage V_i as explained in Section 3.1.

In the case where N_d varies with the distance into the semiconductor the plot of $1/C^2$ against the applied voltage V is not a straight line, but it has been shown[64] that the slope at any point of the characteristic is still given by $2/s^2 q\epsilon_s N_d(W)$, where $N_d(W)$ represents the dopant concentration at the edge of the depletion region and W can be obtained from Eq. (28). Thus, by measuring the slope at given depth W from the surface, the impurity concentration at that point can be determined. This provides a very convenient method of measuring the impurity distribution in a semiconductor.[65]

Equation (27) is based on the assumption that the electron concentration n_0 at each point in the neutral semiconductor equals the donor concentration N_d. When this is the case an increase ΔV_R in the reverse bias increases the depletion region width by ΔW because the electrons move out of this region leaving behind the uncompensated donors. Obviously the condition for point-by-point charge neutrality is not satisfied in the semiconductor bulk when the donor concentration varies with position. However, the difference between n_0 and N_d will not be very significant if the gradient of N_d is not very sharp. In the case where N_d varies by orders of magnitude over a distance equal to the Debye length $L_D = (\epsilon_s kT/q^2 n_0)^{1/2}$, the electron concentration n_0 will be appreciably different from N_d and Eq. (27) will be no longer valid. Thus, impurity distributions in which the dopant concentration changes appreciably over a distance of less than about 0.1 μm cannot be determined using the C–V technique.[66,67]

5.2.2. Effect of Minority Carriers

So far we have neglected the effect of minority carriers (i.e., holes in the above discussion) on the space charge. This neglect is justified in most cases since the band bending is such that the concentration of minority carriers adjacent to the metal is small compared to ionized donor concentration. However, in some cases where the barrier height exceeds $(E_g - \phi_n)$, the hole concentration adjacent to the metal becomes higher than the donor concentration and the space charge in this region is dominated by holes. Since the hole concentration increases exponentially with energy the barrier rises steeply near the surface as shown in Fig. 12. The barrier height deduced from

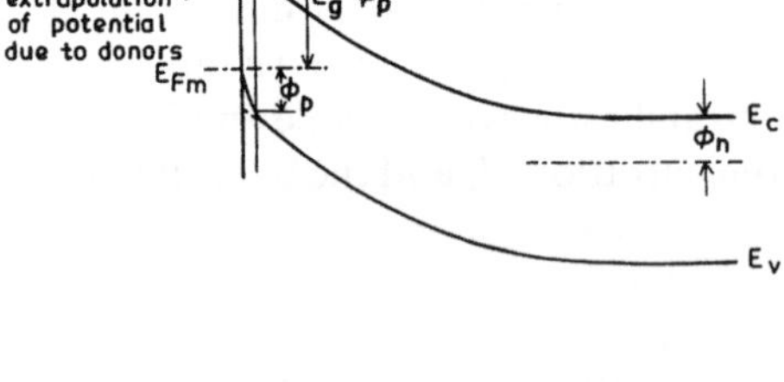

Figure 12. Effect of holes on the shape of potential barrier of Schottky contact on n-type semiconductor (after Rhoderick[9]).

the C–V measurements corresponds to the extrapolated value shown by the dotted curve and is seen to be less than the true value. Thus, when the effect of minority carriers becomes important in the space charge region, a plot of $1/C^2$ against the applied voltage is then no longer linear and if the low bias part of this plot is extrapolated to the V axis the intercept V_0 will be less than $(V_i - kT/q)$. The problem has been discussed by Green.[68]

5.2.3. Effect of Interfacial Layer

If an interfacial oxide layer is present between the metal and the semiconductor a part of the applied voltage appears across this layer and modifies the dependence of the depletion region charge Q_d on the applied voltage. The capacitances of the interfacial layer and the depletion region are effectively in series and the overall capacitance in general may be a complicated function of the interfacial layer parameters and the applied voltage. However, when the interfacial layer is thin (i.e., about 30 Å or less) electrons can tunnel through the layer from the metal to the semiconductor side (or in reverse direction) and the resulting Schottky barrier is nearly ideal. The current voltage characteristics of the nearly ideal diode are described in terms of an ideality factor n given by (see Section 6.2 for details)

$$\frac{1}{n} = \left(1 - \frac{\partial \phi_B}{q \partial V}\right) \tag{29}$$

where $\partial \phi_B / \partial V$ represents the change in the barrier height with applied voltage because of the presence of the interfacial layer. If the states at the oxide–semiconductor interface are uniformly distributed in energy then n is nearly independent of the applied voltage.

In order to obtain the C–V characteristics for the nearly ideal diode we substitute $qV_i = (\phi_B - \phi_n)$ in Eq. (26) and write

$$Q_d = [2\epsilon_s N_d(\phi_B - \phi_n - kT - qV)]^{1/2} \tag{30}$$

and

$$C = S\frac{dQ_d}{dV} = qS\left[\frac{\epsilon_s N_d}{2(\phi_B - \phi_n - kT - qV)}\right]^{1/2}\left(1 - \frac{\partial \phi_B}{q\partial V}\right) \tag{31}$$

which after substituting for $\partial \phi_B / \partial V$ in terms of the diode ideality factor n becomes

$$C = S\frac{dQ_d}{dV} = \frac{qS}{n}\left[\frac{\epsilon_s N_d}{2(\phi_B - \phi_n - kT - qV)}\right]^{1/2} \tag{32}$$

The barrier height ϕ_B for a constant n can be written as

$$\phi_B = \phi_{BO} + \frac{\partial \phi_B}{\partial V} V = \phi_{BO} + \left(1 - \frac{1}{n}\right) qV \tag{33}$$

where ϕ_{BO} represents the zero bias barrier height. Combining Eqs. (32) and (33) one obtains

$$\frac{1}{C^2} = \frac{2nS^{-2}}{q^2 \epsilon_s N_d} [n(\phi_{BO} - \phi_n - kT) - qV] \tag{34}$$

From Eq. (34) one observes that the effect of interfacial layer is to scale up both the slope and the intercept. Thus $1/C^2$ versus voltage V plot underestimates the dopant concentration and results in a larger intercept V_0 over its value in case of an ideal diode without interfacial layer. For a nearly ideal Schottky barrier diode n is 1.1 or less so that the value of N_d is not changed much but V_0 is significantly increased. It is thus clear that if the effect of the nonideal nature of the diode is not taken into account the barrier height deduced from the C–V measurements will be significantly larger than that obtained from the I–V measurements.

5.2.4. Effect of Deep Traps

Deep traps are energy levels of intrinsic lattice defects or impurity atoms that have energy near the center of the band gap. Presence of deep traps in the depletion region of the Schottky barrier makes the junction capacitance a complicated function of the bias voltage and the measuring frequency. The occupancy of the trap level will depend on its position with respect to the electron (or hole) quasi-Fermi level and hence on the junction voltage. If the time constant of traps is so small that they can follow the bias variation as well as the measuring ac signal then the donor concentration N_d in Eq. (27) is replaced by $(N_d + N_t)$, where N_t represents the concentration of deep traps. However, the traps will have no effect on the capacitance if their time constant is so large that they are not able to respond to changes either in the bias voltage or the measuring signal. When the traps can follow the bias variations but not the measuring signal the situation is more complicated.(9) In such cases if the trap density is not negligible compared to the dopant concentration then the impurity distribution obtained from the C–V measurements gets distorted due to the presence of traps.(69) Thus great care should be taken in determining the impurity profiles from the C–V measurements when traps are present.

If the depletion region of a Schottky barrier contains deep traps and the capacitance is changed abruptly, the traps because of their finite response time will not be able to respond to this change instantaneously. Information about the trap energy and the response time can be obtained from the capacitance variation after the transient. As an example consider a Schottky barrier junction on an *n*-type semiconductor containing donorlike traps with energy level below the Fermi level. If the junction is suddenly reverse biased the capacitance will decrease abruptly because of increase in the depletion region width. Initially the depletion region will expand because of increase in the number of ionized donors. However, as time proceeds the traps in the depletion region occupied by the electrons will be emptied. These electrons will be swept out of the depletion region, causing an increase in the positive charge density and thus reducing the depletion region widths. The resulting change in the capacitance can be measured as a function of time, and from these measurements the time constant, concentration and the energy level of the traps can be obtained.(70) Study of deep levels using capacitance transient techniques has developed into an important tool known as deep level transient spectroscopy (DLTS). A review of this technique is given by Miller, Lang, and Kimerling.(71)

6. CURRENT–VOLTAGE CHARACTERISTICS

6.1. Transport Mechanisms

The current flows in a Schottky barrier diode because of charge transport from the semiconductor to the metal or in the reverse direction. There are four different mechanisms by which the carrier transport can occur: (a) thermionic emission over the barrier, (b) tunneling through the barrier, (c) carrier recombination (or generation) in the depletion region, and (d) carrier

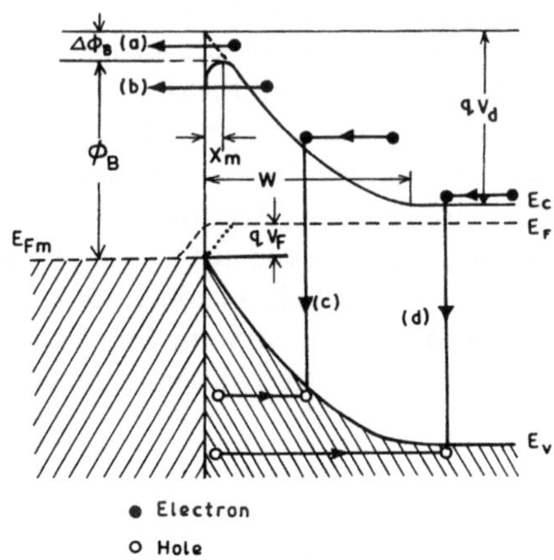

Figure 13. Energy band diagram of a forward biased Schottky barrier junction on *n*-type semiconductor showing different transport processes; (a) thermionic emission, (b) tunneling through the barrier, (c) carrier recombination in the depletion region, and (d) hole injection from the metal into the semiconductor. The dotted line shows the quasi-Fermi-level according to the diffusion theory and the dashed line according to the thermionic emission theory (after Rhoderick(9)).

recombination in the neutral region of the semiconductor which is equivalent to the minority carrier injection. Process (a) is usually the dominant mechanism in Schottky barrier junctions in Si and GaAs and leads to the ideal diode characteristics. The remaining mechanisms mainly (b) and (c) cause departure from the ideal behavior. Figure 13 schematically depicts these processes for a forward biased Schottky barrier made on an *n*-type semiconductor. The inverse processes occur under reverse bias.

6.1.1. Diffusion and Thermionic Emission over the Barrier

Referring to Fig. 13, an electron emitted over the barrier from semiconductor into the metal must move through the high field depletion region. In traversing this region the motion of the electron is governed by the drift and the diffusion processes. The emission of electrons into the metal is controlled by the density of available states in the metal. Thus the two processes, namely, the emission over the barrier and the drift and diffusion in the depletion region are effectively in series and the one which offers the higher resistance determines the current. In their original treatment Wanger,[72] and Schottky and Spenke[73] assumed that the current was limited by the drift and diffusion processes. The diffusion theory leads to the following expression for the diode current[9]:

$$I = qSN_c\mu\mathscr{E}_m \exp\left(-\frac{\phi_B}{kT}\right)\left[\exp\left(\frac{qV}{kT}\right) - 1\right] \tag{35}$$

where S is the diode cross-sectional area, N_c is the effective density of states in the conduction band of the semiconductor, μ is the electron mobility, and all other symbols have their usual meanings. Since the maximum field $\mathscr{E}_m$ in Eq. (35) is voltage dependent, the pre-exponential factor in this equation does not saturate as it should in an ideal Schottky diode. Subsequent work by Bethe[6] showed that the diode current is limited by thermionic emission over the barrier and is not in agreement with Eq. (35). The difference between the two mechanisms is shown by the position of the quasi-Fermi level in the depletion region. According to the diffusion theory the electrons are in equilibrium with the lattice even when the junction is forward biased so that their quasi-Fermi level coincides with the metal Fermi level at the interface as shown by dotted curve in Fig. 13. In the thermionic emission theory on the other hand the electrons entering the metal have energy higher than the metal electrons and their quasi-Fermi level is almost horizontal through the depletion region as shown by the dashed curve.

The effect of drift and diffusion in the depletion region is assumed to be negligible in the thermionic emission theory and the barrier height is assumed to be large compared to kT. From Fig. 13 it is obvious that only those electrons whose kinetic energy exceeds the height of the potential barrier will be able to reach the top of the barrier. Assuming that the electrons have a Maxwellian distribution of velocities, the number of electrons n^* per unit area which have sufficient energy to move over the barrier from the semiconductor into the metal is given by

$$n^* = n_0 \exp\left[\frac{-q(V_i - V)}{kT}\right] \tag{36}$$

where n_0 represents the electron concentration in the neutral semiconductor outside the depletion region and V is the voltage applied to the semiconductor. For a nondegenerate semiconductor

$$n_0 = N_c \exp(-\phi_n/kT)$$

and since $\phi_B = qV_i + \phi_n$ from Eq. (36) we obtain

$$n^* = N_c \exp\left[-\frac{(\phi_B - qV)}{kT}\right] \tag{37}$$

If these electrons are assumed to have an isotropic distribution of velocities then from the kinetic theory the flux of electrons incident on the barrier is $n^*\bar{v}/4$. Supposing that all the incident electrons cross over into the metal and none is reflected back the current I_{SM} due to passing of electrons from the semiconductor to the metal is given by

$$I_{SM} = \frac{qS\bar{v}}{4} N_c \exp\left[-\left(\frac{\phi_B - qV}{kT}\right)\right] \tag{38}$$

where $\bar{v}$ is the average thermal velocity of electrons in the semiconductor. Although the electrons flow from the semiconductor into the metal the current I_{SM} flows from the metal to the semiconductor and is taken to be positive in Fig. 13.

For unbiased junction under thermal equilibrium no net current can flow. Consequently the current given by Eq. (38) must be balanced by an opposite current I_{MS} due to crossing of electrons from the metal into the semiconductor making $I = I_{SM} + I_{MS} = 0$ and

$$I_{MS} = -\frac{qS\bar{v}}{4} N_c \exp\left(-\frac{\phi_B}{kT}\right) \tag{39}$$

In the presence of an applied bias V, the barrier for electron flow from the metal to semiconductor remains practically unchanged at ϕ_B and so is the current $I_{MS} = -I_0$. The current I_{SM}, however, is given by Eq. (38) and combining this equation with Eq. (39) we obtain

$$I = I_0\left[\exp\left(\frac{qV}{kT}\right) - 1\right] \tag{40}$$

For a Maxwellian distribution, the average velocity $\bar{v} = (8kT/\pi m^*)^{1/2}$ and substituting $N_c = 2(2\pi m^* kT/h^2)^{3/2}$ the current I_0 can be written as

$$I_0 = SAT^2 \exp\left(-\frac{\phi_B}{kT}\right) \tag{41}$$

where

$$A = \frac{4\pi m^* q k^2}{h^3}$$

is the Richardson constant for thermionic emission from the metal into the semiconductor with electron effective mass m^*, h is the Planck's constant, and S is the diode area. The question of the correct value of m^* is complicated. For a semiconductor with spherical energy surfaces such as GaAs, m^* is independent of direction. However, for semiconductors like Ge and Si which have anisotropic energy surfaces, m^* is direction dependent.

The thermionic emission theory which predicted Eq. (41) is based on the assumption that the electron collisions within the depletion region are neglected. This is justified only for high-mobility semiconductors. A number of authors have combined the thermionic emission and the diffusion processes in a single theory by considering the two processes in series and by finding quasi-Fermi level at the interface which equalizes the current flowing through each of the two processes. The most complete treatment is that of Crowell and Sze.(74) These authors have taken into account the effect of image force barrier lowering and have defined a recombination velocity v_R at the top of the barrier by equating the electron current into the metal to $v_R(n^* - n_0^*)$. Here n_0^* represents the electron concentration at the top of the barrier at zero bias. Crowell and Sze's analysis leads to the following expression for I_0:

$$I_0 = \frac{SqN_c v_R}{1 + v_R/v_D}\exp(-\phi_B/kT) \tag{42}$$

where v_D is the effective diffusion velocity through the depletion region. In terms of thermionic emission theory the velocity $v_R = \bar{v}/4$. It is evident that for $v_D \gg v_R$ Eq. (42) reduces to Eq. (41) and termionic emission theory applies, while for $v_R \ll v_D$ the diffusion theory is valid and Eq. (42) becomes identical to the preexponential factor of Eq. (35). If we assume the carrier mobility to be independent of the electric field in the barrier then $v_D = \mu\mathscr{E}_m$ and the condition for the validity of thermionic emission theory is given by $\mu\mathscr{E}_m \gg \bar{v}/4$. Substituting the value of $\bar{v}$ and $\mu = q\tau/m^*$ this inequality results in the relation

$$q\mathscr{E}_m\lambda \gg \frac{2kT}{\pi} \tag{43}$$

where the mean free path $\lambda = \bar{v}\tau$ and τ represents the mean free time between the collisions. The left-hand side of Eq. (43) represents the average energy gained by the electron in a mean free path near the top of the barrier. Thus, if this energy is large compared to the thermal energy kT the thermionic emission theory is valid and the effect of electron collisions in the depletion region can be neglected. For Schottky barriers on Si, GaAs, and other high-mobility semiconductors the above inequality is valid except at very large forward bias where the maximum field $\mathscr{E}_m$ becomes small enough to make v_D comparable to v_R. This situation is analogous to high-level injection in a p–n junction.(75)

The thermionic emission diffusion theory of Crowell and Sze(74) assumes that the electron distribution function remains Maxwellian in the barrier depletion region and that the classical drift and diffusion equations can be used throughout this region. Near the top of the barrier the electric field is very high and the distribution function changes considerably within a mean free path. Under these conditions it is no longer possible to split the current into drift and diffusion components. Moreover, we should not expect the distribution function near the top of the barrier to be Maxwellian and isotropic. As the electrons entering the metal are assumed not to return, the distribution function near the boundary must be anisotropic. Monte Carlo calculations have shown(76) that the electron distribution function at the boundary is close to a unidirectional Maxwellian with a mean velocity toward the metal. However, the transport equations derived from the first-order solution of the Boltzmann equation are found to be inadequate near the top of the barrier and a new set of transport equations has been proposed. These considerations show that the problem of hot electrons in a rapidly varying field has not been solved to date and we do not yet have an exact theory of the Schottky barrier current voltage characteristics. It is, therefore, surprising that in spite of the drastic assumptions the existing theory has been so successful in describing the I–V characteristics.

In the original treatment of Bethe(6) it was assumed that once the electrons reach the top of the barrier they are emitted into the metal and do not return back to the semiconductor. Crowell and Sze[74] have refined the theory by including the effects of optical phonon scattering in the metal and quantum mechanical reflections from the barrier. An electron after passing over the barrier into the metal may be scattered back to the semiconductor after the emission or absorption of an optical phonon, with a subsequent reduction in the diode current. Moreover, not all the electrons incident over the barrier will cross over into the metal. According to quantum mechanics there is a finite probability that an electron with kinetic energy larger than the barrier potential energy may be reflected back into the semiconductor. Also an electron with energy less than the barrier energy has a finite probability of tunneling through the barrier. Crowell and Sze[27,74] have calculated the probability f_p of an electron reaching the metal without scattering into the semiconductor and also the probability f_q of its transmission through the barrier in the presence of quantum mechanical reflection and tunneling. They have concluded that the combined effect of these processes and the carrier diffusion effects in the depletion region is to replace the Richardson constant A in Eq. (41) by an effective value A^* given by

$$A^* = \frac{A f_p f_q}{1 + f_p f_q v_R / v_D} \tag{44}$$

The coefficients f_p and f_q vary with the maximum electric field $\mathscr{E}_m$ in the barrier, with temperature T and with the carrier effective mass in the semiconductor. Calculations made on Ge, Si, and GaAs have shown[74] that at 300 K the product $f_p f_q$ is typically of the order of 0.5 so that A^* may be less than A by about 50%. Since the effective mass m^* can vary substantially for different semiconductors, this difference between A and A^* has negligible effect on the barrier height deduced from the $I-V$ measurements. Table 2 below shows the calculated values of A^* for Si and GaAs at 300 K.

Crowell and Sze's ideas of phonon scattering and quantum mechanical reflections are correct conceptually, but whether their calculations yield

TABLE 2. Calculated Values of A* at 300 K

Semiconductor		A^* (A cm^{-2}K^{-2})
Si	n-type	112[a]
	p-type	32[a]
GaAs	n-type	4.4[b]

[a] Reference 94.
[b] Reference 74.

reliable estimates of f_p and f_q is not clear. Due to image force lowering of the barrier, the lower edge of the conduction band slopes steeply between the potential maximum and the metal, and the use of the effective mass approximation in this region in the calculation of f_q is hardly justified.

6.1.2. Tunneling through the Barrier

Besides the diffusion and thermionic emission mechanisms, electrons can also be transported across the barrier by quantum mechanical tunneling. The two ways in which tunneling can occur in a Schottky barrier junction are shown in Fig. 14 for (a) forward bias and (b) for reverse bias. The semiconductor in these figures is assumed to be doped to degeneracy such that the Fermi level lies above the bottom of the conduction band. Because of heavy doping the depletion region is very thin and at low temperatures electrons with energy close to the Fermi level can tunnel from the semiconductor into the metal. This process is known as field emission (FE). At higher temperatures a significant number of electrons are able to rise high above the Fermi level where they see a thinner and lower barrier. These electrons thus can tunnel into the metal before reaching the top of the barrier. This tunneling of thermally excited electrons is known as thermionic field emission (TFE). Since the number of electrons decreases rapidly with energy above the Fermi level whereas the barrier thickness and height also decreases, there exists an energy E_m at which the contribution of TFE becomes maximum. If the temperature is still further raised gradually, a limit is reached at which practically all the

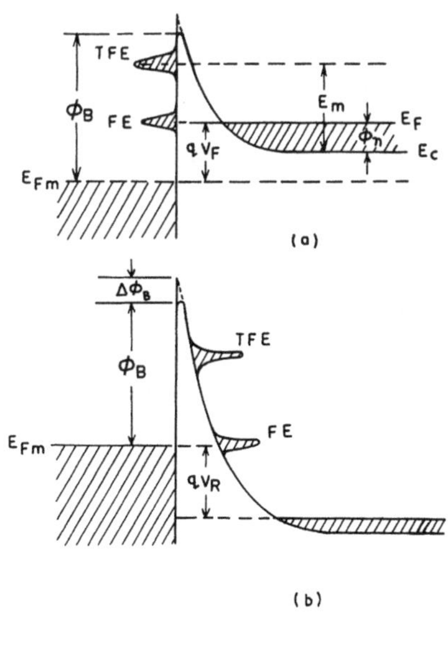

Figure 14. Field emission (FE) and thermionic field emission (TFE) tunneling through a Schottky barrier on n-type semiconductor. (a) Forward bias and (b) reverse bias. Note that tunneling occurs at lower dopings in case of reverse bias (after Padovani and Stratton[77]).

electrons are able to reach the top of the barrier and thermionic emission predominates.

Tunneling through a Schottky barrier has been analyzed theoretically by Padovani and Stratton[77] and by Crowell and Rideout.[78] The main results of their study are described below. Field emission in the forward direction occurs only in degenerate semiconductors, and except for very low forward biases, the $I-V$ characteristic in the presence of tunneling can be described by the relation

$$I = I_s \exp\left(\frac{qV}{E_0}\right) \tag{45}$$

where

$$E_0 = E_{00} \coth\left(\frac{E_{00}}{kT}\right)$$

and

$$E_{00} = \frac{qh}{4\pi}\left(\frac{N_d}{m^* \epsilon_s}\right)^{1/2}$$

where m^* is the electron effective mass and h is Planck's constant. The pre-exponential factor I_S in Eq. (45) is only weakly dependent on voltage and is a complicated function of barrier height, parameters of the semiconductor, and the temperature. The energy E_{00} is an important parameter in tunneling and kT/E_{00} is a measure of the relative importance of TE (thermionic emission) and tunneling.[79] At low temperatures E_{00} may become large compared to kT and we have $E_0 \approx E_{00}$ and the slope of ln I versus V plot is constant independent of T. This is the case for FE. At high temperatures where $E_{00} \ll kT$, we get $E_0 = kT$ and the slope of the ln I versus V plot is q/kT, which corresponds to TE. For intermediate values of temperature the slope can be written as q/nkT with

$$n = \frac{E_{00}}{kT} \coth\left(\frac{E_{00}}{kT}\right) \tag{46}$$

Contribution of TFE to the diode current dominates for $E_{00} \simeq kT$. Figure 15 shows the ranges of temperature and dopant concentration for transition from FE to TFE in n-type GaAs–Au Schottky barriers.[79] The energy E_m at which TFE has its maximum contribution occurs at

$$E_m = \frac{qV_d}{[\cosh(E_{00}/kT)]^2} \tag{47}$$

where V_d is the voltage corresponding to the total band bending and E_m is

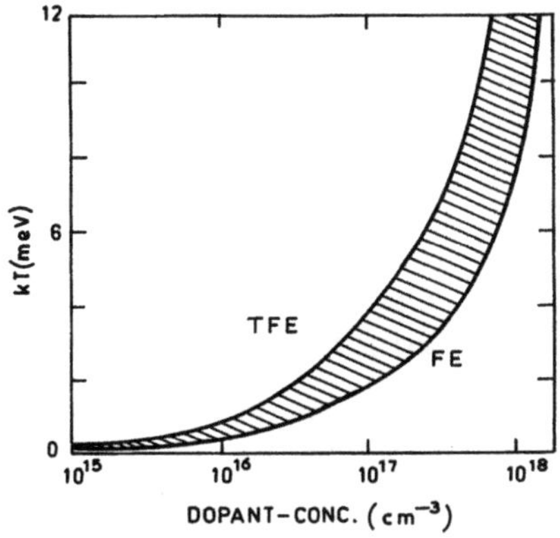

Figure 15. The ranges of temperature and the donor concentration over which FE and TFE occur in Au–n-type GaAs Schottky barriers (after Padovani[79]).

measured from the bottom of the conduction band at the edge of the depletion region. Figure 16 shows the plots of n and E_m/V_d as a function of kT/E_{00} at 300K in the TFE regime.[10] It is seen that n starts increasing above unity and TFE becomes important in Si and GaAs Schottky barriers only for $N_d > 10^{17}\,\text{cm}^{-3}$ at 300 K. At lower dopant concentrations the contribution of TFE becomes negligible and TE dominates.

The analysis of Padovani and Stratton[77] and of Crowell and Rideout[78] on which the above results are based has neglected the image force barrier lowering and the quantum mechanical reflections of electrons from the top of the barrier. Moreover, in both the above works the electron distribution was assumed to be described by Boltzmann statistics. In a further work Rideout and Crowell[80] considered the effect of image force lowering and quantum mechanical reflections using nondegenerate statistics. They observed that inclusion of these effects resulted in a significant change in the current density but only a minor change in the value of n. Chang and Sze[81] in their analysis used degenerate Fermi statistics and took both the image force lowering and quantum mechanical reflections into account. The results of their work, however, are similar to those of Rideout and Crowell[80] in the TFE regime,

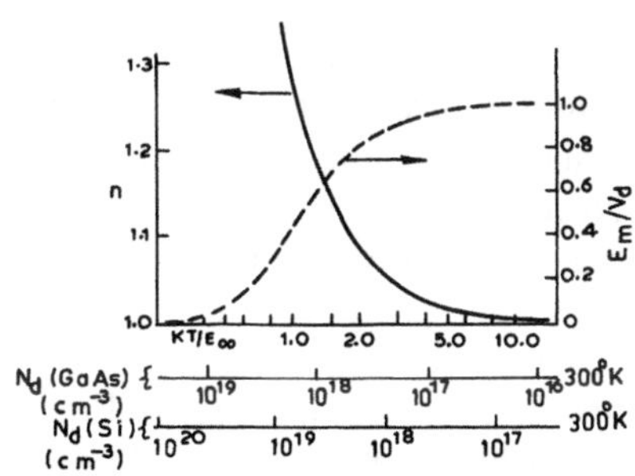

Figure 16. Ideality factor n and position of the maximum of energy distribution E_m of emitted electrons in TFE regime (after Rhoderick[10]).

though they differ from those results in the FE regime. In conclusion we observe that the analyses of Padovani and Stratton[77] and of Crowell and Rideout[78] appear to be reasonably adequate to explain the experimental data on FE and TFE.

6.1.3. Carrier Generation and Recombination in the Junction Depletion Region

At zero bias the depletion region of the Schottky barrier is in thermal equilibrium and the rate of electron–hole pair generation in this region is balanced by the rate of recombination. The electron–hole product is then equal to n_i^2. In presence of an applied voltage the electron–hole product departs from n_i^2 and there will be a net generation or a net recombination of carriers depending upon the polarity of the applied bias. If a reverse voltage is applied to a Schottky barrier junction made on n-type semiconductor, electron–hole pairs in excess of their thermal equilibrium value will be generated in the depletion region. These pairs will be swept out by the electric field of the barrier causing a reverse current. On the other hand, if the junction is forward biased then the electrons will be injected into the depletion region from the neutral bulk semiconductor and holes will be injected from the metal. These excess electron–hole pairs will recombine in the depletion region to give a forward recombination current. A recombination center in a semiconductor is most effective when its energy level is near the center of the band gap. For recombination through these mid-gap deep traps, the current I_{rg} caused by the carrier generation and recombination in the depletion region is given by[82]

$$I_{rg} = I_{RO}[\exp(qV/2kT) - 1] \tag{48}$$

with

$$I_{RO} = \frac{qSn_iW}{2\tau_0}$$

where τ_0 is the minority carrier lifetime in the depletion region and all other symbols have their usual meanings. This current is added to the thermionic emission current and in some cases may be responsible for a value of $n > 1$. Note that the current is a generation current when the junction is reverse biased and is a recombination current when the junction is forward biased.

The existence of depletion region recombination current was demonstrated by Yu and Snow.[83] Note that the exponential in Eq. (48) rises as $(qV/2kT)$ whereas that in Eq. (40) rises as (qV/kT). This shows that the recombination current will be important only at low values of the forward

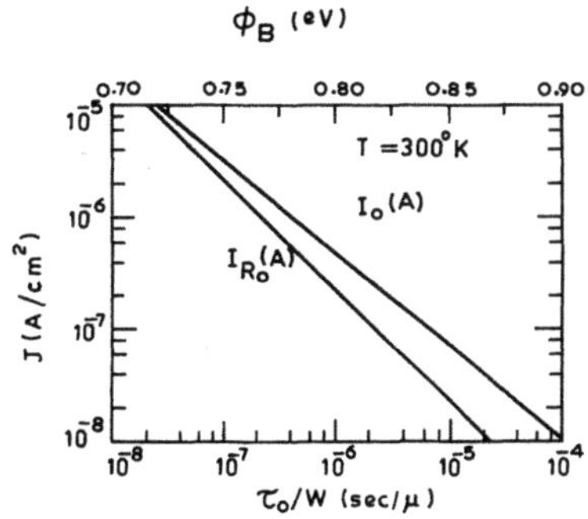

Figure 17. Thermionic emission current (I_0) and the generation current (I_{RO}) as a function of barrier height and the ratio τ_0/W for metal–n-type Si Schottky barriers (after Yu and Snow[83]).

bias. Furthermore, from Eqs. (48) and (41), the ratio I_{RO}/I_0 can be written as

$$\frac{I_{RO}}{I_0} = \frac{qn_i}{A^*T^2}\left(\frac{W}{2\tau_0}\right)\exp\left(\frac{\phi_B}{kT}\right) \tag{49}$$

It is clear from this relation that the depletion region generation recombination current is likely to become important only for large barrier height, low temperature, and lightly doped semiconductor (i.e., large values of W) of low carrier lifetime. Figure 17 shows the relative magnitudes of I_{RO} and I_0 as functions of the barrier height and the ratio τ_0/W. In Al–n-type Si Schottky barriers the ratio I_{RO}/I_0 is nearly unity for $N_d = 6 \times 10^{13}\,\text{cm}^{-3}$ but is about 0.01 for $N_d = 2 \times 10^{15}\,\text{cm}^{-3}$.

6.1.4. Minority Carrier Injection

The Schottky barrier diode is a majority carrier device at least for not too high values of the forward bias. However, very often the barrier height on an n-type semiconductor is substantially higher than half the band gap of the semiconductor. When this is the case, the region of the semiconductor adjacent to the metal has a high concentration of holes and thus becomes p-type. Under a forward bias the electrons from the semiconductor flow into the metal and some of the holes diffuse into the neutral region of the semiconductor causing injection of holes from the metal into the semiconductor. The injected holes gradually disappear by recombination with electrons as they diffuse into the neutral semiconductor [see process (d) in Fig. 13]. Note that the hole injection from the metal is equivalent to flow of electrons from the semiconductor valence band into the metal.

An excellent treatment of minority carrier injection in Schottky barrier contacts on n-type semiconductor is given by Yu and Snow.[84] Using the conventional p–n junction theory[82] the hole current I_p injected into the neutral semiconductor of donor concentration N_d and of thickness large compared to the hole diffusion length L_p can be written as

$$I_p = \frac{qSD_p n_i^2}{N_d L_p}\left[\exp\left(\frac{qV}{kT}\right) - 1\right] \tag{50}$$

Here D_p is the hole diffusion constant and $L_p = (D_p \tau_p)^{1/2}$ where τ_p represents the hole lifetime in the neutral region of the n-type semiconductor. Substituting for L_p and writing $n_i^2 = N_c N_v \exp(-E_g/kT)$, the ratio γ of the hole current I_p to the thermionic emission current I of Eq. (40) is given by

$$\gamma = \frac{q\sqrt{D_p}}{A^* T^2 \sqrt{\tau_p}} \frac{N_c N_v}{N_d} \exp\left[-\frac{(Eq - \phi_B)}{kT}\right] \tag{51}$$

where N_c and N_v represent the effective densities of states in the conduction and the valence band of the semiconductor, respectively. It is evident from Eq. (51) that the ratio γ is higher for larger value of ϕ_B, for higher value of D_p, and for lower values of N_d and τ_p. For Schottky barriers on n-type silicon ϕ_B is typically 0.8 eV and as Yu and Snow(84) have shown, γ is less than 0.1 for $N_d = 10^{14}\,\text{cm}^{-3}$ and above. It has been observed recently(85) that for experimentally obtained values of barrier heights of metal–n-type Ge Schottky barrier contacts with $N_d < 10^{15}\,\text{cm}^{-3}$, γ becomes very large and the dominant current component is due to injection of holes.

The above analysis is valid only for the moderate values of the forward bias. For large forward biases a part of the applied voltage appears across the neutral region of the semiconductor and the electric field in this region causes a drift current component which eventually dominates the minority carrier current. When this is the case the minority carrier injection is substantially altered and the ratio γ can become very large. When γ is plotted as a function of the diode current, it is found to be constant(86) up to a critical current density J_c, and after this increases almost linearly with the current. The current density J_c is given by(86)

$$J_c = \frac{qD_n N_d}{W_n} \tag{52}$$

where W_n is thickness of the quasineutral region of the n-type semiconductor and D_n is the electron diffusion constant.

6.2. Forward Characteristics

The thermionic emission theory predicts the current–voltage characteristics given by Eq. (40). However, a wide variety of practical metal–

semiconductor diodes follow the $I-V$ relation of the form

$$I = I_0\left[\exp\left(\frac{qV}{nkT}\right) - 1\right] \tag{53}$$

where n is often called the "ideality factor," for an ideal Schottky barrier where the barrier height is independent of the bias and current flows only due to thermionic emission $n = 1$. Factors which make n larger than unity are the bias dependence of barrier height, electron tunneling through the barrier, and the carrier recombination within the depletion region. As we have already seen, carrier recombination in the depletion region in Si and GaAs Schottky barrier junctions is important only for concentrations lower than $10^{15}\,\text{cm}^{-3}$. Schottky barriers used in power rectifiers are made on high-resistivity semiconductors. Since these devices operate under high-level injection, deviations from the ideal behavior are also observed because of enhanced minority carrier transport by the drift field in the quasineutral region. For dopant concentrations in excess of $10^{17}\,\text{cm}^{-3}$, n increases above unity because of FE and TFE and also because of image force lowering. A large majority of Schottky barrier junctions are made on n-type Si or GaAs in the dopant concentration range of 10^{15} to $10^{17}\,\text{cm}^{-3}$. The departure of n from unity in these junctions is largely because of the field dependence of barrier height. This field dependence arises either due to the presence of the insulating interfacial layer or due to image force lowering of the barrier. Let us consider the effect of bias-dependent barrier height on the $I-V$ characteristics.

The presence of an insulating interfacial layer in a Schottky barrier junction has three effects[9]:

i. Because of voltage drop across this layer the zero bias barrier height is lower than it would be in the absence of this layer.
ii. Since the electrons have to tunnel through the interfacial layer, the diode current is reduced below its value in an ideal Schottky diode. This corresponds to a lowering in the value of A^*.
iii. Since a part of the applied voltage appears across the interfacial layer the barrier height becomes a function of the applied voltage.

Let us now consider the case of a Schottky barrier junction made on an n-type semiconductor. For the sake of simplicity we assume that the bias dependence of the barrier height ϕ_B can be expressed by the relation

$$\phi_B = \phi_{BO} + \beta q V \tag{54}$$

where ϕ_{BO} is the zero bias barrier height and V is taken to be positive in case of a forward bias. Obviously β is positive since ϕ_B increases with increasing forward bias. Assuming a sufficiently large forward bias so that the unity term in the ideal diode equation (40) can be neglected with respect to the exponential term and making use of Eqs. (41) and (54), the forward current I can be written as

$$I = SA^*T^2 \exp\left(-\frac{\phi_{BO}}{kT}\right)\exp\left[\frac{q}{kT}(1-\beta)V\right] \tag{55}$$

where the Richardson constant A has been replaced by its modified value A^*. Denoting the voltage-independent part of this equation by I_0, a comparison of this equation with Eq. (53), after neglecting the unity term, gives $1/n = (1-\beta)$ and since from Eq. (54) $\beta = (\partial\phi_B/q\partial V)$ we obtain

$$\frac{1}{n} = \left(1 - \frac{\partial\phi_B}{q\partial V}\right) \tag{56}$$

Thus if $\partial\phi_B/\partial V$ is known as a function of the bias the ideality factor n can be calculated theoretically.

In the case where no interfacial layer is present and the barrier lowering occurs only due to the image force, Eq. (9) gives

$$\frac{\partial\phi_B}{\partial V} = \frac{q}{4}\left(\frac{q^3 N_d}{8\pi^2\epsilon_s^3}\right)^{1/4}\left(V_i - \frac{kT}{q} - V\right)^{-3/4} \tag{57}$$

and n also varies with the bias. However, it remains almost constant if qV is less than $\phi_B/4$. For Si and GaAs Schottky barriers with $N_d \leq 10^{17}\,\text{cm}^{-3}$ n has a value of about 1.02 so that the effect of image force barrier lowering is unimportant in the case of a forward bias but may become important in the case of a reverse bias.[9]

The ideality factor n is a constant independent of temperature in some diodes while in others it is found to increase with temperature. In diodes where the current components caused by FE, TFE, or due to carrier recombination in the depletion region become comparable with the thermionic emission current, n undoubtedly is expected to vary with temperature. However, it has been observed that even in devices where these processes are unimportant n may become temperature dependent and in some cases this temperature dependence can be expressed by the relation

$$n = \left(1 + \frac{T_0}{T}\right) \tag{58}$$

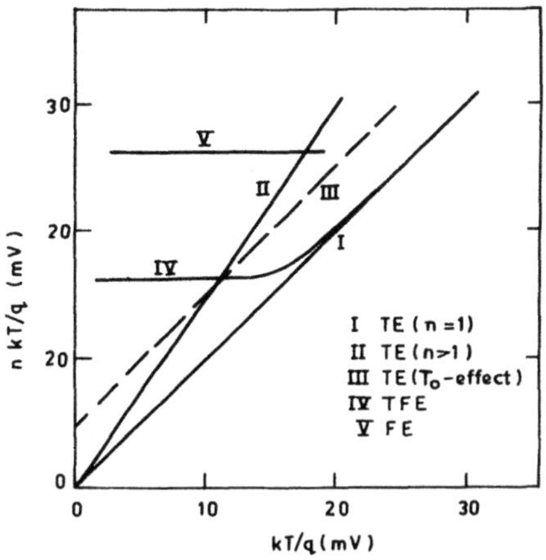

Figure 18. Plots of nkT/q as a function of kT/q for identifying the different current transport mechanisms (after Saxena[87]).

where T_0 is a constant independent of temperature. When n varies according to Eq. (58) the diode is said to exhibit the "T_0 effect." It has been proposed[87] that measurement of $I-V$ characteristics at different temperatures can be used to determine the mechanism of current transport by observing the temperature dependence of n. It is clear that for $V > 2kT/q$ Eq. (53) can be written as

$$I = I_0 \exp(qV/nkT) \tag{59}$$

If I is measured as a function of V at a number of temperatures and nT is plotted as a function of T, different current transport mechanisms can be identified as shown in Fig. 18. Curves I, II, and III in this figure represent the cases where TE is the dominant mechanisms with $n = 1$, with $n > 1$, and with T_0 effect, respectively. Curve IV correspond to TFE and V to FE. It has been pointed out recently[88] that the above-mentioned assignment for the mechanisms of current transport is possible only in the case of homogeneous diodes having no fluctuations in the geometrical and material parameters over the diode area. In a large area diode where T_0 fluctuates over the diode area, the nT versus T plot may indicate the transport mechanism to be FE or TFE even when the current flow is only due to thermionic emission.

Various attempts have been made to explain the temperature dependence of n in Eq. (58). Levine[89] observed that in many Schottky barrier diodes T_0 is bias dependent at constant temperature and is temperature independent only at a constant current. He deduced that T_0 anomaly uniquely defines an exponential distribution of surface states resulting in a charge $Q_{ss} = Q_c e^{-\phi_B/E^*}$ per unit area of the surface. Here Q_c is a constant of integration and E^* is characteristic energy associated with the surface state distribution. This expression for Q_{ss} does not admit a physically meaningful density of states distribution since the charge Q_{ss} does not vanish at flat band. A more appropriate form for Q_{ss} is proposed by Crowell[90]:

$$Q_{ss} = Q_f \left\{ \exp\left[-\frac{(\phi_B - \phi_0)}{kT} \right] - 1 \right\} \tag{60}$$

where ϕ_0 represents the neutral level at the surface and $-Q_f$ is the interface charge that neutralizes Q_{ss} at flat band. Levine's analysis has assumed the charge Q_m at the surface of the metal to be zero. However, inclusion of the charge on the metal surface and of an interfacial oxide layer between the metal and the semiconductor in the model do not alter the conclusions reached in Levine's work.[91]

Crowell[90] has pointed out that the T_0 effect is not uniquely related to an exponential energy distribution of surface states. He has analyzed Levine's parameters in terms of more detailed models like the Bardeen interface model and a doped interface model. Crowell concluded that the Bardeen interface model with a parabolic density of states distribution leads to results equivalent to Levine's model for $E^* > kT$. However, for $E^* < kT$ the Bardeen model is found to be physically untenable though the T_0 anomaly is still observed. The doped interface model in which the n-type semiconductor under the metal contact gets converted to p-type results in an energy band diagram similar to that of Fig. 10 and is a possible mechanism to explain T_0 anomaly even for $E^* > kT$. The T_0 effect can also be obtained when the surface states arise due to tunneling of metal electrons into the forbidden gap of the semiconductor. This situation leads to the same potential profile as the doped interface model, i.e., Fig. 10. It is very likely that the T_0 anomaly is related to the unavoidable interfacial disorder at the metal–semiconductor boundary. According to Levine,[92] an exponential distribution of interface states suggests that there is a slightly amorphous or disordered semiconductor layer at or near the metal–semiconductor junction. We have already seen that the energy band diagram of Fig. 10 can result if the energy states in the interfacial region of a metal n-type semiconductor junction can have a net negative charge. Assuming that the T_0 anomaly is related to the interfacial disorder one would expect the energy E^* to be sensitive not only to the choice of metal and semiconductor but also to the method of surface preparation and amount of interfacial oxide and contaminants on the surface.

The thermionic emission process leads to a saturation current density J_0 of the order of 10^{-7} A/cm^2 in Si and GaAs. This is about 4 orders of magnitude higher than the current density in p–n junctions of these semiconductors. Thus for the same forward current the Schottky barrier diode will have about 0.2 V less voltage drop than the p–n junction diodes. Also the ideality factor n of the Schottky barrier diode is usually smaller than that of a

p–n junction diode on the same semiconductor. These features make the Schottky barrier diode a preferred device in low-voltage high-current rectifiers.

6.3. Reverse Characteristics

According to thermionic emission theory the reverse current of a Schottky barrier junction should saturate at the value $I_0 = SA^*T^2 \times \exp(-\phi_B/kT)$. This saturation has not been observed in practical diodes and there are many causes which make the current increase with the reverse bias. These are discussed below.

It is obvious that the reverse current will not saturate if the barrier height ϕ_B depends upon the reverse bias. As we have already noted ϕ_B is a decreasing function of the electric field inside the barrier. Since this field increases with increasing reverse bias the barrier height decreases with the reverse bias making the reverse current to increase as $\exp(\Delta\phi_B/kT)$, where $\Delta\phi_B$ is the barrier lowering at reverse bias. A primary and unavoidable cause of barrier lowering is the image force on the electron emitted from the metal into the semiconductor. The image force barrier lowering is given by Eq. (9) with V replaced by $-V_R$. Thus when $\Delta\phi_B$ results from the image force lowering alone a plot of log I against $V_R^{1/4}$ should give a straight line. Although devices have been reported[93] in which the reverse current can be completely explained in terms of image force barrier lowering, such perfect diodes are exceptions rather than a rule. In a large majority of devices the barrier lowering necessary to explain the reverse I–V characteristics is far too much in excess of that which can be accounted by the image force lowering alone. Type 4 contacts which have a thin oxide layer between the metal and the semiconductor show a much larger barrier lowering as we have already seen and the nonsaturation of reverse current in such diodes is obvious. However, Andrews and Lepselter[94]

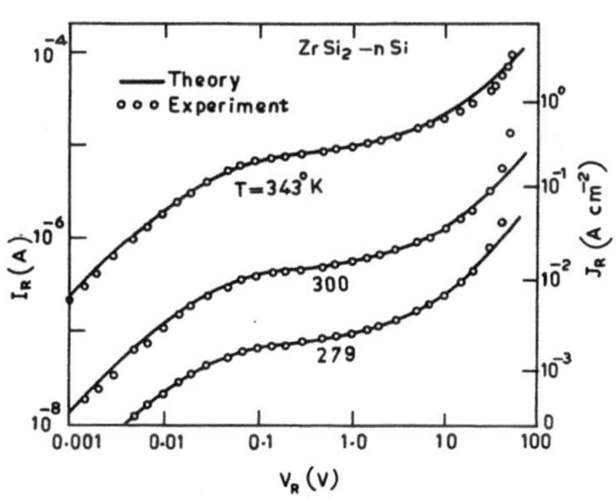

Figure 19. The reverse I–V characteristics of a $ZrSi_2$–Si Schottky barrier diode formed in n-type silicon at different temperatures (after Andrews and Lepselter[94]).

have observed a large barrier lowering even in metal–silicide diodes having no interfacial oxide layer (type 3 contact). The reverse current of these diodes shown in Fig. 19 can be explained by assuming a barrier lowering of the form $\Delta\phi_B = \alpha \mathscr{E}m$ with values of α lying in the range of 15 to 35 Å. This simple model is also found to explain the reverse current in Al–Si Schottky diodes made on relatively high resistivity *n*-type silicon.[10] In type 4 contacts the $\alpha\mathscr{E}m$ dependence arises due to the presence of an interfacial layer. But in clean contacts and type 3 contacts such a relation may result either due to Heine tails[28] of metal electrons into forbidden gap of the semiconductor or due to interdiffusion effects at the metal–semiconductor boundary.[10]

Besides the barrier lowering, other mechanisms which cause the reverse current to increase with bias are tunneling of electrons from the metal into the semiconductor conduction band through the barrier and the electron–hole pair generation inside the depletion region. Tunneling can occur either as FE or as TFE and enters at lower doping levels for reverse bias case than for foward bias case. This is because a moderate reverse bias causes the potential barrier to become sufficiently thin near the top so that the electrons before reaching the top can be emitted into the semiconductor by TFE as shown in Fig. 14b. Near room temperature TFE in reverse biased Schottky barrier becomes important for $N_d > 10^{17}$ cm^{-3} in Si and for $N_d > 2 \times 10^{16}$ cm^{-3} in GaAs. The corresponding figures for forward bias are[10] $N_d > 5 \times 10^{17}$ cm^{-3} in the case of Si and $N_d > 10^{17}$ cm^{-3} in the case of GaAs. At higher concentrations in excess of 10^{18} cm^{-3} FE may occur at moderate values of reverse bias. In the range where TFE becomes important the reverse current can be written as[77]

$$I_R = I_s \exp(qV_R/E')$$

where

$$E' = E_0\left[\left(\frac{E_{00}}{kT}\right) - \tanh\frac{E_{00}}{kT}\right]^{-1} \tag{61}$$

In many devices TFE becomes particularly important near the edge of the metal contact where the electric field is enhanced because of junction curvature which reduces the barrier height in this region. A large current thus flows through the edges. This edge leakage is a common cause of soft I–V characteristics in Schottky barrier diodes. Edge leakage can be eliminated by using a guard ring.[95]

Electron–hole pair generation in the depletion region contributes a generation current directly proportional to the width of the depletion region W. Since W varies as $(V_i + V_R)^{1/2}$, the generation current increases as the square root of the reverse bias. Like the recombination current under forward bias, the generation current becomes important for large barrier heights, low temperatures, low dopant concentrations, and low values of carrier lifetime. Carrier generation in the depletion region is a common cause of lack of saturation of reverse current in GaAs and Si Schottky barrier diodes with $N_d < 10^{15}\,\mathrm{cm}^{-3}$ at room temperatures and below.

For large barrier heights well in excess of $E_g/2$ and low dopant concentrations the injection of minority carriers from the semiconductor into the metal may become important, as has been recently observed in metal–n-type Ge Schottky barriers.[85] However, this minority carrier current saturates to a constant value for reverse biases in excess of about $2kT/q$.

7. TRANSIENT BEHAVIOR

The diode current–voltage characteristics discussed in the previous section are obtained under static condition. This assumes that the voltage across the diode is varied slowly or one waits long enough for the carrier distributions in the various regions of the diode to reach steady state before making the measurements. When the diode voltage is changed suddenly the carriers require a finite time for adjusting to this change, and during this time the behavior of the device is different from its static behavior. The response time of the Schottky barrier to an applied voltage is much shorter than that of a p–n junction diode.

When a p–n junction diode carrying a steady forward current is reverse biased suddenly, the excess minority carriers injected in the neutral regions of semiconductor must be removed before the diode can respond to the reverse voltage. A forward biased Schottky barrier contains two types of excess carriers. Firstly, there are excess electrons injected from the semiconductor into the metal which have kinetic energy higher than the metal electrons and hence are "hot." If the diode is suddenly reverse biased these electrons can return back to the semiconductor. This, however, does not happen because on the average a hot electron after arriving into the metal loses about half of its excess kinetic energy during a mean free time which is less than 10^{-13} sec. After losing this energy the electron is no longer able to surmount the barrier. Thus, the hot electron storage into the metal does not limit the response time of a Schottky barrier.

The other form of excess carriers are the holes injected from the metal into the neutral region of the semiconductor. For Schottky barriers made on high-resistivity n-type Ge, the hole current dominates the electron current.[85] One would, therefore, expect the response time of Ge Schottky barrier diodes to be determined by the hole storage in the neutral semiconductor. However, for Si and GaAs Schottky barrier diodes the hole current forms a negligibly small fraction (of the order of 10^{-3}) of the total current. Consequently, the effect of hole storage on the transient response of the diode will be unimportant except for high-level injection.[86]

In practical Schottky barrier diodes the *RC*-time constant which is the product of the depletion region capacitance and the diode series resistance determines the response time. For Schottky barrier diodes on Si and GaAs this time constant is typically a few picoseconds, which is orders of magnitude lower than the switching time of *p–n* junction diodes. Thus, Schottky barrier diodes are capable of operation at much higher frequencies than the *p–n* junction diodes, and this is an important advantage of these devices.

8. LOW-RESISTANCE SCHOTTKY BARRIER CONTACTS

In order to measure the forward and the reverse *I–V* characteristics of a rectifying junction the contact on the other end of the semiconductor must be of low resistance. In fact, the contact resistance R_c should be small compared to the resistance of the semiconductor sample on which the contact is made. Such low-resistance contacts have been conventionally classified as ohmic contacts although they are nonohmic in the sense that they have nonlinear *I–V* characteristics.

A low-resistance metal contact to a semiconductor results when the barrier height is zero or is small compared to the thermal energy *kT*. If one could make Schottky contacts in which the difference between the metal work function and the semiconductor electron affinity alone determines the barrier height then it would have been possible to obtain a low barrier height by

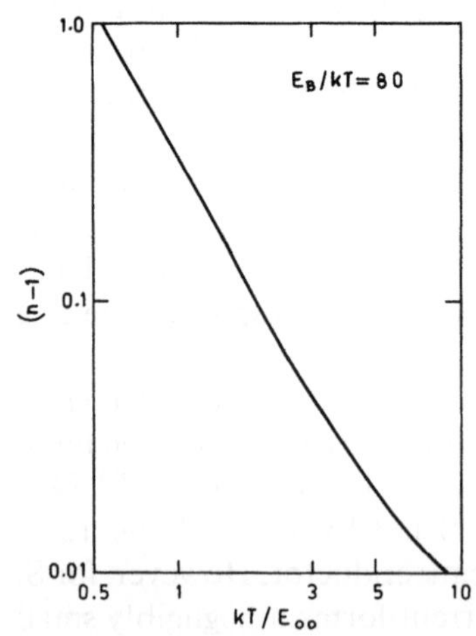

Figure 20. The predicted deviation of ideality factor *n* from unity versus kT/E_{00} (after Rideout[96]).

choosing a metal with appropriate work function. This concept is of hardly any use since, as we have seen, the barrier heights on most of the semiconductors are not controlled by the metal work function.

Low-resistance contact on a semiconductor most often occurs when the depletion layer contact barrier is reduced in thickness (and also in height) by heavy doping of the semiconductor. A sufficiently thin barrier (about 100 Å or less) is transparent to electrons in either direction and FE and TFE dominate the current transport. Thus, to make low-resistance contacts on a semiconductor, the semiconductor region just underneath the metal contact is doped so heavily that the depletion layer width is sufficiently reduced for the FE and TFE to dominate the current flow. Although the contact on heavily doped semiconductor is of low resistance its $I-V$ characteristic is still nonohmic. The tunneling theory of Section 6.1.2 applies to this type of contacts and deviation of the ideality factor n from unity can be used as a relative contribution of tunneling as shown in Fig. 20. Here it is assumed that the deviation of n from unity is due to tunneling and the effect of the image force, interfacial layer, and surface states, etc., on n is negligible. It has been pointed out[96] that in tunneling dominated metal–semiconductor contacts the ideality factor should also appear in the reverse current term and the current–voltage relation can be expressed as[80]

$$I = I_s\left\{\exp\left(\frac{qV}{nkT}\right) - \exp\left[\left(\frac{1}{n} - 1\right)\frac{qV}{kT}\right]\right\} \tag{62}$$

The $I-V$ characteristics predicted by this equation are shown in Fig. 21. For $n = 1$ Eq. (62) reduces to the ideal diode equation while for $n - 2$ the diode conducts better in the reverse direction than in the forward direction. Recent measurements on Schottky diodes made on amorphous Si have shown[97] that

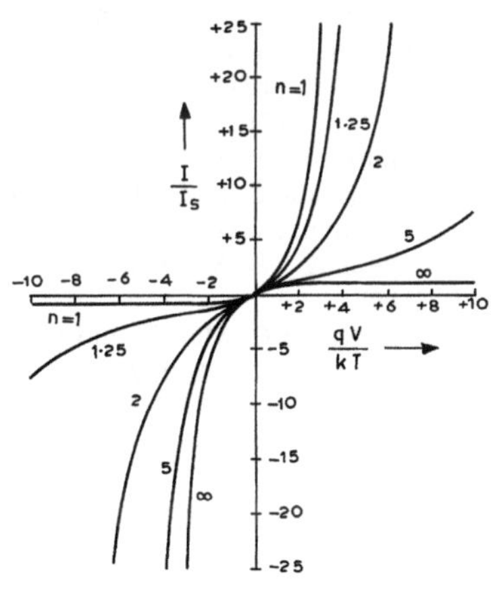

Figure 21. Normalized current voltage characteristics of low resistance depletion region contact predicted by Eq. (62) (after Rideout[96]).

Eq. (62) can be used to fit the experimental data only when n obtained under reverse bias differs from that obtained under forward bias.

A contact is generally characterized by its specific contact resistance which is the product of the contact resistance and the contact area. Contacts having a specific contact resistance of $10^{-3} \Omega \, cm^2$ are found to have reverse characteristics that are usually masked by the bulk resistance of the semiconductor.

ACKNOWLEDGMENTS. I would like to thank my students Mr. A.K. Gupta and Mr. A.V. Chaturvedi for many helpful discussions and for their help in the preparation of the manuscript. Thanks are also due to Mr. Rakesh Lal and to Dr. J. Narain for critical reading of the manuscript.

REFERENCES

1. F. Braun, Über die Stromleitung durch Schwefelmetalle, *Ann. Phys. Chem.* **153**, 556 (1874).
2. H.C. Torrey and C.A. Whitmer, *Crystal Rectifiers*, McGraw-Hill, New York (1948).
3. W. Schottky, R. Strömer, and F. Waibel, *Hochfrequenztechnik* **37**, 162–165 (1931).
4. W. Schottky, *Naturwissenschaften* **26**, 843 (1938).
5. N.F. Mott, Note on the contact between a metal and an insulator or semiconductor, *Proc. Camb. Phil. Soc.* **34**, 568–572 (1938).
6. H.A. Bethe, Theory of the boundary layer of crystal rectifiers, MIT Radiation Laboratory, Report 43–12 (1942).
7. H.K. Henisch, Rectifying *Semiconductor Contacts*, Clarendon Press, Oxford (1957).
8. A.G. Milnes and D.L. Feucht, *Heterojunctions and Metal–Semiconductor Junctions* (Chap. 6 and 7), Academic Press New York (1972).
9. E.H. Rhoderick, *Metal–Semiconductor Contacts*, Clarendon Press, Oxford (1978).
10. E.H. Rhoderick, Metal–semiconductor contacts, *IEEE Proc.* **129**(1), 1–14 (1982).
11. V.L. Rideout, A review of the theory, technology and applications of metal–semiconductor rectifiers, *Thin Solid Films* **48**, 261–291 (1978).
12. B.L. Sharma and S.C. Gupta, Metal–semiconductor Schottky barrier junctions and their applications, *Solid State Technol.* **23**(5), 97 (1980); and **23**(6), 90 (1980).
13. J. Bardeen, Surface states and rectification at a metal–semiconductor contact, *Phys. Rev.* **71**, 717–727 (1947).
14. C.A. Mead, Metal–semiconductor surface barriers, *Solid-State Electron.* **9**, 1023–1032 (1966).
15. M. Aven and C.A. Mead, Electrical transport and contact properties of low-resitivity *n*-type zinc sulphide crystals, *Appl. Phys. Lett.* **7**, 8–10 (1965).
16. L. Pauling, *The Nature of the Chemical Bond*, Cornell University Press, Ithaca, New York (1960).
17. S. Kurtin, T.C. McGill, and C.A. Mead, Fundamental transition in electronic nature of solids, *Phys. Rev. Lett.* **22**, 1433–1436 (1969).
18. J.M. Andrews and J.C. Phillips, Chemical bonding and structure of metal–semiconductor interfaces, *Phys. Rev. Lett.* **35**, 56–59 (1975).
19. G. Ottaviani, K.N. Tu, and J.W. Mayer, Interfacial reaction and Schottky barrier in metal–silicon systems, *Phys. Rev. Lett.* **44**, 284–287 (1980).

20. P.S. Ho, P.E. Schmid, and H. Föll, Stoichiometric and structural origin of electronic states at the Pd_2 Si–Si interfaces, *Phys. Rev. Lett.* **46**, 782–785 (1981).
21. J.L. Freeouf, G.W. Rubloff, P.S. Ho, and T.S. Kuan, Reactive Schottky-barrier formation: The Pd–Si interfaces, *J. Vac. Sci. Technol.* **17**, 916–925 (1980).
22. L.J. Brillson, Transition in Schottky barrier formation with chemical reactivity, *Phys. Rev. Lett.* **40**, 260–263 (1978).
23. D.C. Northrop and E.H. Rhoderick, The physics of Schottky barriers, in *Variable Impedance Devices* (M.J. Howes and D.V. Morgan, ed.), Chap. 2, pp. 37–73, John Wiley and Sons, New York (1978).
24. A.M. Cowley and S.M. Sze, Surface states and barrier height of metal–semiconductor systems, *J. Appl. Phys.* **36**, 3212–3220 (1965).
25. O. Wada, A. Majerfeld, and P.N. Robson, InP Schottky contacts with increased barrier height, *Solid-State Electron.* **25**, 381–387 (1982).
26. A.N. Daw, A.K. Dutta, and M.C. Ash, On the determination of the neutral level and charge density in the interfacial layer of a MIS diode, *Solid-State Electron.* **25**, 431–432 (1982).
27. S.M. Sze, *Physics of Semiconductor Devices*, John Wiley and Sons, New York (1981).
28. V. Heine, Theory of surface states, *Phys. Rev. A* **138**, 1689–1696 (1965).
29. W. Schockley, On the surface states associated with a periodic potential, *Phys. Rev.* **56**, 317–323 (1939).
30. A.J. Bennet and C.B. Duke, Self-consistent model of bimetallic interfaces I. Dipole effects, *Phys. Rev. B* **160**, 541–553 (1967); Metallic interfaces II., Influence of the exchange correlation and lattice potentials, *Phys. Rev. B* **162**, 578–590 (1967).
31. B. Pellegrini, A detailed analysis of the metal–semiconductor contact, *Solid-State Electron.* **17**, 217–237 (1974).
32. J.C. Inkson, Many-body effects at metal–semiconductor junctions II; The self-energy and band structure distortion, *J. Phys. C: Solid-State Phys.* **6**, 1350–1362 (1973).
33. J.C. Phillips, *J. Vac. Sci. Technol.* **11**, 947 (1974).
34. W.A. Harrison, Schottky barriers without mid-gap states, *Phys. Rev. Lett.* **37**, 312–313 (1976).
35. S.G. Louie and M.L. Cohen, Electronic structure of a metal semiconductor interface, *Phys. Rev. B* **13**, 2461–2469 (1976).
36. S.G. Louie, J.R. Chelikowsky, and M.L. Cohen, Theory of semiconductor surface states and metal–semiconductor interfaces, *J. Vac. Sci. Technol.* **13**, 790–797 (1976).
37. J.E. Rowe, S.B. Christman, and G. Margaritondo, Metal-induced surface states during Schottky-barrier formation on Si, Ge and GaAs, *Phys. Rev. Lett.* **35**, 1471–1474 (1975).
38. W.E. Spicer, P.W. Chye, P. Skeath, C.Y. Su, and I. Lindau, *J. Vac. Sci. Technol.* **15**, 1422 (1979).
39. L.J. Brillson, R.Z. Bachrach, R.S. Bauer, and J. McEnamin, Chemically-induced charge distribution at Al–GaAs interfaces, *Phys. Rev. Lett.* **42**, 397–401 (1979).
40. J. Van Laar and A. Huijser, Contact potential differences for III–V compound surfaces, *J. Vac. Sci. Technol.* **13**, 769–772 (1976).
41. P.W. Chye, T. Sikegawa, I.A. Babalola, H. Sunami, P.E. Gregory, and W.E. Spicer, Surface and interface states of GaSb: A photo-emission study, *Phys. Rev. B* **15**, 2118–2126 (1977).
42. W.E. Spicer, I. Lindau, P. Skeath, C.Y. Su, and P. Chye, Unified mechanism for Schottky-barrier formation and III–V oxide interface states, *Phys. Rev. Lett.* **44**, 420–423 (1980).
43. W.E. Spicer, *J. Vac. Sci. Technol.* **17**, 1019 (1980).
44. W. Mönch and H. Gant, Chemisorption-induced defects on GaAs (110) surfaces, *Phys. Rev. Lett.* **48**, 512–515 (1982).
45. A. Hiracki, A model for the mechanism of room temperature interfacial intermixing reactions in various metal–semiconductor couples, *J. Electrochem. Soc.* **127**, 2662–2665 (1980).

46. L.J. Brillson, R.S. Bauer, R.Z. Bachrach, and G. Hansson, Atomic interdiffusion at Au–AlGaAs interfaces, *Appl. Phys. Lett.* **36**, 326–328 (1980).
47. J.M. Palau, E. Testemale, A. Ismail, and L. Lessabatere, Silver Schottky-diodes on Kelvin, AES and LED characterized (100) surfaces of GaAs cleaned by ion bombardment, *Solid-State Electron.* **25**, 285–294 (1982).
48. A.M. Goodman, Metal–semiconductor barrier height measurement by differential capacitance method—One carrier system, *J. Appl. Phys.* **34**, 329–338 (1963).
49. R.H. Fowler, The analysis of photoelectric sensitivity curves for clean metals at various temperatures, *Phys. Rev.* **38**, 45–56 (1931).
50. R. Hackam and P. Harrop, Temperature dependence of the Schottky barrier height in gallium arsenide, *Solid State Commun.* **11**, 669–672 (1972).
51. M.S. Tyagi, Electrical properties of metal–GaAs Schottky barriers, *Surf. Sci.* **64**, 323–333 (1977).
52. A.M. Cowley, Depletion capacitance and diffusion potential of GaP Schottky barrier diodes, *J. Appl. Phys.* **37**, 3024–3032 (1966).
53. R. Mach, H. Treptow, and W. Ludwig, Physical properties of Au–ZnSe metal–semiconductor contacts, *Phys. Status Solidi (a)* **25**, 567–573 (1974).
54. M.S. Tyagi and S.N. Arora, Metal–ZnSe Schottky barriers, *Phys. Status Solidi (a)* **32**, 165–172 (1975).
55. M.J. Turner and E.H. Rhoderick, Metal–silicon Schottky barriers, *Solid State Electron.* **11**, 291–309 (1968).
56. A. Thanailakis, Contacts between simple metals and atomically clean silicon, *J. Phys. C: Solid State Phys.* **8**, 655–668 (1975).
57. A. Thanailakis and A. Rasul, Transition-metal contacts to atomically clean silicon, *J. Phys. C: Solid State Phys.* **9**, 337–343 (1976).
58. W.G. Spitzer and C.A. Mead, Barrier height studies on metal semiconductor systems, *J. Appl. Phys.* **34**, 3061–3069 (1963).
59. R.K. Swank, M. Aven, and J.Z. Devine, Barrier heights and contact properties of *n*-type ZnSe crystals, *J. Appl. Phys.* **40**, 89–97 (1969).
60. L.J. Brillson, G. Margaritondo, and N.G. Stoffel, Atomic modulation of interdiffusion at Au–GaAs interfaces, *Phys. Rev. Lett.* **44**, 667–670 (1980).
61. J.D. McCaldin, T.C. McGill, and C.A. Mead, Correlation for III–V and II–VI semiconductors of the Au Schottky barrier energy with anion electronegativity, *Phys. Rev. Lett.* **36**, 56–61 (1976).
62. M.L. Cohen, Electrons at interfaces, in *Advances in Electronics and Electron Physics* (L. Marton and C. Marton, Eds.), Vol. 51, pp. 1–62, Academic Press, New York (1980).
63. M. Schlüter, Chemical trends in metal–semiconductor barrier heights, *Phys. Rev. B* **17**, 5044–5047 (1978).
64. J. Hilibrand and R.D. Gold, Determination of impurity distribution in junction diode from capacitance voltage measurements, *RCA Rev.* **21**, 245–252 (1960).
65. P.J. Baxandall, D.J. Colliver, and A.F. Fray, An instrument for the rapid determination of semiconductor impurity profiles, *J. Phys. E: Sci. Inst.* **4**, 213–221 (1971).
66. D.P. Kennedy, P.C. Murley, and W. Kleinfelder, On the measurement of impurity atom distributions in silicon by the differential capacitance technique, *IBM J. Res. Dev.* **12**, 399–409 (1968).
67. M. Nishida, Depletion approximation analysis of the differential capacitance–voltage characteristics of an MOS structure with nonuniformly opened semiconductors, *IEEE Trans. Electron Devices* **ED-26**, 1081–1085 (1979).
68. M.A. Green, The capacitance of large barrier Schottky diodes, *Solid-State Electron.* **19**, 421–422 (1976).

69. L.C. Kimerling, Influence of deep traps on the measurement of free-carrier distributions in semiconductors by junction capacitance techniques, *J. Appl. Phys.* **45**, 1839–1845 (1974).
70. D.V. Lang, Deep level transient spectroscopy: A new method to characterize traps in semiconductor, *J. Appl. Phys.* **45**, 3023–3032 (1974).
71. G.L. Miller, D.V. Lang, and L.C. Kimerling, Capacitance transient spectroscopy, *Ann. Rev. Mater. Sci.* 377–448 (1977).
72. C. Wagner, Theory of current rectifiers, *Phys. Z.* **32**, 641–645 (1931).
73. W. Schottky and E. Spenke, Quantitative treatment of the space charge and boundary-layer theory of the crystal rectifier, *Wiss. Veroff. Siemens—Werken* **18**, 225–291 (1939).
74. C.R. Crowell and S.M. Sze, Current transport in metal–semiconductor barriers, *Solid-State Electron.* **9**, 1035–1048 (1966).
75. J.M. Wilkinson, J.D. Wilcock, and M.E. Brinson, Theory and experiment for silicon Schottky-barrier diode at high-current density, *Solid-State Electron.* **20**, 45–50 (1977).
76. G. Baccarani, Current transport in Schottky barrier diodes, *J. Appl. Phys.* **47**, 4122–4126 (1976).
77. F.A. Padovani and R. Stratton, Field and thermionic-field emission in Schottky barriers, *Solid-State Electron.* **9**, 695–707 (1966).
78. C.R. Crowell and V.L. Rideout, Normalized thermionic-field emission in metal–semiconductor barriers, *Solid-State Electron.* **12**, 89–105 (1969).
79. F.A. Padovani, The voltage–current characteristic of metal–semiconductor contacts, in *Semiconductors and Semimetals*, Academic Press, New York, Vol. 7, pp. 75–146 (1971).
80. V.L. Rideout and C.R. Crowell, Effects of image force and tunneling on current transport in metal–semiconductor (Schottky barrier) contacts, *Solid-State Electron.* **13**, 993–1009 (1970).
81. C.Y. Chang and S.M. Sze, Carrier transport across metal–semiconductor barriers, *Solid-State Electron.* **13**, 727–740 (1970).
82. A.S. Grove, *Physics and Technology of Semiconductor Devices*, John Wiley and Sons, New York (1967).
83. A.Y.C. Yu and E.H. Snow, Surface effects on metal–silicon contacts, *J. Appl. Phys.* **39**, 3008–3016 (1968).
84. A.Y.C. Yu and E.H. Snow, Minority carrier injection of metal–silicon contacts, *Solid-State Electron.* **12**, 155–160 (1969).
85. D.A. Buchanan and H.C. Card, On the dark current in germanium Schottky-barrier photodetectors, *IEEE Trans. Electron Devices* **ED-29**, 154–157 (1982).
86. D.L. Scharfetter, Minority carrier injection and charge storage in epitaxial Schottky barrier diodes, *Solid-State Electron.* **8**, 299–311 (1965).
87. A.N. Saxena, Forward current–voltage characteristics of Schottky barriers on *n*-type silicon, *Sur. Sci.* **13**, 151–171 (1969).
88. G.S. Visweswaran and R. Sharan, Current transport in large area Schottky diodes, *Proc. IEEE* **67**, 436–437 (1979).
89. J.D. Levine, Schottky barrier anomalies and interface states, *J. Appl. Phys.* **42**, 3991–3999 (1971).
90. C.R. Crowell, The physical significance of the T_0 anomalies in Schottky barriers, *Solid-State Electron.* **20**, 171–175 (1977).
91. E.H. Rhoderick, A note on Levine's model of Schottky barriers, *J. Appl. Phys.* **46**, 2809 (1975).
92. J.D. Levine, Power Law reverse current–voltage characteristic in Schottky barriers, *Solid-State Electron.* **17**, 1083–1086 (1974).
93. M. Hirose, N. Altaf, and T. Arizumi, Contact properties of metal–silicon Schottky barriers, *Jpn. J. Appl. Phys.* **9**, 260–264 (1970).
94. J.M. Andrews and M.P. Lepselter, Reverse current–voltage characteristics of metal–silicide Schottky diodes, *Solid-State Electron.* **13**, 1011–1023 (1970).

95. M.P. Lepselter and S.M. Sze, Silicon Schottky barrier diode with near-ideal $I-V$ characteristics, *Bell Syst. Tech. J.* **47**, 195–208 (1968).
96. V.L. Rideout, A review of the theory and technology for ohmic contacts to group III–V compound semiconductors, *Solid-State Electron.* **18**, 541–550 (1975).
97. M.P. Shaw, Metal–Semiconductor junctions, in *Handbook on Semiconductors*, Vol. 4, Chap. 1, pp. 50–85, North-Holland Publishing Company, Amsterdam (1981).
98. E. Testemale, J.M. Palau, A. Ismail, and L. Lassabatere, Properties of the contact on ion-cleaned *n*- and *p*-type silicon surfaces, *Solid-State Electron.* **26**, 325–331 (1983).

2

Interface Chemistry and Structure of Schottky Barrier Formation

R.Z. Bachrach

1. INTRODUCTION

The current microscopic understanding of the interface chemistry and structure of Schottky barrier formation is described in this chapter. This chapter builds upon the physics of Schottky barriers presented in Chapter 1 and provides some background to the fabrication of Schottky barriers described in Chapter 3. The more recent investigations are discussed and the current status of the detailed picture we now have for metal–semiconductor barriers on group IV, III–V, and II–VI semiconductors is reviewed. In particular, this chapter emphasizes the pervasive importance of interfacial interdiffusion and interactions in metal–semiconductor systems. As will become apparent from the discussion below, many intrinsic aspects of Schottky barrier formation have been elucidated, but much detail remains to be unraveled in specific metal–semiconductor systems.

Rectification at metal–semiconductor interfaces has been a subject of much interest over the last few decades and significant work has gone into forming a microscopic understanding of the origin of what is commonly referred to as a Schottky barrier. The occurrence of these phenomena has been known for a long time and the initial insight into the nature of the electrical barrier is due to Schottky[1] and Mott,[2] who considered the role of a single phase of metal joined to a single phase of semiconductor. In most cases, this

R.Z. Bachrach • Xerox Palo Alto Research Center, 3333 Coyote Hill Road, Palo Alto, California 94304.

description was not found to apply and Bardeen[3] proposed the presence of interface states as the explanation. The essential insight provided by Bardeen was that real states at the interface residing in the gap can determine the resultant barrier height. In 1947, the microscopic nature of these states was not known and they were generally referred to as surface states, but in current terminology they are termed interface states.

Over the years, great effort has gone into providing a systematic understanding of this seemingly simple system. Complete determination of the detailed mechanisms responsible for Schottky barriers has proven quite elusive. Perhaps the best explanation for this is that there is not one single mechanism which will account for all the different material situations encountered. Although there are unifying themes to the microscopic description of Schottky barrier formation, there are in fact a number of different mechanisms which come into play at the interface which determine the macroscopic electrical barrier when all the scope of metals and semiconductors are considered.

Much of the impetus for renewed research work on Schottky barriers in the last ten years resulted from the application of advanced vacuum preparation techniques and the application of surface sensitive spectroscopies both of which created new opportunities. The application of these new tools over the last decade has helped to elucidate the nature and extent of the interfacial regions. Even during room-temperature formation of the metal–semiconductor interface, interdiffusion and interfacial reactions are important and get accentuated at elevated temperatures. The two phenomena of metal–semiconductor interdiffusion and interfacial reactions are pervasive in most systems, and therefore specifying the extent and nature of this interfacial region is essential. An understanding of the details of the interfacial region is the major feature that was missing from the early work.

In addition to the advances in measurement techniques and the need for technological control, the landmark paper of Kurtin, Mead, and McGill[4] in 1969 sparked significant interest in discovering the unifying principles underlying the characteristics of the interfaces and specification of the systematics which characterizes much of the work carried out in the 1970s.

Many groups have contributed to the detailed understanding of Schottky barriers that can be set forth today and that will be described below. This work built upon extensive studies during the fifties and sixties. A good overview of this prior work can be obtained from several books.[5–8] Brillson[9] has presented a broad review of more recent work on metal–semiconductor interfaces. In this chapter, the discussion is restricted primarily to specific aspects of work that has been done investigating the intrinsic microscopics of Schottky barrier formation.

2. PERSPECTIVES ON SCHOTTKY BARRIER FORMATION

2.1. Introduction

A Schottky barrier is a limiting situation which can be described as two infinite half planes of material, one a metal and the other a semiconductor, brought into contact. Distinguishing traditional technical barriers from ideal barriers is useful. A typical technical barrier would result from contacting metal on the semiconductor after a series of in-ambient preparations. An ideal barrier results from depositing the metal in a carefully controlled way where precautions are taken to keep the interface atomically clean—i.e., the only atoms present are those of the initial semiconductor and the desired metal. The procedure for cleaning the surface may strongly influence the outcome by altering the surface structure stoichiometry or by introducing surface defects.

In making a Schottky barrier interface, the metal is typically deposited onto the semiconductor. Two regimes are

a. building up the metal layer by layer by evaporative deposition;
b. pressing two bulk pieces together such as in forming a point contact.

In both these cases, the technical interface would have some amount of extraneous material trapped at the interface. In actual situations, this extraneous material might consist of a native oxide of 10–15 Å thickness with some carbonaceous overlayer of a couple of monolayers. Nicollian and Sinha[10] have considered in detail the special aspects of technical barriers with, for example, thin interfacial oxide layers, and the reader is referred to their review for more detail.

With the widespread use of MBE growth techniques[11,12] or general *in situ* cleaning, the general distinction between technical and ideal barriers is disappearing and this chapter is only addressing aspects of ideal barriers. Ideal barriers were often called intimate contacts in previous literature. In general, Schottky barrier interfaces are not ordered, but with silicide regrowth techniques or molecular beam epitaxy it is possible to create interfaces which are both atomically clean and ordered. Examples are aluminum–GaAs(100)[13] and $NiSi_2$-Silicon[14] treated in Section 6. These interfaces present particularly interesting opportunities for separating the mechanisms controlling the barrier height.

Although the principal conception of the Schottky barrier is a two-component contact, Fig. 1, reproduced from Ref. 5, with a generalized interfacial region represents that actual system more completely and defines a

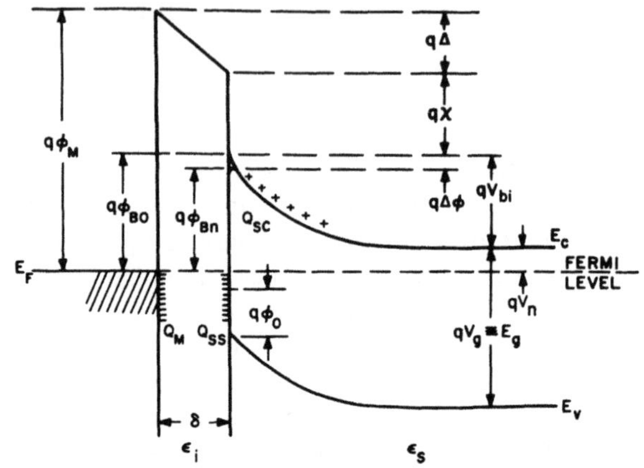

Figure 1. Schematic representation of Schottky barrier with symbols defined for an *n*-type semiconductor (after Sze, Ref. 5). Key; ϕ_M, work function of metal; ϕ_{bn}, barrier height of metal–semiconductor barrier, ϕ_{b0}, asymptotic value of ϕ_{bn} at zero electric field; ϕ_0, energy level at surface; $\Delta\phi$, image force barrier lowering; Δ, potential across interfacial layer; χ, election affinity of semiconductor, V_{bi}, built-in potential; ϵ_s, permittivity of semiconductor; ϵ_i, permittivity of interfacial layer; δ, thickness of interfacial layer; Q_{sc}, space–charge density in semiconductor; Q_{ss}, surface-state density on semiconductor; Q_M, surface-charge density on metal.

number of parameters needed to characterize the barrier. As outlined below, the interfacial layer has a number of components. The presence of the interface component phases can arise from a number of effects, but even for ideal barriers where the preparation technique provides an atomically clean interface, the inherent interactions can create new phases. The application over the last decade of the new tools discussed in Section 3 has helped to elucidate the nature and extent of the interfacial regions.

The diagram of Fig. 1 can in fact be used to characterize any of the material–semiconductor contacts such as semiconductor–semiconductor and metal–insulator–semiconductor. Using this approach, the difference in these cases depends upon how much freedom exists in specifying the electric field at the interface.

In addition to the intrinsic aspects of this interface region it is possible to tailor barriers to desired characteristics with modern techniques such as MBE. Variation of composition or doping can be made on an atomic scale in ways

TABLE 1. Interface Formation Regimes

a. Metal deposition onto cleaved surfaces
b. Metal deposition onto sputtered surfaces
c. Metal deposition onto sputtered and reannealed surfaces
d. Single crystal metal growth on *in situ* grown semiconductors

that would not be possible with diffusion profiles. Full discussion of the range of specially prepared profiles is beyond the scope of this chapter, but the concepts will be understandable from the discussion presented here. These modified barrier structures are the subject of an extensive literature and have significant applications (See Ref 9, Section 6.4).

The various interface formation regimes pertinent to the discussion of ideal barriers in this chapter are listed in Table 1.

2.2. Brief Review of Phenomenological Schottky Barrier Data

In order to set the background for the discussion of this chapter, we introduce some of the systematics that have been obtained in studying Schottky barrier heights. We are focusing on data that are likely to represent the ideal barrier case, i.e., presumably two-phase interfaces created in an atomically clean manner. More general situations and an extensive review of data are presented elsewhere in the book.

The largest amount of pertinent data is that produced by deposition onto cleaved surfaces.[15,16] Even in these cases, there are interfacial aspects particular to the cleavage process which are important. The second class of emerging data results from metal deposition and *in situ* grown semiconductor surfaces which provide examples where intrinsic aspects are exhibited.

Figure 2 shows barrier height data obtained from the literature plotted versus metal electronegativity. Shown are representative data for Si, GaAs, ZnSe, and ZnS.[15–17] In well-controlled situations, one sees that for a given semiconductor, the barrier heights reasonably follow a linear relationship with respect to the metal electronegativity, X.

Note that for the metals shown Si and GaAs are strongly in the Bardeen

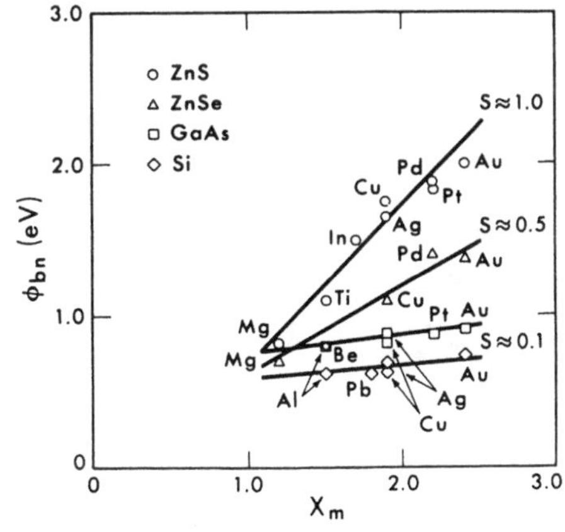

Figure 2. Barrier height plotted versus metal electronegativity for Si, GaAs, ZnSe, and ZnS. The corresponding slope S is indicated (after Louie *et al.*, Ref. 17).

limit. Rowe *et al.* for Si[18] (Fig. 3a) and Lindau *et al.*[19] for GaAs (Fig. 3b) have reproduced the trend. As originally found by Scheer and van Laar[20] for GaAs, these works showed that the Fermi level pinning position is essentially but not completely established at the 0.1–1 monolayer level.

In a paper in 1965, Crowley and Sze[21] derived a model expression for Schottky barriers based on the representation of Fig. 1 which reproduces the Schottky and Bardeen limits. This expression is given by

$$\phi_b(m, s) = S(s)^* X_m + \phi_0(s) \tag{1}$$

with

$$S(s) = \epsilon_i/(\epsilon_i + q^2 \delta D_s) \tag{2}$$

$$\phi_0(s) = \frac{E_g}{q} - \frac{S(s)X + c_3 + \Delta\phi}{1 - S(s)} \tag{3}$$

From these equations, D_s, the density of interface states, is given by

$$D_s = [1 - S(s)]\epsilon_i/S(s)\delta q^2 \quad \text{states cm}^{-2}\,\text{eV}^{-1} \tag{4}$$

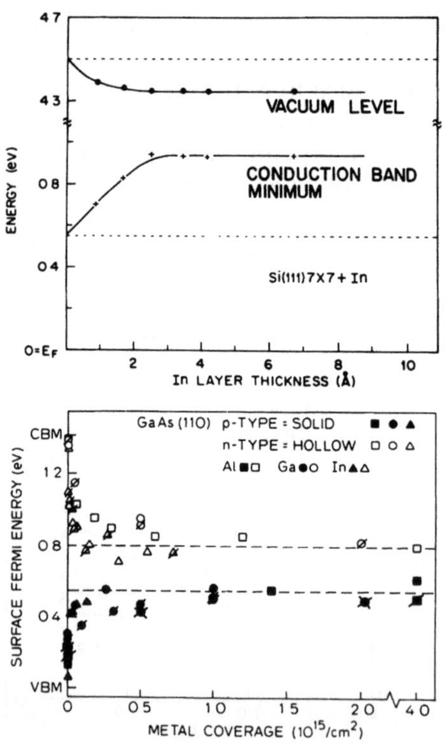

Figure 3. Fermi level position versus metal coverage for (top) Si (after Rowe, Ref. 18) and (bottom) GaAs (after Lindau *et al.*, Ref. 19).

Note that D_s does not necessarily refer to a descrete level and in many cases a continuous distribution will occur. Also note that the metal electronegativity is linearly related to the metal work function.[17,22] The lumped parameter $\phi_0(s)$ represents the center of gravity of the state distribution. As an example, ϕ_0 for Au Schottky barriers is usually found to be one third of the band gap.

The model reasonably describes much of the data shown, and Crowley and Sze derived the parameters for several cases from available data, including the number of states likely to be present at the interface. These results based upon early data are shown in Table 2.

The results of this phenomenological model depicted in Fig. 1 reduce the problem of understanding Schottky barriers to describing the mechanisms that produce an interfacial layer and the mechanisms that produce real states Q_{ss} either in the interfacial layer or in the semiconductor. Real situations are then in general combinations of these two cases. The states Q_{ss} essentially provide an interface dipole that leverages the semiconductor space-charge region and makes it independent of the metal. The width of the dipole is not as important as the changes it induces in the space-charge region. In his paper, Bardeen provided considerable perspective on a variety of cases that arise.

With respect to Fig. 1, the minimum interfacial region is determined by the layers from which the interface is derived. For interface states derived from intrinsic surface states, this would be approximately the dimension of the outermost layer on the semiconductor. Based on current work, one can say that Q_{ss} is comprised of several components.

How many states are required to have a significant effect? One finds in general that $q^2\delta/\epsilon_i$ is on the order of 10^{+13}cm^{-2}. Thus to be in the Schottky limit, $D_s \leqq 10^{+12}$. A typical semiconductor surface monolayer consists of 5–10 × 10^{+14}cm^{-2} so that very few states relative to the surface are necessary to determine the barrier height, and even if the number goes to one per surface unit cell, no additional change results. If this interface density were carried

TABLE 2. Schottky Barrier Parameters Deduced by Cowley and Sze[a]: Summary of Barrier Height Data and Calculation for Si, GaP, GaAs, and Cds

Semiconductor	c_2	c_3(V)	χ(V)	$D_s \times 10^{-13}$ (eV^{-1}cm^{-2})	$q\phi_0$(eV)	$q\phi_0/E_g$
Si	0.27 ± 0.05	−0.55 ± 0.22	4.05	2.7 ± 0.7	0.30 ± 0.36	0.27
GaP	0.27 ± 0.03	−0.01 ± 0.13	4.0	2.7 ± 0.4	0.66 ± 0.2	0.294
GaAs	0.07 ± 0.05	+0.49 ± 0.24	4.07	12.5 ± 10.0	0.53 ± 0.33	0.38
CdS	0.38 ± 0.16	−1.20 ± 0.77	4.8	1.6 ± 1.1	1.5 ± 1.5	0.6

[a] After Sze, Ref. 5.

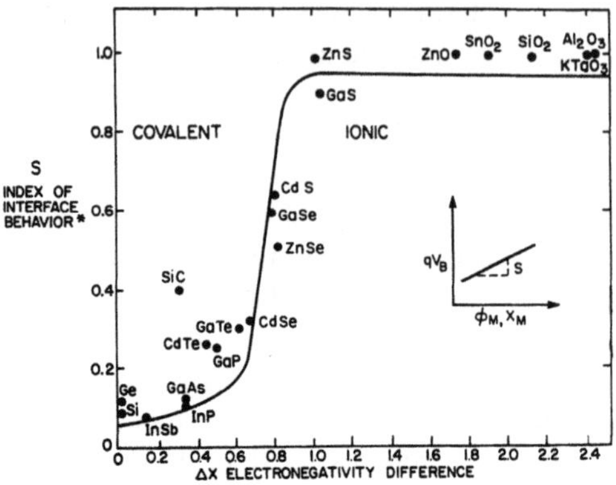

Figure 4. Plot of slope index S versus electronegativity difference for various semiconductors. The inset shows the derivation from the original data such as in Fig. 2 (after Kurtin *et al.*, Ref. 4).

over to a bulk density, the effective doping would be $10^{+20}\,\text{cm}^{-3}$, or well into the degenerate regime for all semiconductors.

The phenomenological model of Crowley and Sze[21] provides a good basis for discussion of Schottky barriers and a basis for understanding the transition between the various regimes. Given this background, the challenge in deriving a further microscopic understanding is to develop the materials systematics. These systematics have often been cast in a number of ways as described below.

Using a linear formulation of Schottky barrier height equivalent to Eq. (1), Kurtin, McGill, and Mead[4] showed a further relationship for the slope parameter, S. Figure 4 plots S versus ΔX, the semiconductor constituent electronegativity difference which has the aspect of a transition region. Brillson[23] has shown that the same form results if instead one plots versus ΔH_f, the relative heat of formation, as used by Andrews and Phillips[24] to characterize silicide–Si barriers.

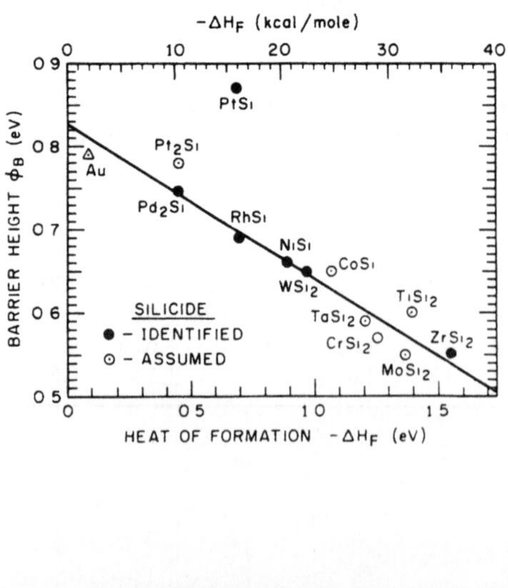

Figure 5. Metal silicide–Schottky barrier height versus relative heat of formation for a variety of silicide forming metals on silicon (after Andrews and Phillips, Ref. 24).

A number of reevaluations of the extant barrier height data have been performed which have confirmed the general trend as originally presented. In a reevaluation of the data, Schluter[25] has shown that in fact the Schottky limit is not reached and that more likely the data for more electronegative semiconductors continues to rise after the transition rather than reaching the Schottky limit at $S = 1$. Cohen[22] has shown that the likely limit for the Schottky case would be $S = 1.5$.

The primary conclusion that can be drawn from the relationship shown in Fig. 4 is that metal reactivity has less impact on the barrier height as the ionicity increases. Another way to state this is that the covalent semiconductors are less stable against interreaction with metals than ionic semiconductors.

In addition to the phenomenological trends seen with this presentation, Andrews and Phillips showed that a plot of silicide–Si Schottky barrier

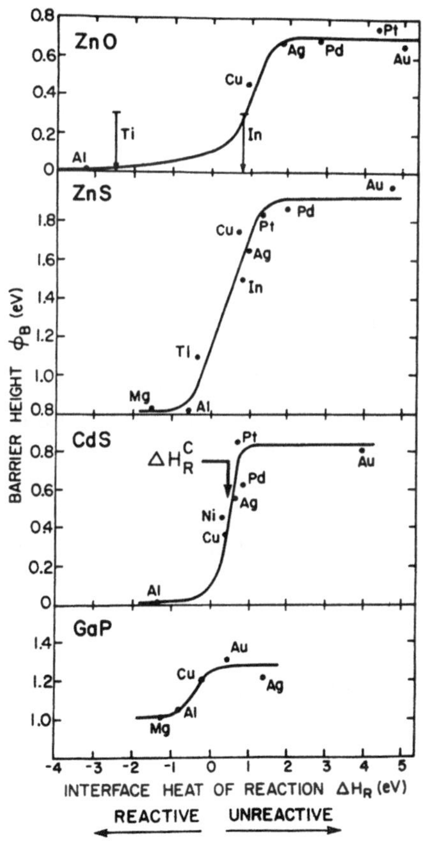

Figure 6. Barrier height versus interface heat of reaction plotted for different metals on specific semiconductors (after Brillson, Ref. 27).

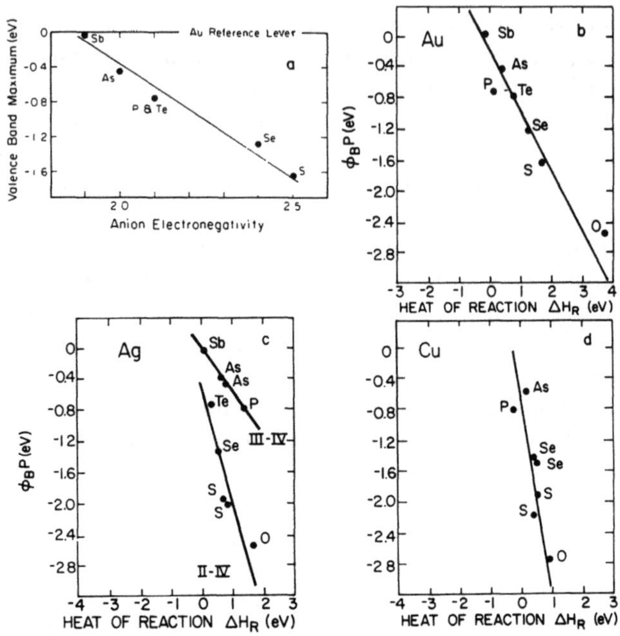

Figure 7. The common anion rule showing (a) barrier height for gold Schottky barriers versus anion electronegativity (after McCauldin *et al.*, Ref. 28) and (b)–(d) plotted for specific metals versus semiconductors with common anions (after Brillson, Ref. 27).

heights followed an approximately linear relationship with respect to the relative heat of formation as shown in Fig. 5. If one used the eutectic temperature as done by Ottiavani,[26] one achieved a similar presentation. Brillson[9,27] pursued these type of relationships for Schottky barriers on III–V and II–VI compound semiconductors and found the results summarized in Fig. 6 and 7b–d. These data for metals on *p*-type semiconductors showed trends in barrier height which depended on the interface heat of reaction. In the case of Fig. 7a, these trends are similar to those deduced by McCaldin *et al.* regarding the relationship to the anion electronegativity.[28] The major conclusion one draws from these data is the importance of interfacial interactions.

3. THE CHEMISTRY AND STRUCTURE OF THE INTERFACIAL LAYER

From the previous discussion we have seen that the real focus for the discussion of the microscopics of Schottky barriers is understanding the mechanisms which determine the interfacial layer. The interplay of a variety of effects then determines the interfacial composition, interfacial width, and the number of interfacial states. The complexity of the general problem arises from the diversity of concurrent effects that can be at work in specific material situations.

Contemporary Schottky barriers are fabricated with the layer-by-layer approach, or the deposition of the metal upon the semiconductor, and this will be the focus of the remaining discussion. In the discussion in this chapter, we are not considering interfacial phenomena due to extraneous or intentional additional material at the interface. Rather, the intrinsic aspects of the interfacial layer formation in an initially two-phase system and the formation of localized states are elucidated.

The microscopic probes that will be described utilize the control that can be achieved in performing the layer-by-layer fabrication, interrupting the development at a given point, and then studying the interface before depositing more material. A second type of approach to studying Schottky barriers results from the application of probes where material is sliced away layers at a time and then studied. These reverse layer-by-layer studies can have the problem that the etching techniques often mix the material on a microscopic scale.

The problem one faces in interpreting these layer-by-layer studies is that the Schottky barrier is a macroscopic property and the intermediate observations do not always lead one to the description of the end point. Understanding the thickness-dependent evolution is interesting and the knowledge of the evolution can also lead to intervention for the purposes of achieving better control. One should keep in mind, however, the metal thickness dependence of the barrier evolution and the subtantial changes that take place at various stages of the interface formation.[29]

Before treating the details of the barrier evolution stages, it is worth describing the overall process to motivate the division into the stages enumerated in Table 3 into which the evolution can be decomposed. The tools

TABLE 3. Enumeration of the Layer-by-Layer Stages of Schottky Barrier Formation

0. Clean surface	2. Metal nucleation
a. Ordered	a. Eutectic formation
b. Disordered	b. Compound formation
1. Dilute limit	c. Interdiffusion and interaction
a. Chemisorption	3. Asymptotic overlayer
b. Reaction	

that have been used to understand the stages are then briefly described. Following expansion about the details of the stages, a case study approach will fill in further details.

3.1. Synopsis of the Layer-by-Layer Evolution

Following the preparation of the clean surface, the initial state of the semiconductor surface can have many different configurations which will affect the outcome of the Schottky barrier that forms. The state can consist of either structural or compositional disorder reflected in structural defects such as steps, surface vacancies, or antisite arrangements. In addition to aspects of structural ordering, surface states can be generated by the structural defects.[30,31] Many examples exist, however, where the clean surface can be prepared in a well-ordered state relatively free from defects.[32–34]

In the initial metal deposition, the density of surface atoms, which can be determined accurately, is low enough that they do not interact. In this case, one can identify site-specific chemisorption as well as reaction or replacement. Energy released at this stage such as the heat of condensation or reaction may dislodge other surface atoms from their equilibrium sites. The outcome of this stage may depend upon the surface temperature either through surface diffusion or through activated processes.

As the surface concentration grows, the probability of metal–metal interactions becomes a significant aspect. Depending upon the surface mobility and the strength of the interaction, the metal may have a tendency to form clusters. For many metals, the cluster formation or metal film nucleation is exothermic, so this stage can disrupt the underlying lattice.

Beyond the development of the initial metal overlayer, depending on the configuration of the metal, the metal film may be under a significant stress (either tensile or compressional), which can be a driving force for interdiffusion and at some point dislocation formation. Stress, dipole formation, and other phenomena drive interdiffusion which can proceed over tens of angstroms even at room temperature. The final interface may be compositionally and spatially inhomogeneous so that there is significant dimension to the situations that need to be evaluated in detail. These effects can be accentuated with the thermal processing that usually accompanies device processing.

3.2. Some Techniques for Studying the Stages of Interface Formation

A number of techniques have evolved which can provide significant information about the microscopics of interface formation during the layer-by-layer evolution. The primary ones are either electron or ion spectroscopies that rely on the short range to enhance surface contrast. Surface photovoltage, Kelvin probe, and Raman scattering measurements have also provided significant information. Full treatment of these measurement techniques is beyond the scope of this chapter, but the brief description will lead the reader to references for more detail. Reference 9 provides an expanded description of many of these techniques as applied to metal–semiconductor interfaces.

The techniques of low-energy electron diffraction and reflection high-energy diffraction LEED[35] and RHEED[36] provide information on surface ordering. Surface extended x-ray absorption fine structure, EXAFS,[37] provides information on local bonding and coordination. Electron energy loss[38] provides information on local vibrations and transitions to unoccupied states. Rutherford backscattering,[39] ion scattering,[40] and small-angle x-ray diffraction[41] provide significant information about both the surface and the overall interfacial layer.

High-resolution transmission electron microscopy, TEM, with its various forms such as lattice imaging has allowed the local aspects of the interface to be investigated directly.[42] In many cases, one can see directly the abruptness of interfaces as well as the extent of inhomegenieties.

Depth profiling Auger and Auger spectroscopy[43] as well as direct image mass analysis has allowed the compositional distribution of the interfaces to be explored.

Various capacitive and photovoltage spectroscopies are quite useful in probing the states in the gap directly.[44] These techniques can be coupled with Kelvin probe measurements to determine the surface work function as the interface develops. They have been developed to the point that surface images can be formed and spatial effects related to structural damage.

A particularly effective technique for exploring interface development has been the application of core and valence band synchrotron radiation excited photoemission spectroscopy.[45] Much of this work has been enabled by the development of a high-performance soft x-ray monochromator first implemented at the Stanford Synchrotron Radiation Laboratory by F.C. Brown *et al.*[46] This monochromator allowed the experiments to be performed where the electron escape depths were tuned to be a minimum, thus enhancing the surface and interface contrast. By varying the escape depth with photon energy, distribution effects could also be elucidated. This technique allowed both bonding and interdiffusion to be explored. Many of the studies described below used this technique.

4. EVOLUTION OF THE INTERFACIAL LAYER

The details of the formation of the interfacial layer are focused upon in this section without concurrently emphasizing the processes which generate

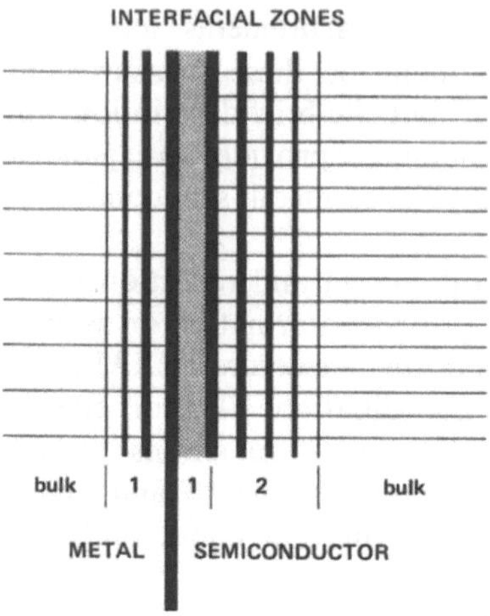

Figure 8. Schematic depiction of metal–semiconductor interfacial zones discussed in text. All of the zones do not necessarily develop in specific situations or with sharp boundaries.

interface states. This separation better allows the individual aspects to be identified, but should not imply a decoupling of the two phenomena.

The outcome of the processes discussed here is an interfacial layer of some total width. The interface layer can have a multiplicity of zones with a division between metallic and semiconducting regions as depicted schematically in Fig. 8.

On the semiconducting side, the interfacial layer has two zones. The first zone lies within the evanescent tail of the metal as originally defined by Heine.(47) The second zone is the remaining region where due to interdiffusion a composition or doping different from the original bulk semiconductor exists.

A similar description can be characterized on the metal side and the new alloyed metal zone may be of sufficient width to become the metal forming the barrier. This new interfacial metal can have different characteristics from the originally deposited metal.

4.1. State 0: The Clean Semiconductor Surface

The layer-by-layer evolution starts from the clean semiconductor surface. The literature bearing on this subject spans the full spectrum of surface physics experimental and theoretical technique, and this subject is still one of active investigation. The limited review presented here should put forth the essential aspects of the current models for these semiconductor surface structures. By choice, aspects of the structure of clean metal surfaces have not been included in the discussion since its importance to the barrier height problem has not been established.

The clear semiconductor surfaces can be divided into cases of ordered and disordered regimes. In the case of cleaved surfaces, disordering usually results from structural damage induced by the cleaving process employed. Huijer and van Laar found that the pinning of the Fermi level on GaAs after cleavage was dependent upon the number of steps.[48]

The noncleavage faces are typically cleaned by a sputtering technique[49] which usually leaves the surface disordered and in the case of compound semiconductors, nonstoichiometric. Thermal annealing will usually restore the ordering. The most perfect noncleavage surfaces are those grown *in situ* with MBE techniques.[11]

Some work has been performed investigating the effect of reordering on barrier height following sputter cleaning. Amith and Mark found for gold Schottky barriers on GaAs(110) that the same barrier height resulted for a wide variety of preparations whether or not the initial surface was reordered.[50] The properties under bias were very sensitive to the details. These conclusions were confirmed by Palau *et al.*, who showed that defects present before the metal deposition did not strongly influence the barrier height.[51] They found, however, that states introduced by sputtering would play a significant role.

Most vacuum–semiconductor interfaces reconstruct or relax so that they show lateral periodicities or coordinates different from a strict truncation of the bulk lattice.[52–54] A variety of driving forces for these reconstructions have been identified, but in general they can be summed up as deriving from the surface lowering its energy by remaining semiconducting.[53] The reconstruction has a strong influence on the spectrum of intrinsic surface states and their energy position. The character of surface states observed with electron

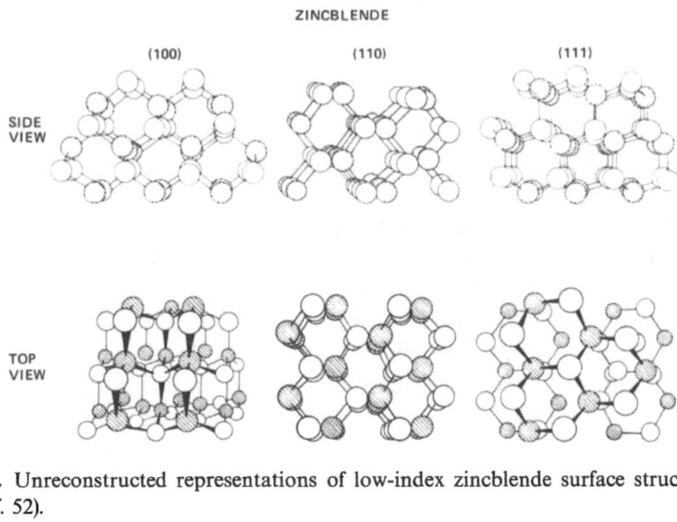

Figure 9. Unreconstructed representations of low-index zincblende surface structures (after Duke, Ref. 52).

spectroscopies is very sensitive to the presence of adsorbates or overlayers. Trends in the surface reconstructions studied with LEED have been discussed by Duke *et al.*[55] Figure 9 shows the unreconstructed bulk terminated structures for three low index planes of zincblende lattices. One can see that different behavior is likely for adatoms being deposited onto these surfaces.

In most cases, the detailed dependence of Schottky barrier heights upon crystal surface orientation has not been established. There is good evidence, however, that aspects like noise figure do improve if fully ordered interfaces are established, and these can often be more readily achieved on specific surface orientations.[56] Below, brief descriptions will be given of the current understanding of selected surfaces of Si and GaAs. Establishing a structural model that will represent all the data is a very challenging, complicated undertaking so that these models will continue to evolve.

4.1.1. Silicon (100) and (111) Surfaces

The silicon surfaces have been the subject of intense investigation over the last 30 years, but many issues remain.[54] Much of the work focuses on the (100) and (111) faces.

The reconstructions of the (100) surface are felt to be accurately understood and are dominated by dimerization of the outermost layer. The model proposed by Chadi to describe the 2×1 and 4×2 reconstructions has become the most generally accepted one for the surface.[57] The most energetically optimal geometries correspond to asymmetric and partially ionic dimers. With reference to Fig. 9, one can picture this as the atoms in the outermost layer pulling closer to form molecularlike bonds.

The details of the (111) cleavage surface still remain in dispute even though it is one of the most studied. There are two primary reconstructions: the 2×1 obtained directly upon cleaving, and the 7×7 obtained after thermal annealing. The transition temperature between the two is typically 250–500°C, but is a strong function of the number of cleavage steps present.[58] Many models have been proposed, but over the years none has been found to completely explain all the available data.

The Haneman buckling model[59] of the 2×1 in which alternate atoms are displaced in and out of the surface seemed good for close to 20 years, but recent photoemission data have not been found to be fully consistent with this geometry. Recently proposed surface chain and molecular models such as put forth by Pandey[60] and by Chadi[61] can better account for most data, although coordinates have not been generated which give a good LEED fit.

The 7×7 surface has not been explained, although evolving experimental evidence now favors rough surface models.[62] The 7×7 surface is one of the few semiconductor surfaces which exhibit a metallic character. Probably much more work is needed before this surface is considered solved.

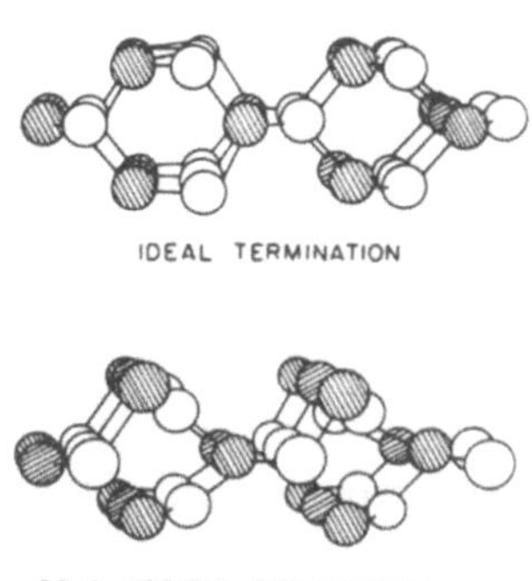

Figure 10. Representation of the unrelaxed and relaxed GaAs(110) surface (after Duke, Ref. 52).

4.1.2. GaAs(110) and GaAs(100) Surfaces

Additional aspects are introduced by the compound semiconductors primarily due to the ionic aspects. In the case of the (110) cleavage face, the general trends seem well determined.[55]

The GaAs(110) surface is considered to be the best understood semiconductor with coordinates of the surface atoms well located.[55,63] The vacuum–semiconductor interface has been found to have no states in the gap[64] so that it has proven to be a particularly fertile vehicle for surface studies. As with other cleavage faces, great care needs to be exercised in preparation to minimize steps.

The (110) surface is thought to be accurately described by the model shown in Fig. 10, where the As atoms rotate out and the Ga atoms rotate in while preserving the 1 × 1 unit cell. This relaxation has the effect of removing

TABLE 4. GaAs Surface Phases for Various Ordered Reconstruction[a]

		Our data[a]		Previous data[b]	
Rec.	$R = \text{Ga/As}$	As coverage	As atoms per Ga atoms in the unit cell	As coverage	As atoms per Ga atoms in the unit cell
$c(4 \times 4)$	0.65	1.00	8/8	0.86	7/8
$c(2 \times 8)$	0.72	0.89	7/8	0.61	5/8
$c(8 \times 2)$	0.98	0.52	4/8	0.22	2/8
1×6	1.07	0.42	2.5/6	0.52	3/6
4×6	1.17	0.31	7.5/24	0.27	6.5/24

[a] After Bachrach *et al.*[67]
[b] Drathen, Ranke, and Jacobi, *Surf. Sci.* 77, 162 (1978).

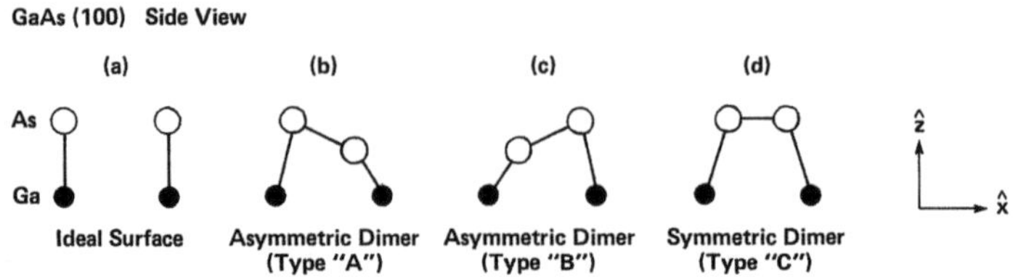

Figure 11. Representation of surface As dimerization involved in GaAs(100)$c(4 \times 4)$ reconstruction (after Chadi *et al.*, Ref. 68).

states from the gap so that one can typically achieve flat band conditions at the surface.

The GaAs(100) surface has the most technological importance since it is the one upon which most epitaxial growth is performed for device applications. A wide variety of complex surface phenomena driven by changes in the surface stoichiometry characterize this surface.(65,66) The reconstructions of the (100) face are listed in Table 4 along with estimates of the As/Ga ratio.(67) Two general regimes exist which fall into As-rich and Ga-rich classes. Some experiments indicate a 0.1 eV difference in barrier height prepared on these two surface composition regimes.(56)

The As-rich reconstructions are dominated by As dimerization. In the case of the $c(4 \times 4)$ Chadi has shown that if this surface is a complete monolayer, then one can describe the reconstruction as arising from a set of covalent and ionic dimers as shown in Fig. 11.(68) Sequences of the A, B, and C dimers can then arranged in the following patterns:

C C C C C		C C C C C C
A B A B A B	or	A B A B A B
C C C C C		C C C C C C
B A B A B A		B A B A B A

The detailed structures for the other structures have not been established but are under active investigation. A transition occurs as the surface goes Ga rich, indicating a fundamental change in the surface bonding. A special situation exists for the (100) face with aluminum. Single-crystal interfaces result even with deposition at room temperature.(13,56) This is discussed further below.

4.2. Stage 1: The Dilute Limit (< 1/2 Monolayer)

The initial stage of interface development constitutes the dilute limit. The dilute limit is characterized by a density of metal adatoms small enough that a continuous metal film does not form. In its most basic form, the major dilute limit issue can be stated as: "Where does a metal atom sit when deposited on an empty semiconductor surface?" This seemingly simple question is in fact quite complicated, for although a metal atom may be strongly bound to the surface, for many cases the atoms even at room temperature have considerable lateral mobility. This mobility may lead to metal cluster formation as the concentration grows, but can occur well below the 1/2 monolayer used for the demarcation of this section. An additional dimension to this stage is that the atoms may find sites beneath the surface.

The structural aspects of the dilute limit are complex and depend strongly on the semiconductor during deposition. Changes in the surface structure at this stage of deposition are known as impurity stabilized. Deposition rate effects are not well explored but have been demonstrated with respect to nucleation of single-crystal metals.(69) An important aspect of the dilute limit is that the heat of condensation of the metal atom may be sufficient to dislodge semiconductor surface atoms. If there is a tendency to surface reaction, then this also strongly influences this stage.

A number of detailed insights have been obtained about this stage and will be expanded upon in Sect. 6. An example is the study of the Al–Ga–As system. The work of Chadi and Bachrach(70) and of Ihm and Joananopolous(71) among others have shed considerable light on this problem. Figure 12 depicts the twofold site deduced for an Al or Ga adatom on GaAs(110).(70) Figure 13 presents the full surface total energy contours calculated for this situation. The calculation establishes that the twofold coordinated site is the lowest energy, but there are other sites of competing energy.(71) One of these is the Al–Ga single coordinated site proposed by Swartz *et al.*(72) An interesting aspect is that the intuitive single coordinated Al–As site which would be the position of the next metal in an epitaxial growth is not the most favored site on the free surface.

Zunger(73) has investigated the transition from the ordered chemisorption regime to cluster formation. The importance of clusters to the Schottky barrier problem is not clear. It is one of many intermediate metal nucleation phenomena. Examples exist, however, where the cluster stage does not appear to be predominant.

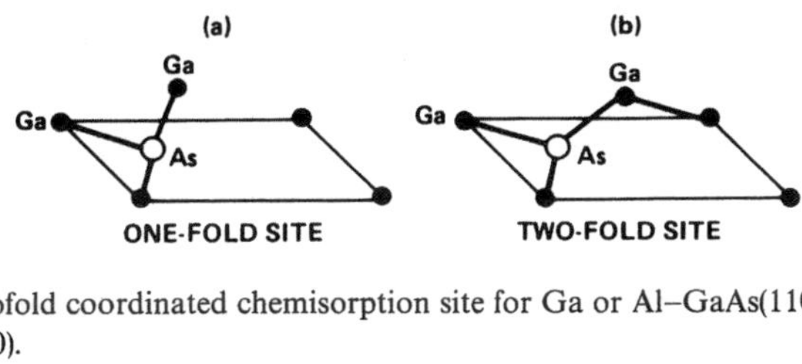

Figure 12. Twofold coordinated chemisorption site for Ga or Al–GaAs(110) (after Chadi and Bachrach, Ref. 70).

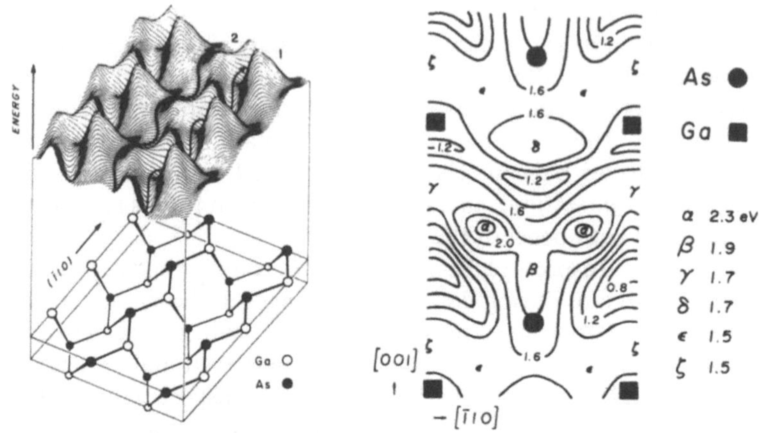

Figure 13. (a) Al adsorption energy contour plot on GaAs(110) surface. The substrate surface lies in the plane of the figure and the adsorbed Al atom lies above the plane. (b) Surface total energy for chemisorption of Al on GaAs(110) determined. The arrows indicate paths of easy surface diffusion (after Ihm and Joannopoulos, Ref. 71).

The dilute limit is very informative for a detailed understanding of the evolution of the chemistry of Schottky barriers because significant Fermi level pinning is found at this early stage[18,19,74] and interdiffusion is initiated.[75,76] The resulting dipole which is also connected with the eventual barrier height can provide an interdiffusion driving force during the later stages. This will be discussed further in Sections 5.4 and 6.2.

Figure 14 gives an example of the evolution of Fermi level pinning for Al on GaAs as a function of coverage obtained by Skeath *et al.*[77] Figure 15 shows similar information obtained in a different manner.[76] From both sets of data one sees that the barrier height is mostly established to within ± 0.1 eV at the end of the dilute range. The continued variation during monolayer formation most likely signals a significant change in the states being added.

What is not yet clear about the dilute limit is the extent to which the zone 1 discrete interface levels maintain their character through the subsequent stages. This is discussed later but the essential conclusion is that a transformation in the nature of the states takes place as additional metal is added.

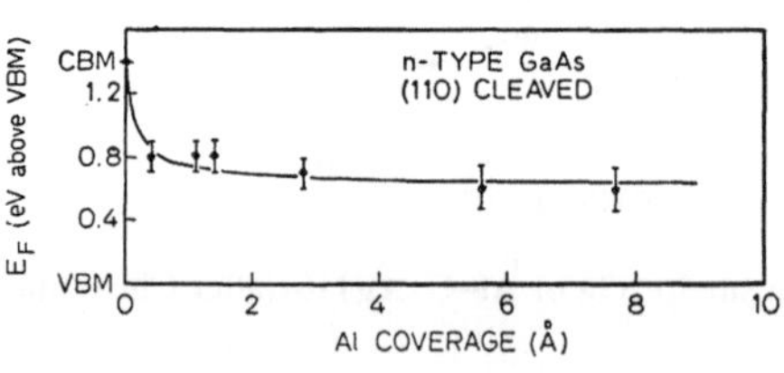

Figure 14. Evolution of Fermi level pinning as function of Al coverage on GaAs(110) (after Skeath *et al.*, Ref. 77).

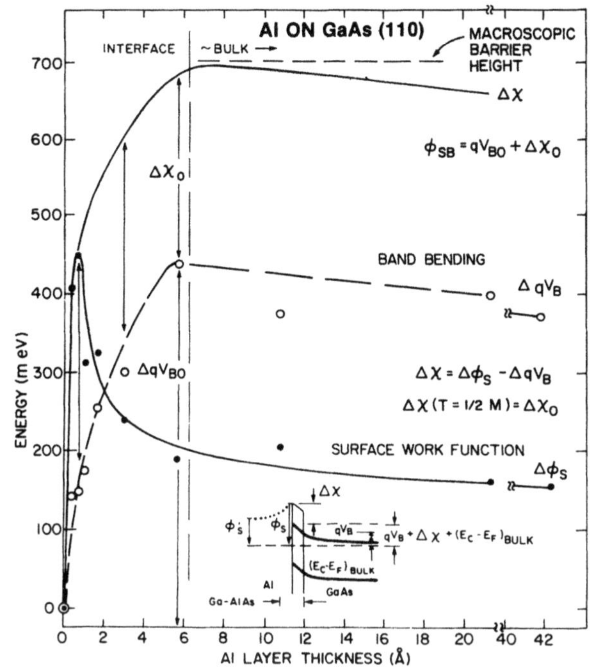

Figure 15. Evolution of Schottky barrier height versus Al coverage determined from surface photovoltage and Kelvin probe measurements. The variation of surface work function φ and band bending qV_b are shown. The inset illustrates interface dipole $\Delta\chi$ schematically (after Brillson *et al.*, Ref. 76).

4.3. Stage 2: Monolayer Formation—Metal Film Nucleation

The monolayer formation stage plays the most important part in determining the characteristic of the interface. As the density of atoms increases, metal nuclei begin to form.[78] The nucleation of the metal film is a very complex subject.

Characteristic of the monolayer formation stage is the production of interdiffusion due to the heat released as the metal nucleates. This heat augments electromigration due to the presence of the dipole.[74,79] In some cases, the release of this heat can drive other reactions over their activation barrier. The approach to monolayer formation may have several substages at which ordered metal overlayers may result. The structural aspects of the monolayer stage are strongly dependent upon the deposition temperature and subsequent annealing history.

A wide number of observations have shown that in most systems the Fermi level pinning is mostly complete once the first monolayer of metal is established. The increase in metal density during this stage is also likely to effect the character of interfacial defects created during stage 1.

Quite a bit of theoretical effort has gone into studying the nature of the one half to one monolayer region of metal adsorbates since it is a more tractable problem than for the case of a thick metal overlayer. Examples of this are the work of Chelikowski and Cohen,[80] Chadi and Bachrach,[70] Ihm and Joanopolous,[71] and Zhang and Schluter.[81] These studies have shown that many aspects of the interface are sensitive to the ordering of the metal at this stage. To some extent this reflects the fact that the full metallic character of the overlayer has not developed. Studies by Kleinman *et al.*[82] have found that three monolayers are generally required to converge to metallic like properties. Batra and Herman[83] have performed calculations which explore this transition region for the case of Al–Ge.

4.4. Stage 3: Additional Monolayers and Interdiffusion

The final characteristics of the interface can continue to be driven as additional monolayers are added. Strain fields due to lattice mismatch and grain boundary phenomena can play an important role. Usually the interface reaches a stable configuration within 3–10 monolayers for most metals on GaAs.

Interfaces are often metastable, however, and the interdiffusion is accelerated at increased temperatures or under biased conditions. In the case of silicides, the interaction and phase stabilization often proceeds for quite long distances.[84]

The application of core level photoemission to this problem confirmed what had already been deduced with Auger spectroscopy,[85] but on a finer dimensional scale. An overriding fact of metal–semiconductor interfaces is the intrinsic aspects of interfacial interdiffusion and interaction. These two aspects of the chemistry often tend to overwhelm issues that might be deduced from the clean separate surfaces. The interdiffusion can result in interface layers with radically different characteristics from the supposed components.

For example, in the case of Al–GaAs, an interfacial layer of AlAs or GaAlAs can result so that a heterojunction interface is included in the barrier.[86,87] As shown by Bachrach *et al.*, the interdiffusing As is always chemically trapped by Al–As bonding. Although the Ga tends to segregate to the surface of aluminum, the zone 1 metal probably is an Al–Ga alloy. Such alloy effects are also seen with Au and Pd Schottky barriers.[88,89] In the case of silicides, the interdiffusion which creates the metallic silicide moves the barrier away from the interface with the initial metal. In many cases, these reactions will proceed in the presence of a thin oxide layer.

One way of determining the interfacial width is by use of core level

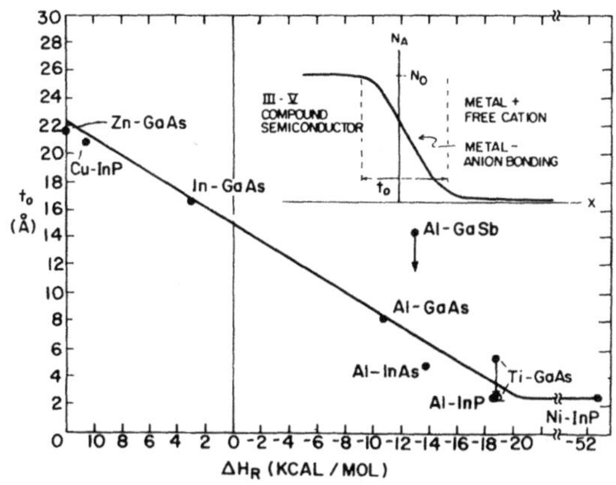

Figure 16. Interface widths determined from core level photoemission measurements with marker atoms. The interface width versus interface heat of reaction is shown for metal/III–V compound semiconductor junctions. The inset shows the anion profile schematically (after Brillson *et al.*, Ref. 91).

interface markers in conjunction with photoemission studies.[90] Figure 16 shows an example of interface widths deduced with this technique for room temperature deposition with a variety of metals.[91] This shows a good correlation of the intrinsic width increasing with the reactivity.

The asymptotic interface results when additional deposition does not appreciably change the composition or doping of either the deposited metal or the semiconductor. The asymptotic interface may be either an homogeneous or inhomogeneous interface layer. Cases exist where an interfacial layer will spontaneously phase segregate in the plane of the interface and result in an

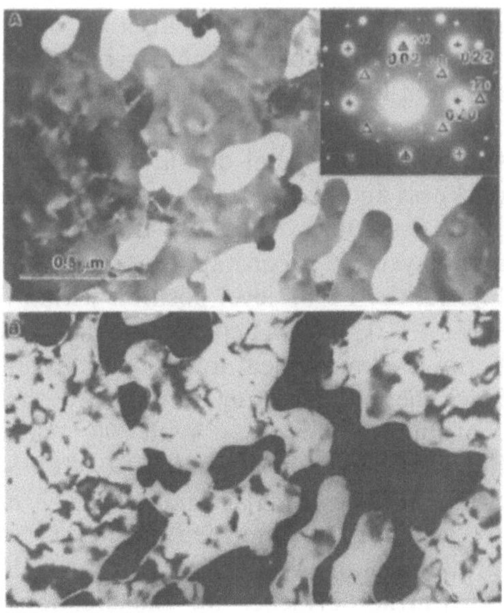

Figure 17. Single-crystal aluminum growth on GaAs(100) seen with high-resolution TEM. Specific orientation regimes appear as contrasting areas (after Petroff *et al.*, Ref. 92).

inhomogeneous situation. The application of high-resolution electron microscopy has shown examples of this. For example, in growing single-crystal Al on GaAs(100), depending on the starting surface, one can get crystallites of mixed orientation. Figure 17 shows some results of Petroff *et al.*[92] The mixed orientation domains produce the pronounced contrast. More detail is provided in Section 6.

The asymptotic interface may also depend upon the subsequent annealing history. Time versus temperature is important both from the point of view of diffusion and from activation energies for reaction. Auger profiling measurements of RBS show that diffusion at elevated temperature can spread the interface over thousands of angstroms.

An important aspect of the asymptotic layer often is that it provide a diffusion barrier to other metals used in subsequent processing. This secondary property is the subject of significant literature and is beyond the scope of this chapter. Often multiple metal films are employed to establish the overall compatibility.[93]

4.5. Some Specific Characteristics of the Interfacial Layers

Certain aspects of the interfacial layers have been identified with a number of studies. Examination of combined Kelvin probe and surface photovoltage experiments have shown that the electron affinity is modified as the interface develops. As shown by Brillson, the dielectric aspects of the interfacial layer may show new characteristics.[94] Such behavior had been discussed by Phillips in connection with interfacial plasmons.[95]

An example of the evolution of the barrier height with metal coverage was shown in Fig. 15. The evolution of the electron affinity change is plotted and can be seen to be a strong function of the metal coverage. Assuming the semiconductor–metal dipole $\Delta\chi$ equals $(\Delta\phi - \Delta_q V_b)$ below a critical metal thickness and $(\Delta\phi - \Delta_q V_b)_{\max}$ for coverages above that thickness, then for Al on GaAs(110), $\nabla\chi$ increases with Al coverage to a maximum value at one half monolayer and is constant for larger aluminium thickness. From the inset in Fig. 15, the Fermi level position within the band gap is $E_c - E_f = {}_qV_b + \Delta\chi + (E_c - E_f)_{\text{bulk}}$. This dependence of ${}_qV_b$ and $\Delta\chi$ is in close agreement with the Fermi level movement found by photoemission techniques shown in Fig. 14. Since the interfacial region has finite thickness, the $\Delta\chi$ voltage represents a contribution to ϕ_b. With ϕ_b defined as $E_c - E_f - {}_qV_b + \Delta\chi_0$ in Fig. 15, a barrier height of 0.7 ± 0.08 eV is obtained, in good agreement with electrical values. Similar studies have been performed by Palau *et al.*[96] which are generally in agreement with this analysis and have split the barrier formation into two parts.

Further evidence for the creation of the interface phases has been

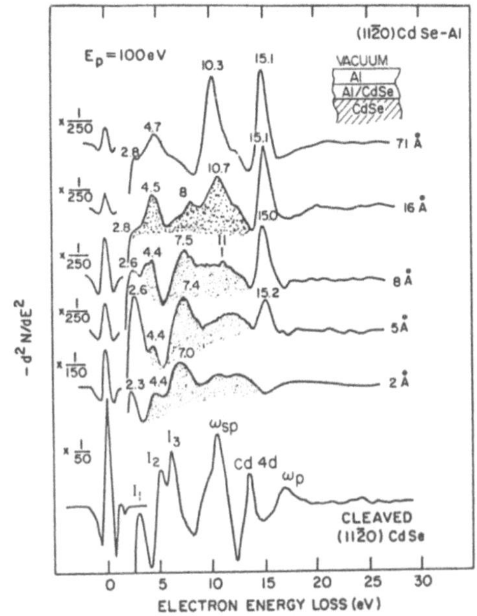

Figure 18. Example of dielectric function modification from electron energy loss spectroscopy (after Brillson *et al.*, Ref. 94).

established with low-energy electron loss measurements. The signature of interfacial reactions is provided by new dielectric aspects. Figure 18 shows an example for Al–CdS interfaces measured by Brillson.[94] The electron energy loss is shown as a function of aluminum coverage. The plasma modes for the clean surface are labeled. As the interface forms, a series of new interfacial plasmons can be identified. The resultant interface dielectric constants deduced from the measurements are lower than those of the semiconductor, consistent with the formation of Al chalcogenides at the interface. This interfacial behavior is analogous to that observed by Rowe *et al.*[97] for Al–Si.

5. FORMATION OF INTERFACE STATES

During the stages discussed above which create the interfacial layer, simultaneous mechanisms proceed which produce states in the band gap of the

TABLE 5. Types of Interface States

a. Intrinsic surface states
b. Interfacial relaxation induced states
c. Metal-induced gap states
d. Vacancy, anti-site, or defect related states
e. Exchange reaction doping

semiconductor. The character of the states is one of the key issues in establishing a microscopic understanding of the barrier height. This section describes aspects of interface states summarized in Table 5 and relates them to the stages of interface formation. The table is organized as a progression from intrinsic related to defect and impurity related.

5.1. Intrinsic Interface States Derived from the Metal and Semiconductor

As discussed above, the clean surfaces of both semiconductors and metals have surface states and resonances. Silicon intrinsically has surface states in the band gap,[30,98] while for many of the compound semiconductors the surface states lie outside the gap.[32,99] For some time, it was thought that these surface states would play an important role in determining Schottky barrier properties.

Even in cases where surface states are not initially present in the band gap such as on some compound semiconductors, it is possible that the change in surface reconstruction upon interface formation can pull states into the band gap.[70] Aspects of this are related in Section 6 in discussing GaAs(110). Detailed calculations have shown that the surface often responds to an adsorbate by changing its reconstruction. The interfacial relaxation thus provides a new mechanism by which states are reintroduced into the gap region.

A crucial issue is the evolution of these states in the presence of the asymptotic metal. Heine[47] found using a model calculation that the screening of the semiconductor derived states by the evanancent field from the metal changes the nature of semiconductor surface states appreciably. This screening effect has an impact on any form of interface states, but in the case of states derived initially from the clean surface states, the states are transformed into interface resonances. Experiments by Rowe have indicated that in the case of silicon, such a transformation of the character of the states has been observed.[18]

The idea of Heine was elaborated upon by Bennet and Duke,[100] by Pelligrini,[101] and by Louie *et al.*,[17] who showed in detail the development of metal-induced gap states. Pursuing a somewhat different approach, Phillips[102] had argued that interface bonding would provide an intrinsic mechanism determining the barrier height that would reflect the ionicity trends observed.

The importance of Louie *et al.*'s calculation is that it shows how the localized states persist in the presence of the screening field although their character is changed. The main aspect of these theories is that the density of metal-induced gap states is spread throughout the band gap whereas the surface states at the vacuum interface can be highly localized in the band gap.

Because these calculations probably underlie much of the important interfacial physics whereby intrinsic electronic phenomena determine the Schottky barrier height once the asymptotic interface is established, we will expand on their calculation below.

5.2. Localized Defect and Impurity Related States

Classes of localized states arise from the processes which disrupt the surface before or during interface formation. For example, the various processes discussed in Section 4 can possibly lead to either vacancy formation[71,103] or antisite defects[104] at the interface. If these defects have charge states associated with them, then they can potentially have a strong influence on the barrier. A body of suggestive evidence for such states has accumulated from core level photoemission measurements.[105,106]

A similar class of states arises from situations where the metal when incorporated in the semiconductor lattice forms a donor or acceptor. Examples of this are Au–GaAs or Al–ZnSe. In extreme cases, if the incorporated metal atoms convert the type of the semiconductor, then a *p–n* junction would form and strongly influence the properties of the barrier. This doping case is analgous to the heterojunction case described in Section 4 for example with Al–GaAs exchange reactions and interdiffusion.

The presence of surface defect states has been deduced from the observation of interdiffusion of cations and anions. Some of the first evidence for the generation of interfacial defects was provided by Bachrach and Bianconi in a study of Ga–GaAs interfaces.[75] Their demonstration with core

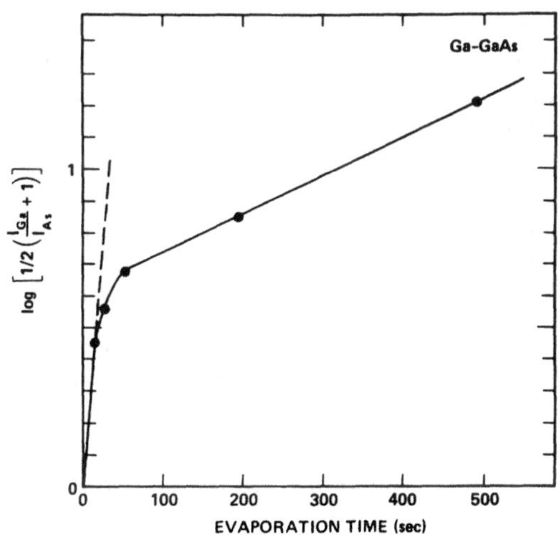

Figure 19. Change in Ga and As 3*d* core level intensities as a function of Ga layer deposition. The deposition rate was approximately one monolayer per minute (after Bachrach and Bianconi, Ref. 75).

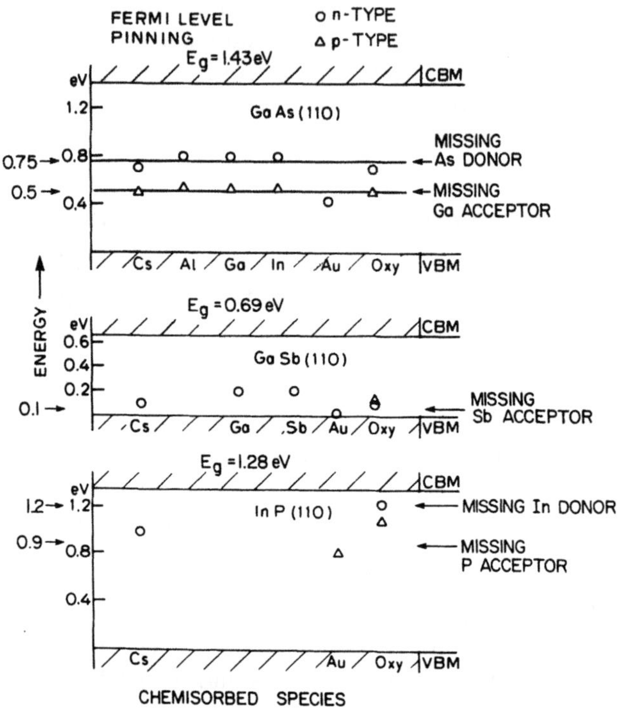

Figure 20. Defect level assignments based on Fermi level pinning observations (after Spicer *et al.*, Ref. 106).

level photoemission of As outdiffusion shown in Fig. 19 was interpreted as resulting in interfacial vacancies. The data are plotted as an exponential rise. Above the monolayer level, however, significant deviation from the expected dashed line is found. Comparison with the attenuation of valance band spectra indicates that the film is continuous and that outdiffusion is occurring.

Anion and cation outdiffusion has been found to be a persistent aspect of interface formation on III–V's. Lindau and Spicer have discussed similar types of work in a review.[105] Spicer *et al.* have proposed that these defects are the primary mechanism for Fermi level pinning based upon data such as shown in Fig. 3b. From these ideas, the possibility of defect levels as shown in Fig. 20 was postulated to determine barrier formation in III–V compound semiconductors.[106] The data, however, are not fully consistent with a set of energy levels independent of the metal. Note that the appearance of a single position determined for monolayer coverage does not hold at thicker coverage. Such an assertion is also not consistent with the higher-resolution barrier height determination from thick barriers.

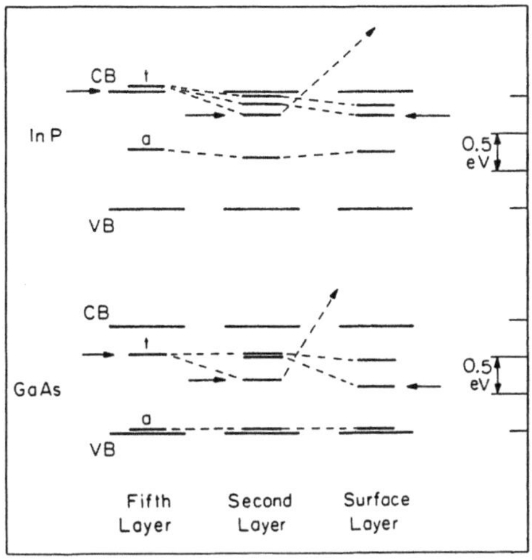

Figure 21. Surface vacancy energy levels for GaAs (after Daw and Smith, Ref. 103).

Allen and Dow[(104)] have argued that it is more likely for the important localized state to be an antisite defect. Figure 21 shows the expected vacancy related energy levels determined by Daw and Smith[(103)] which Allen and Dow compare to their results. Allen and Dow have produced a predictive result for antisite defects on compound semiconductor surfaces shown in Fig. 22.[(104)]

Although it is possible for vacancies to remain at the interface, theoretical arguments have been put forward that in many cases these would be associated with metal. In the case of antisite defects, the properties of isolated defects on a free surface are clear, but if the density increases, then the layer is converted into a zone of the metal. An anion antisite defect, for example a Ga on an As site, then has the aspects of a metal cluster when additional monolayers develop. One can anticipate that continued TEM studies as well as local probes such as surface EXAFS may help to unravel some of these issues.

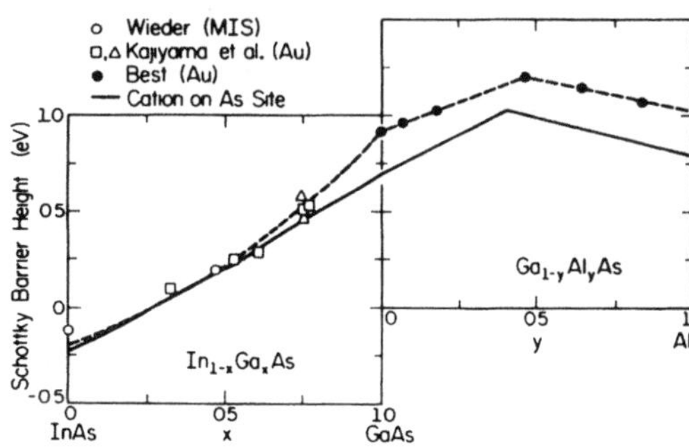

Figure 22. Prediction of barrier heights from antisite defect energy levels for compound semiconductor alloy sequence (after Allen and Dow, Ref. 104).

5.3. Interface States and the Stages of Interface Formation

The interface states are distributed within the two zones of the interface characterized above, one being derived from the evanescent tail of the metal wave function into the semiconductor and the remaining interfacial layer being the second zone. The interface states in zone 1 in many cases have the most significant effect on determining the barrier height. The first zone is at most 10 Å thick from estimates available in the literature.[17,47] The second zone can be very large depending on the thermal history of the sample and can also significantly affect the applied bias performance of the barriers.

Considerable discussion has occurred in recent years on the contribution of semiconductor-derived interfacial states due either to interdiffusion doping or to interfacial defects. No conclusive evidence for these states playing the dominant role has been established for asymptotic barriers, although they clearly are the important contributor in the dilute limit.

Observing the localized states involved with Schottky barrier formation in the Bardeen limit directly with spectroscopic techniques is a challenging problem. A possible reason for this is that they are often smeared out and not resolved. The photoemission band bending technique is indirect and does not have the necessary energy resolution to be definitive (± 0.1 eV at present).

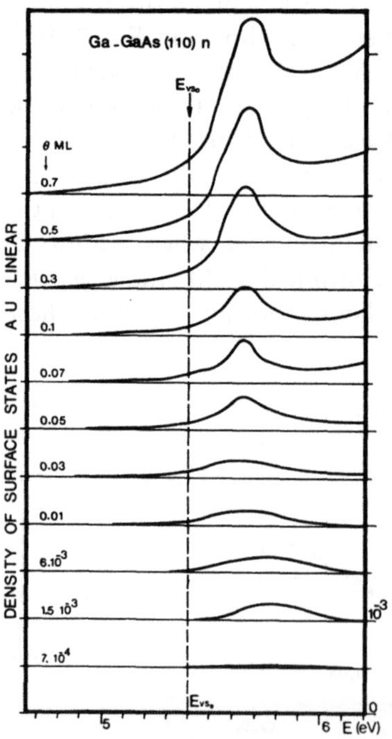

Figure 23. Photoemission yield for initial metal overlayers on GaAs indicating the development of interfacial states (after Bolmont *et al.*, Ref. 107).

Surface photovoltage spectroscopy or other capacitive spectroscopies do reveal aspects of these states,[96] but these would reside primarily in zone 2. Photoemission yield spectroscopy has shown distributions of states in the gap.[107] Specific examples are provided by the work of Bolmont *et al.* as shown in Fig. 23. One sees a continuous development of occupied states within the band gap.

A number of theoretical studies have been carried out on the aspects of surface defects such as surface vacancies and surface antisite defects. It is interesting that these studies show levels at energies appropriate for pinning the Fermi level when the defects lie on the free surface. When these defect or impurity related states are incorporated into the interface and lie in zone 2, their attributes can be well defined. For those lying is zone 1, however, the same screening effects on their position needs to be considered as with localized states derived from intrinsic surface states. Theoretical studies have not been performed of the properties of both surface vacancies and antisite defects for the case where these defects are within the evanescent tail of the metal such as discussed above for the transformation of intrinsic surface states into metal-induced gap states. More work needs to be done to fully understand the superposition.

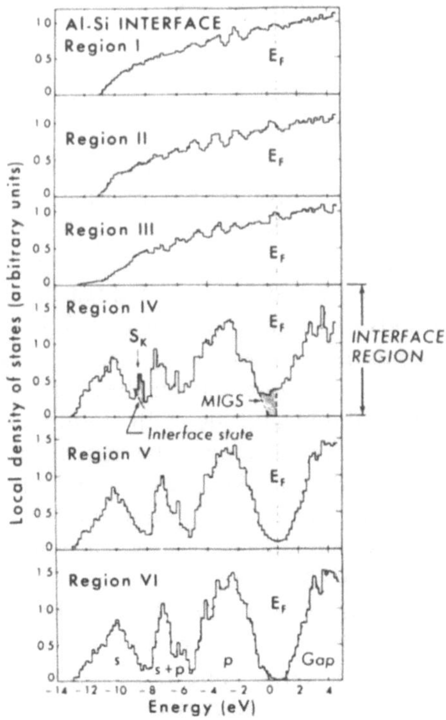

Figure 24. Layer density of states for the Al–Si interface with the metal treated within a jellium model. An interface state as well as the metal-induced gap states are marked. The layers shown would extend about 10 Å in real space (after Louie *et al.*, Ref. 17).

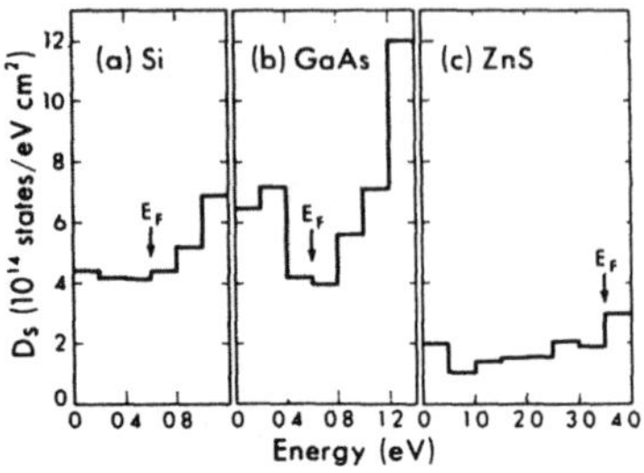

Figure 25. Density of metal-induced gap states for Si, GaAs, and ZnS show the change with ionicity. Note the rise in density of states at 0.4 and 0.8 eV in the region of postulated states shown in Fig. 20 (after Louie *et al.*, Ref. 17).

This section is concluded by expanding the discussion of metal-induced gap states which become most important at the asymptotic overlayer stage. The principal results to be described are those of Louie and Cohen's[17,108] studies of Al interfaces with Si, GaAs, and ZnS. One should note that the results of Pelligrini are consistent with this although not presented in as graphical form.

The concept of metal-induced gap states was introduced above as one of the mechanisms which produces states in the interfacial region. As this is an intrinsic mechanism, one needs to understand whether or not other effects could override this density of states.

Louie, Chelikowski, and Cohen have considered in detail the electronic structure of Al–Si, and their results are informative. Figure 24 shows their curves of layer density of states for this interface. The physical extent of the layers would be approximately 10 Å. Note how rapidly the convergence to a metallic density of states occurs. The feature S_k is indicative of the formation of interface states. These states have a truly interfacial character and decay in both directions from the interface. Consideration of the effect of the interface states labeled MIGS on the barrier resulted in determining a barrier height in good agreement with the experimentally observed values.

Louie and Cohen extended their calculation to GaAs and ZnS so that trends with ionicity were exposed. The principal result is their density of states comparison shown in Fig. 25. The magnitude of states is sufficient to explain the barriers observed on these materials. Note also the trend with ionicity where the magnitude of the number of states decreases substantially.

Figure 26 shows the extent of localization of these states for the semiconductors Si, GaAs, ZnSe, and ZnS. The states thus become more confined as the ionicity increases.

A consequence of these calculations is that any complete explanation of Schottky barrier heights needs to include these zone 1 states. Unfortunately this type of calculation has not been extended to other metals. One would expect that the primary variable affecting these states would be the metallic density of states. Thus the interdiffusion modification of the metal adjacent to semiconductor discussed in Section 4 is an important aspect for the full elucidation of this issue.

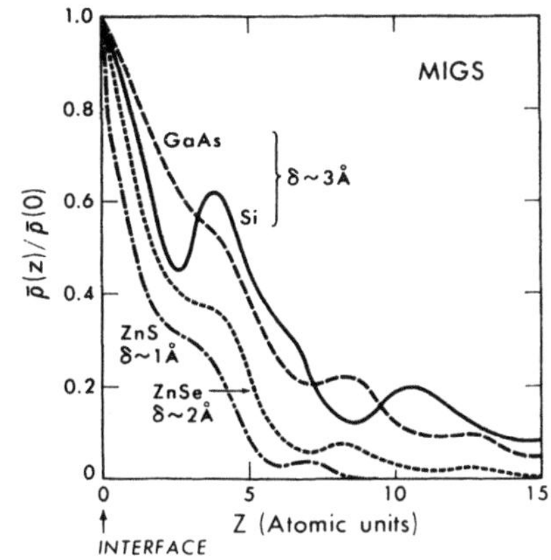

Figures 26. Localization of metal-induced gap states for Si, GaAs, and ZnS (after Louie *et al.*, Ref. 17).

Batra and Herman[85] have explored the evolution of metal-related interface states in the transition region from the dilute limit to several monolayers. Although not yet complete, their result is consistent with the idea of a transformation in the nature of the interface states as the metal thickness develops.

6. CASE STUDIES OF THE CHEMISTRY AND STRUCTURE OF SCHOTTKY BARRIER FORMATION

This chapter on the chemistry and structure of Schottky barrier formation is concluded with some selected case studies for specific semiconductors with emphasis on the features that determine the Schottky barrier. The cases presented build upon the general discussions of the preceding sections. The cases are separated into Si, and selected III–V and II–VI compound semiconductors and look at specific aspects which were not addressed previously in as much detail.

6.1. Case Studies of Silicon Schottky Barriers

6.1.1. Al, Ag, Cu, and Au Schottky Barriers

Extensive studies have been performed on the formation of silicon Schottky barriers because of their technological applications, and device detail is presented elsewhere in the book. Of the various Schottky barriers, the silicide–Si are most commonly used and have replaced Al–Si, which are often problematical for LSI because of the thinness of the layers.

Silicon Schottky barriers fall into two classes as seen from Fig. 2 and 5.

Metals such as Al, Ag, Cu, and Au have very similar barrier heights ranging from 0.6 to 0.7 eV,[16] while those metals forming silicides, such as Ni, Pt, Cr, etc. show a wider range[24] and appear to be closer to being in the Schottky limit if the silicide work function is considered.

As discussed earlier, the vacuum silicon interface has intrinsic surface states in the bandgap, and the Fermi level at the surface is pinned. Rowe *et al.* studied the evolution of these states with the initial deposition of In metal and found that the sharp surface states were removed but were replaced by a broad distribution.[18,97] Thus, the nature of the interface states pinning the Fermi level and determining the barrier height had changed. Figure 27a shows the electron energy loss data from which this was deduced.

Detailed studies have been performed of the dilute and monolayer level

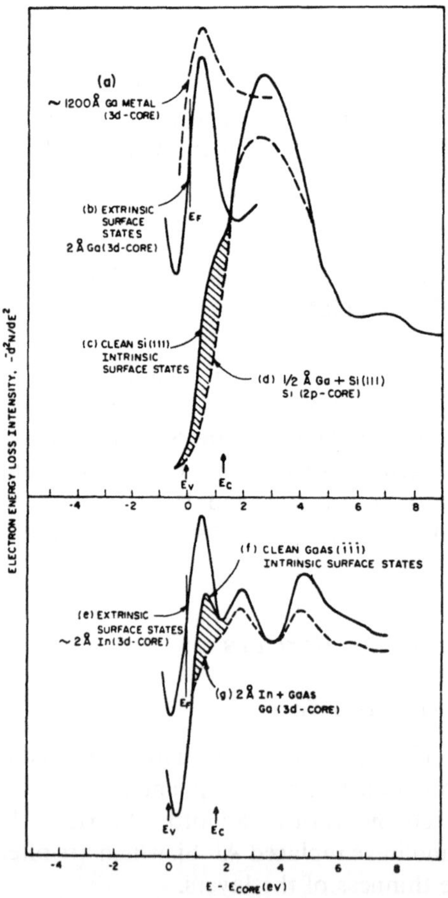

Figure 27. Evolution of the Si surface states with Ga coverage seen with electron energy loss. Ga atoms produce new states localized on the adatoms (b) which are different from a Ga metal film (a) and which change the localized Si surface states (d) and (c). (After Rowe *et al.*, Ref. 18.)

TABLE 6. Thermal Annealing Behavior to Obtain Ordered Overlayers on Si(111)[a]

Structure	Metal	Evaporated coverage (ML)	Proposed stoichiometric (ML)	Anneal temp. (°C)	Anneal time (min.)
$\sqrt{3} \times \sqrt{3}$R30°	Al	0.35	1/3	700	0.25
$\sqrt{7} \times \sqrt{7}$R19.1°	Al	0.50	3/7	700	0.25
$\sqrt{3} \times \sqrt{3}$R30°	Ag	1.0	1	450	1
$\sqrt{19} \times \sqrt{19}$R33.9°	Ni	0.09	1/19	800	15

[a]After Hansson *et al.*, Ref. 111.

formation stages on ordered Si surfaces.(109) Typically for deposition at room temperature, the metals are disordered on the surface and often coexist with the underlying substrate reconstruction.(110) Thermal annealing can convert the surface to ordered structures as outlined in Table 6, but the specific metal reconstruction depends upon the coverage.

The development of Schottky barriers with aluminum depends very much on the temperature of the substrate. Because aluminum is very reactive with oxygen, deposition must be performed in good quality ultrahigh-vacuum systems. The initial deposition of aluminum on Si at room temperature leads to a disordered overlayer which thermal annealing will convert to an ordered layer, the nature of which depends on the coverage. Once the surface coverage exceeds one monolayer, the surface appears disordered.

For aluminum, the reconstruction sequence is $\sqrt{3} \times \sqrt{3}$ at 1/3 monolayer and $\sqrt{7} \times \sqrt{7}$ at 1/2 monolayer. At one monolayer and above, the metal does not find specific registry, and ordered overlayers are not found. The thermal annealing that creates the ordered structures promotes interdiffusion, and there appears to be a second ordered $\sqrt{3} \times \sqrt{3}$ which includes aluminum diffused beneath the surface. Figure 28 shows this trend.(111) Comparing the emission from all the surfaces, one sees that the 7×7 surface has the strongest emission close to the Fermi level, i.e., stronger metallic character than the metal covered surfaces. In the energy region where the dangling bond occurs for the clean surface, only a weak shoulder is seen on the metal–silicon surfaces. One would expect the dangling bond to be changed by some orbital bonding to the metal atoms.

Al has a strong tendency to sit beneath the silicon surface and thermal annealing tends to drive it into the bulk so that the surface remains silicon rich.(112) In fact, it is possible to proceed with annealing so that Auger spectroscopy would see no aluminum on the surface. The interdiffusion, however, is an important aspect of Al–Si Schottky barriers and strongly

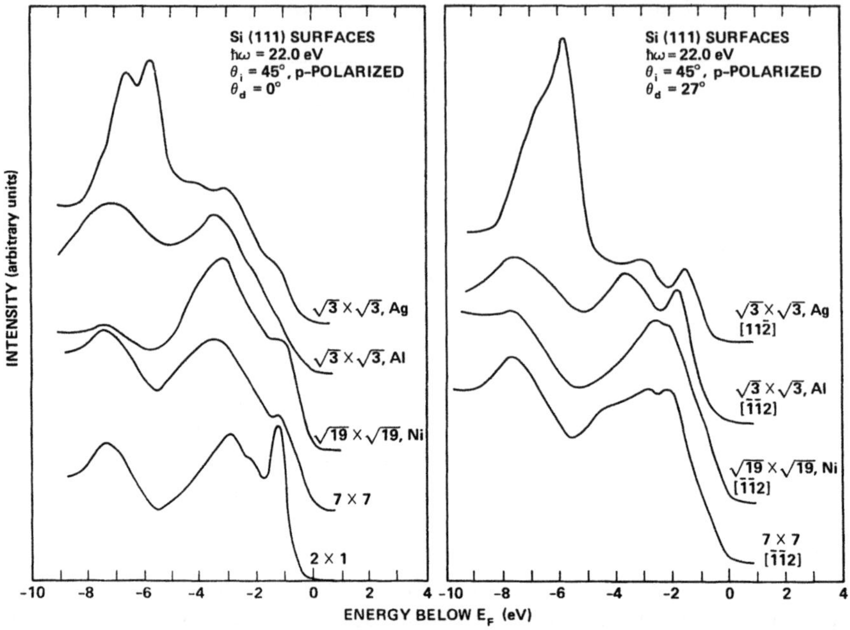

Figure 28. Photoemission study of metals on silicon comparing different reconstructions designated. (a) Normal emission and (b) various polar and azimuthal spectra from the various Si(111) surfaces selected to emphasize the physics discussed in this paper (after Hansson *et al.*, Ref. 111).

influences their usefulness. For thin planar structures, this interdiffusion often leads to shorts.

The detail structures for the reconstructions have not been solved, and the effect on the barrier height of proceeding from a reordered deposition has not been determined. Several calculations have, however, investigated the states that would arise for aluminum situated in several geometries. These works are an extension of the jellium calculation of Louire and Cohen, but need to be extended to thicker metal films before they are fully relevant to a deeper understanding of Schottky barrier.

6.1.2. Silicide–Silicon Interfaces

Many metals form silicides upon contact with silicon or at elevated temperatures. The interfacial reactions are a dominant aspect of this system.[12] Of the wide variety of silicides that occur, the examples in this section will be based upon Pt, Pd, or Ni. As with aluminum, the metals which form silicides are disordered when deposited at room temperature.

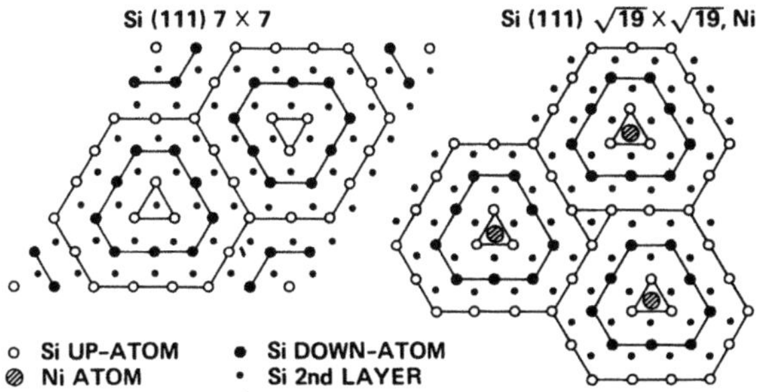

Figure 29. Model from impurity stabilized $\sqrt{19} \times \sqrt{19}$ Si(111) surface (after Hansson *et al.*, Ref. 111).

Thermal annealing converts them to ordered structures. In many cases, thick epitaxial silicides can form.

When deposited on Si(111), Ni at only 5% of a monolayer will with thermal annealing convert the 7×7 reconstruction to a $\sqrt{19} \times \sqrt{19}$.[110] For many years the impurity-stabilized aspect of this surface was not recognized, but it has been now definitely proved.[111] This impurity-stabilized surface is a precursor of the growth of $NiSi_2$. Hansson *et al.* have proposed the following model shown in Fig. 29 for this reconstruction based upon analogy with a reconstruction proposed for the 7×7 surface by Chadi *et al.*[113] The similarity between the spectra seen in Fig. 28 suggests that the two reconstructions are quite similar.

With continued Ni deposition under MBE type conditions, epitaxial $NiSi_2$ can be grown.[14] In addition to creating Schottky barriers, metal–semiconductor–metal structures have been grown and metal base transistors are being explored.

With room-temperature deposition and then thermal annealing, a

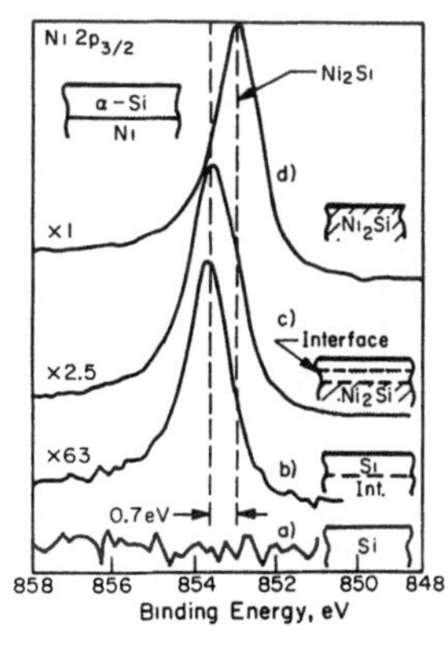

Figure 30. XPS study of Ni–Si interfacial reaction (after Cheung *et al.*, Ref. 114).

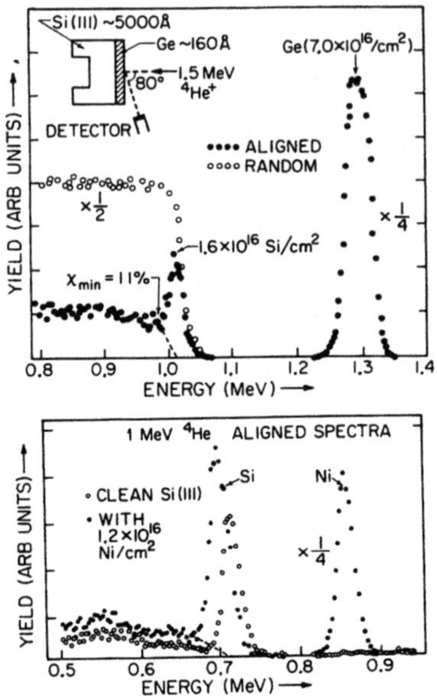

Figure 31. RBS study of Ni–Si interfacial reactions (after Cheung *et al.*, Ref. 114).

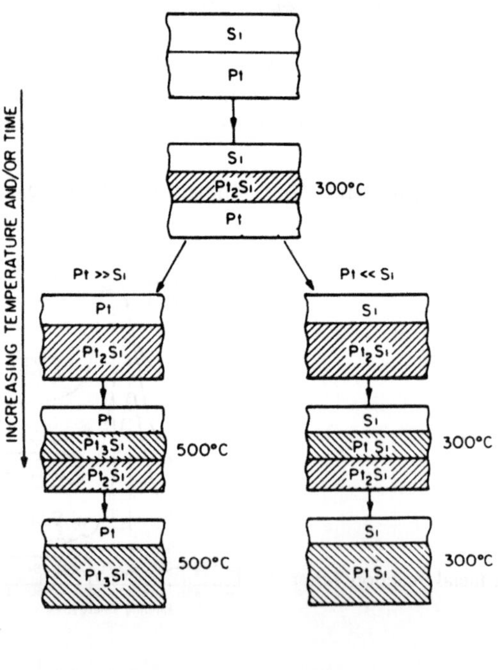

Figure 32. Silicide evolution paths depending upon metal-to-Si ratio (after Otiavani *et al.*, Ref. 26).

progressive conversion to silicide occurs. This silicide has been studied extensively by RBS and by XPS. Examples of the investigation of aspects of the conversion for Ni silicides are shown in Figs. 30 and 31.[114] Figure 30 shows examples of Ni $2p$ XPS which can be used to deduce the interface composition. Figure 31 compares RBS from a nonreactive interface of Si–Ge with Si–Ni. The reaction has significantly moved Si atoms so that a large peak is observed. The Schottky barriers with metals forming silicides are dominated by the reactions and these move the junction away from the initial surface.

Figure 32 gives an example of the possible evolution of platinum silicide phases depending on the material available.[115] This suggests the metallic silicide phase is silicon rich. Consideration of this aspect led Freouff to propose that the silicide–Si Schottky barriers were in the Schottky limit if a Si-rich phase existed at the interface.[116] Deducing the work function for these phases, Freouff prepared the graph shown in Fig. 33. Although suggestive, the existence of the required interfacial phase compositions has not been established, and this analysis is still to be proved.

In the case of Pd–Si interfaces, it is likely that the Si-rich phase does not exist, and interface states still play an important role. Figure 34 shows a TEM lattice image of the Pd–Si interface.[117] One can see that the interface is ordered but not fully abrupt. Analysis of photoemission data was consistent with the interface silicide being Pd_2Si.

A wide range of other silicides has been studied, and a clearer picture of the interface states should emerge. A recent interesting study by Schmid *et al.* looked for the effect of morphology and composition with Ni silicides but in fact found that little variation was observed. They thus concluded that the local interfacial environment was relatively insensitive to these details.

Silicides are discussed further in the chapter by Nemanich *et al.* in connection with Schottky barriers on amorphous silicon.

6.2. Case Studies of III–V and II–VI Compound Semiconductor Schottky Barriers

The barriers on compound semiconductors are strongly influenced by interdiffusion and interaction. This was originally seen on a gross scale with Auger profiling, but photoemission proved the inherent microscopic aspects of this even for low-temperature depositions.

It was also clear from the early work that GaAs was strongly in the Bardeen limit, and a clearer picture of this evolved from the photoemission studies. One of the interesting cases is from the Ga–Al–As system.

6.2.1. The Ga–Al–As System

The Ga–Al–As system is a particularly interesting one for the study of Schottky barriers. $Ga_xAl_{1-x}As$ forms a ternary alloy over the whole

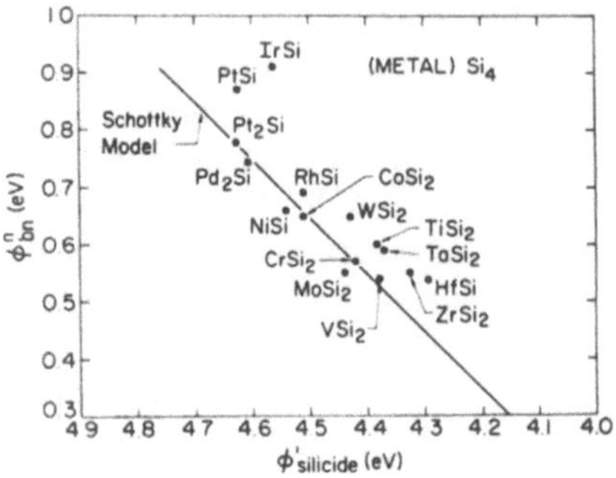

Figure 33. Plot of silicide barrier height versus deduced Si-rich silicide work function (after Freouff, Ref. 116).

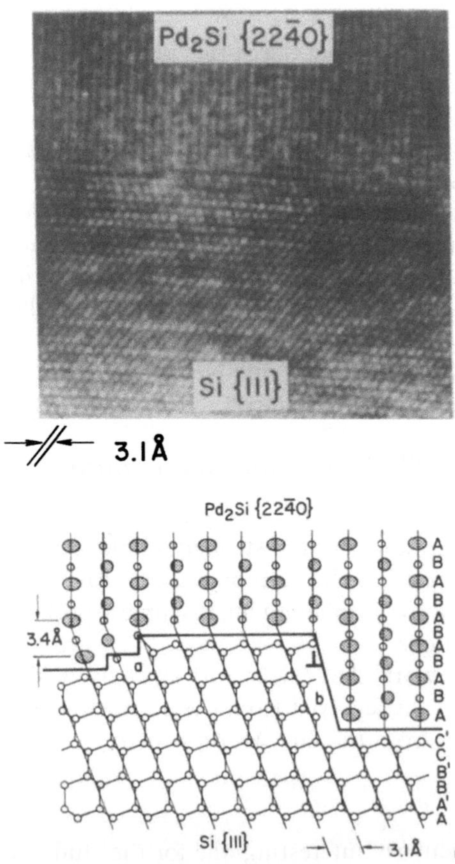

Figure 34. TEM study of Pd–Si interface (after Schmid *et al.*, Ref. 117).

composition range and in addition, the ternaries are closely lattice matched. Deposition of Ga or Al metal will form a Schottky barrier so this is an isoelectronic system. The barriers for Al are well established, but Ga shows more variability. Bachrach and Bauer investigated the electrical characteristics of an Ga–GaAs (*p*-type) interface.[72] In fact they found that the outdiffusion resulted in an ohmic interface for *p*-type samples at thick overlayers in contrast to the low coverage result of Skeath *et al.*[118] Woodall has reported observing ohmic behavior for Ga on *n*-type GaAs but that the system with time converted to a barrier.[119]

Extensive information exists on GaAs(100) and the cleavage face GaAs(110). The cleavage face has provided a fertile laboratory for the investigation of surface and interface effects. A key feature of GaAs(110) is that the intrinsic surface can be prepared by cleavage so that few states remain in the gap. This property was apparently first identified by Huijer and van Laar.[63] Subsequent studies have confirmed their result. Gobelli and Allen[120] first identified the surface pinning position that does arise due to many circumstances and has been subsequently confirmed with photoemission studies.

We have seen in Section 2 that the Schottky barrier height on GaAs(110) is essentially only weakly dependent on the metal work function and therefore is strongly in the Bardeen limit. This effect was verified with photoemission by Lindau *et al.*,[19] who showed that the surface band bending is substantially complete at the submonolayer level for most of the metals deposited on GaAs(110). As discussed earlier, the barrier formation is not fully complete at this stage.[70] What the studies of GaAs thus show is that the essential localized states are formed in the initial metal deposition of stage 1 and that the further stages have only a small altering effect on the asymptotic pinning position.

Investigation of Ga–GaAs(110) interface formation by Bachrach and Bianconi established some of the first evidence that the metal deposition resulted in the outdiffusion of As.[75] They further deduced that the resulting

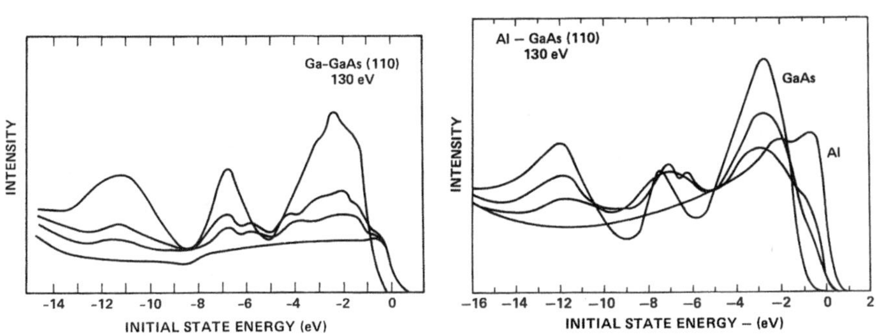

Figure 35. Evolution of interface states (after Bachrach *et al.*, Ref. 86).

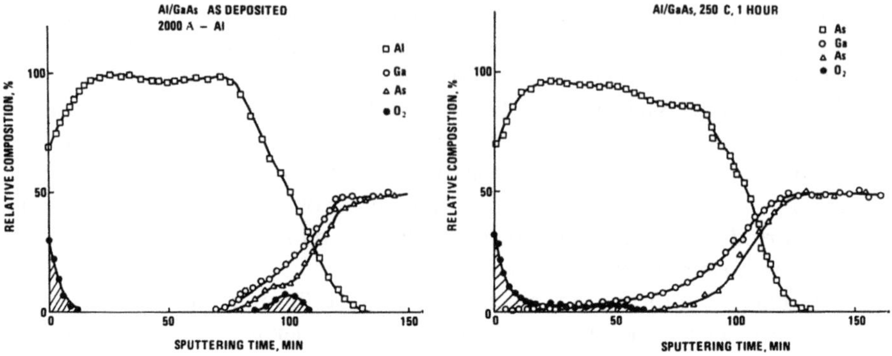

Figure 36. Interdiffusion (after Christou and Day, Ref. 85).

interfacial vacancies most likely would have an important connection with barrier formation. This work also showed the evolution of interface states as seen in Fig. 35a. The prominent changes that occur in the spectra at one-half and one-monolayer Ga coverage show the development of two new peaks at −4.2 and −5.8 eV. These are indicative of interfacial bonding and are analogous to the interfacial states shown in the case of Si in Fig. 24. Bolmont *et al.* examined Ga–GaAs interfaces with photoemission yield and showed the evolution of states in the gap.[85]

A key aspect of the evolution of Schottky barriers on GaAs that is seen with photoemission studies is the interdiffusion of Ga and As into the metal. An example was shown in Fig. 19. Figure 36 shows a further example of interdiffusion established with auger sputter profiling. In this case the distance is quite large with elevated temperatures. The observation of the interdiffusion of Ga and As occurs under almost all circumstances for metal–GaAs interfaces. This makes it likely that a Ga or As deficiency can occur at the interface.

Interdiffusion data such as this led Spicer and his group to conclude that these defects determined the barriers on all III–V's.[106] Figure 21 shows an example of their result. Williams has discussed issues surrounding the role of defects in a review paper.[121]

The identification of the interfacial defects stimulated a theoretical investigation of the electronic states that might be associated with both vacancies and antisite defects discussed in Section 5. No direct evidence has been obtained showing these are actually present in the asymptotic interface.

Investigation of Al–GaAs interface formation revealed a number of important features already discussed. An example of the microscopic behavior of this system is shown in Fig. 37 which shows the evolution of the Al 2*p* and Ga 3*d* core lines. The initial deposition in this case was chemically shifted

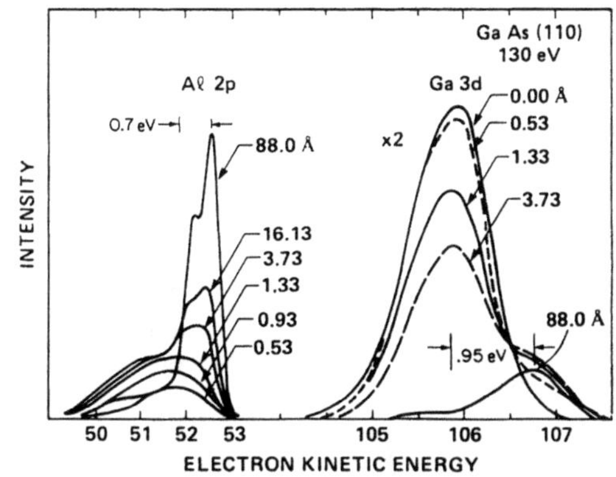

Figure 37. Identification of Al–Ga exchange reaction from core level shifts (after Bachrach *et al.*, Ref. 122).

indicative of the Al replacing the Ga in the surface. This argued in favor of the Al–Ga exchange reaction as exemplified in Fig. 37.[122,123] The metallic Ga thus shows up shifted to lower binding energy. A number of different regimes for this system have been established depending upon the number of initial surface defects and the deposition rate.[124–126] This system is still under active investigation.

In order to gain more insight into the behavior of this system, Bachrach and Bauer investigated the two complementary situations of Al–GaAs and Ga–AlAs.[122] They found, in fact, that the interfacial chemistry behaved differently. Figure 38 shows examples of valance band spectra. The case for Al–GaAs is similar to that shown in Fig. 35b while Ga–AlAs did not substantially modify the substrate. In the case of Al–GaAs(100), the surface remains ordered even during the dilute stage.

A number of workers have extensively studied the low-temperature growth of thick single-crystal aluminum layers on GaAs(100).[13,56,127–129]

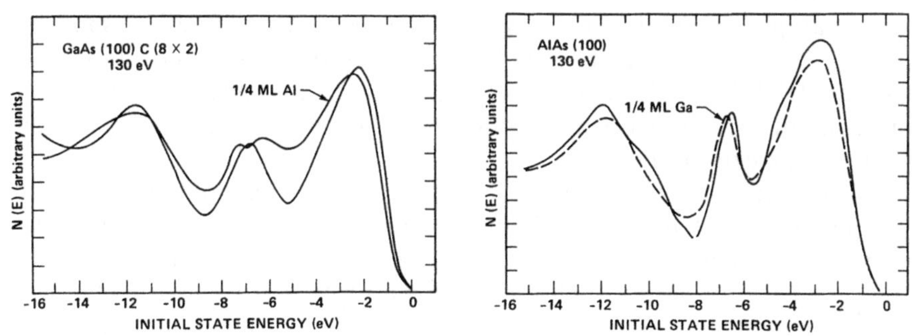

Figure 38. Comparison of Al–GaAs and Ga–AlAs showing the effect of the Al–Ga exchange reaction (after Bachrach *et al.*, Ref. 122).

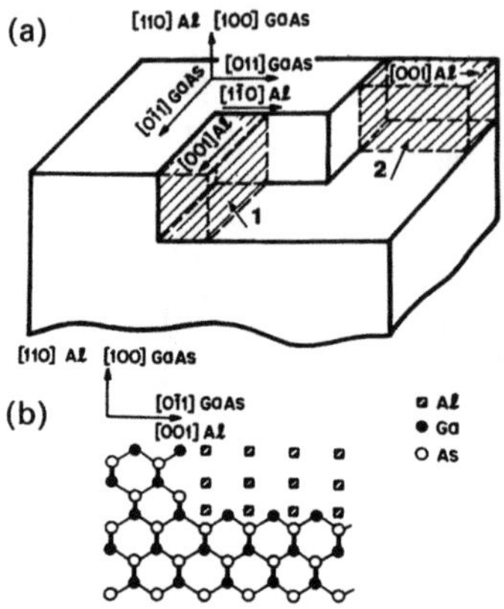

Figure 39. Possible crystallographic origin of Al surface domains for example shown in Fig. 17. (a) Schematic of two possible configurations of [110] oriented Al domains epitaxially grown at steps on a [100] substrate. (b) Schematic of the possible atomic configuration of a [110] Al film epitaxially grown on a [100] surface. Projection on a (011) GaAs plane (after Petroff *et al.*, Ref. 92.)

Prinz *et al.* have achieved single-crystal growth on the (110) face using rapid deposition with the substrate slightly cooled.[67] Whereas the (100) face provies a two-dimensional match, the (110) surface only provides a direct match in one direction. There are a number of different regimes which pertain, depending upon the stoichiometry of the starting surface and the deposition rate. In general, dislocations in the metal film do not start to appear until the film is thicker than 600 Å.

Figure 39 shows an example of the origin of the different orientations deduced by Petroff which could nucleate at steps.[92]

6.2.2. The GaAlAs Ternary System with Au Schottky Barriers

In addition to the isoelectronic Schottky barrier system discussed above, Schottky barriers with gold have been extensively studied. Gold is relatively easy to evaporate and chemically inert so that it is a stable system to investigate.

Examples of detailed investigation are the work of Best[130] shown in Fig. 22. The variation in barrier height across the whole alloy range was measured.[130] This alloy system in fact deviates from the common anion rule discussed in Section 2 since in this case one would expect little variation as one changed the cation ratio.

The microscopics of interface development with gold has been investigated and a characteristic feature is strong interdiffusion. The properties of Au Schottky barriers are examples of another class of phenomena where eutectic formation seems to play an important role. A particularly interesting probe is monitoring the gold spin–orbit splitting as a function of deposited material. Figure 40 shows some data from Chye *et al.*,[131] while Fig. 41 shows similar data obtained by Bauer *et al.*[123] displayed as an evolution plot. Note

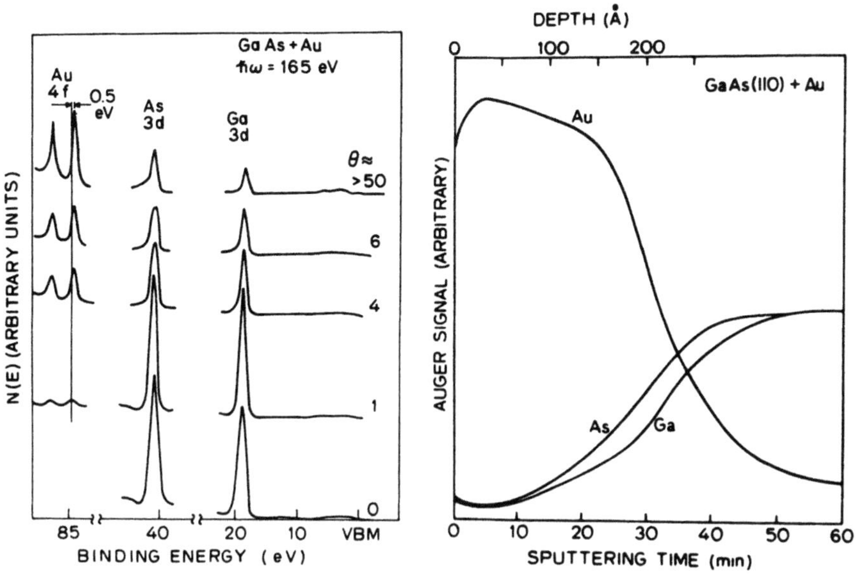

Figure 40. (a) Evolution of gold spin–orbit splitting as a function of gold coverage. (b) Sputter profile showing the interdiffusion of Au and GaAs (after Chye *et al.*, Ref. 131).

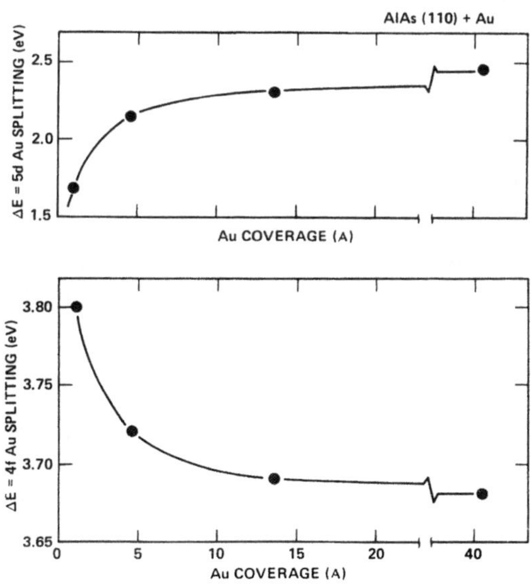

Figure 41. Gold spin–orbit splitting convergence as a function of deposition (after Bauer *et al.*, Ref. 123).

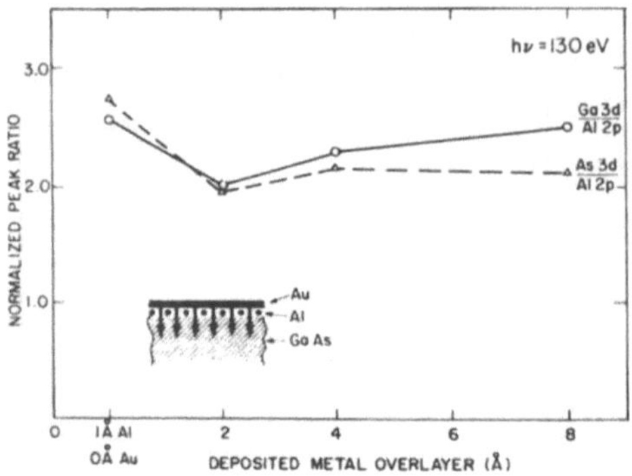

Figure 42. Example of determination of gold interdiffusion relative to an aluminum surface marker (after Brillson *et al.*, Ref. 90).

that the equivalent of many monolayers of gold are deposited before the film displays the characteristic of a metal film.

Brillson, Bauer, and Bachrach studied the interdiffusion with the use of a diffusion marker.[(90)] By putting a one-half monolayer of aluminum at the interface, they showed that one could deduce the direction of diffusion of Au into the surface. Figure 42 characterizes data obtained for anion/marker and cation/marker. The decrease by 25% with the initial Au deposition is consistent with Au diffusion past the Al into GaAs. With further deposition, however, some outdiffusion is observed. A strong driving force for the interdiffusion seems to be gold eutectic alloy formation.[(88,89)] Comparison of the behavior of GaAs and AlAs resulted in a similar picture being developed.

Brillson extended these measurements and used a range of different marker elements such that the systematics of the barrier width were determined as shown in Fig. 16.[(91)] Brillson showed that the chemical trapping deduced for Al–GaAs was an important aspect determining the width of the interfacial region.

6.2.3. InP

InP has a number of unique aspects because its surface is particularly unstable against loss of phosphorous. Wieder,[(132)] Williams,[(133)] and Brillson[(134)] have produced many interesting studies of barrier formation on this semiconductor.

InP is an interesting example of the influence of surface preparation on the barrier height. Williams[(133)] has shown that InP has a tendency to lose surface P. Metals which form large barriers on the stoichiometric surface will form ohmic contacts on the P depleted surfaces. Figure 43 shows an example of this behavior. The inset shows the data which for the reactive metals have a negligible width. The main figure shows the very different interdiffusion behavior for the reactive and unreactive metals.

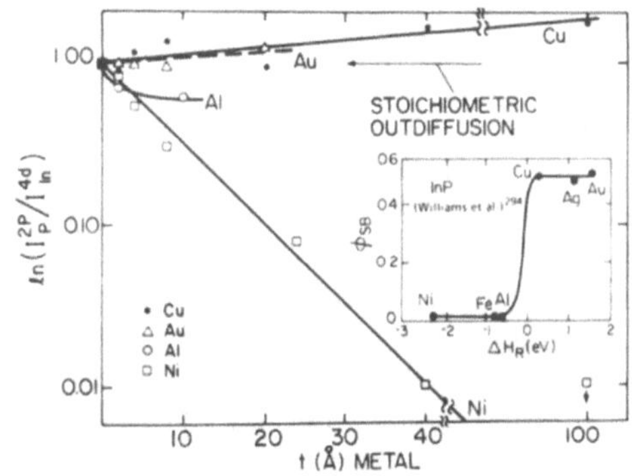

Figure 43. Schottky barrier behavior on InP (after Williams *et al.*, Ref. 133).

6.2.4. Some II–VI Examples

The II–VI's have been intensively investigated for they provide the transition from the covalent to ionic regime. Particularly interesting studies have been performed on II–VI alloys.[135] Significant analogies exist with the behavior of the III–V's, and these are discussed in Ref. 9. In general, the

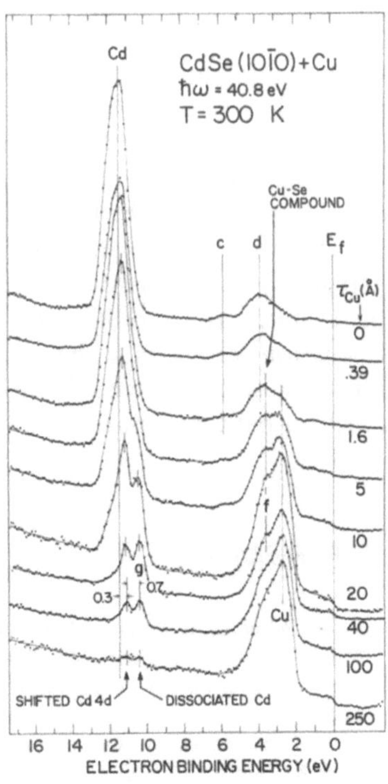

Figure 44. Photoemission core level and valance band spectra for Cu deposited on CdSe (after Brucker and Brillson, Ref. 137).

diffusion constants are larger for the II–VI's, so that the interdiffusion distances are much larger. In general, the mechanisms for Schottky barrier formation decompose into the same components as for III–V's, but with lower numbers of interface states.

Figure 44 gives a UPS example of evidence for compound formation in the case of Cu on CdSe.[136] The Cd chemical shifts are indicative of disassociated metal. Examination with energy-dependent variation of the photoemission suggests that the Cd remains beneath the metal surface and near the interface.

Interestingly, the sign of the interface dipole in the II–V's is reversed with respect to the III–V's and this gets reflected in the interdiffusion that occurs. An example is Al–CdS which can be compared with Al–GaAs. Investigation of barrier evolution for Al on CdS with the surface photovoltage technique showed the dipole had the reverse sign.[137]

7. SUMMARY

This chapter has described the current microscopic understanding that has been developed of the chemistry and structure of Schottky barrier formation. The discussion has been restricted to ideal barriers created from what would initially be a two-phase system. Presentation of the general aspects has been followed with a set of case studies. The discussion has described how the formation of Schottky barriers is dominated by interfacial reactions and interdiffusion which create new phases of interfacial layers and charged states which determine the barrier that forms. The origins of the charge states were related to the two principal mechanisms of metal-induced gap states and interfacial defect states. The barrier height depends both on these states and modification of the electron affinity at the interface. The discussion provides a background for understanding why most metal–semiconductor systems deviate from the original conception of an abrupt junction between phases.

REFERENCES

1. W. Schottky, *Z. Phys.* **113**, 367 (1939).
2. N.F. Mott, *Proc. Cambridge Phil. Soc.* **34**, 568 (1938).
3. J. Bardeen, *Phys. Rev.* **71**, 717 (1947).
4. S. Kurtin, T.C. McGill, and C.A. Mead, *Phys. Rev. Lett.* **22**, 1433 (1970).
5. S.M. Sze, *Physics of Semiconductor Devices*, 2nd ed., Wiley–Interscience, New York (1981).
6. H.K. Henisch, *Rectifying Metal Contacts*, Clarendon Press, Oxford (1957).
7. D.R. Frankl, *Physics of Semiconductor Devices*, Pergamon Press, New York (1967).

8. E.H. Rhoderick, *Metal Semiconductor Contacts*, Clarendon Press, Oxford (1980).
9. L.J. Brillson, Surface Science Reports, Vol. 2 (1982). This review contains over 1000 references.
10. E.H. Nicollian and A.K. Sinha, in *Thin Films—Interdiffusion and Reactions*, Chap. 13, (J.M. Poate, K.N. Tu, and J.W. Mayer, eds.), John Wiley and Sons, New York (1978).
11. A.Y. Cho and J.R. Authur, *Prog. Solid State Chem.* **10**, 157 (1975).
12. R.Z. Bachrach, *Crystal Growth*, 2nd ed. Chap. 6, (Brian Pamplin, ed.), Pergamon Press, New York (1980).
13. R. Ludeke, L.L. Chang, and L. Esaki, *Appl. Phys. Lett.* **23**, 202 (1973).
14. J.C. Bean and J.M. Poate, *Appl. Phys. Lett.* **37**, 643 (1980).
15. C.A. Mead, *Solid State Electron.* **9**, 1023 (1966).
16. A. Thanailakis, *J. Phys. C: Solid State Phys.* **8**, 655 (1975).
17. S.G. Louie, J.R. Chelikowsky, and M.L. Cohen, *Phys. Rev. B* 15, 2154 (1977).
18. J.E. Rowe, *J. Vac. Sci. Technol.* **13**, 798 (1976).
19. I. Lindau, P.W. Chye, C.M. Garner, P. Pianetta, C.Y. Su, and W.E. Spicer, *J. Vac. Sci. Technol.* **15**, 1332 (1978).
20. J.J. Scheer and J. van Laar, *Surf. Sci.* **18**, 130 (1969).
21. A.M. Cowley and S.M. Sze, *J. Appl. Phys.* **36**, 3212 (1965).
22. M.L. Cohen, *J. Vac. Sci. Technol.* **16**, 1135 (1979).
23. L.J. Brillson, *Phys. Rev. Lett.* **40**, 260 (1978).
24. J.M. Andrews and J.C. Phillips, *Phys. Rev. Lett.* **35**, 56 (1975).
25. M. Schluter, *Phys. Rev.* **B 15**, 5044 (1978).
26. G. Ottaviani, *J. Vac. Sci. Technol.* **18**, 924 (1981).
27. L.J. Brillson, *Phys. Rev. Lett.* **40**, 260 (1978).
28. J.O. McCaldin, T.C. McGill, and C.A. Mead, *Phys. Rev. Lett.* **36**, 56 (1976); *J. Vac. Sci. Technol.* **13**, 802 (1976).
29. R.Z. Bachrach, *J. Vac. Sci. Technol.* **15**, 1340 (1978).
30. J. van Laar and J.J. Scheer, *Surf. Sci.* **8**, 342 (1967).
31. M. Henzler, *Surf. Sci.* **36**, 109 (1973).
32. P.W. Chye, I.A. Babalola, T. Sukegawa, and W.E. Spicer, *Phys. Rev. Lett.* **35**, 1602 (1975).
33. R.H. Williams, R.R. Varma, and A. McKinley, *J. Phys. C: Solid State Phys.* **10**, 4545 (1977).
34. L.J. Brillson, *Surf. Sci.* **69**, 62 (1977).
35. J.B. Pendry, *Low-Energy Electron Diffraction*, Academic Press, London (1974); G.A. Somorjai and M.A. van Hove, in *Structure and Bonding*, Springer-Verlag, Berlin (1979); J.J. Lander, *Prog. Solid State Chem.* **2**, 26 (1965).
36. D.W. Pashley, *Epitaxial Growth*, Part A (J.W. Mathews, ed.), Academic Press, New York (1975).
37. A. Bianconi, *Appl. Surf. Sci.* **6**, 392 (1981); T.H. Hayes and J.B. Boyce, *Solid State Physics*, pp. 173–351, Academic Press, New York (1982).
38. See Ref. 9 Section 3.3.2 for a good discussion.
39. J.W. Mayer and J.M. Poate, Depth profiling techniques, *Thin Films—Interdiffusion and Reactions*, Chap 6, (J.M. Poate, K.N. Tu, and J.W. Mayer, eds.), John Wiley and Sons, New York (1978).
40. L.C. Feldman and P. Eisenberger, *Science V* **214**, N4518, 300 (1981).
41. W.C. Marra, P. Eisenberger, and A.Y. Cho, *J. Appl. Phys.* **50**, 6927 (1979).
42. P.M. Petroff, *J. Vac. Sci. Technol.* **14**, 973 (1977).
43. C.C. Chang, in *Characterization of Solid Surfaces* (P.F. Kane and G.B. Larabee, eds.), Plenum Press, New York (1974).
44. H.C. Gatos and J. Lagowski, *J. Vac. Sci. Technol.* **10**, 130 (1973). See also Ref. 9, Section 3.3.1; J.M. Palau, E. Testemale, and L. Lassabatere, *J. Vac. Sci. Technol.* **19**, 192 (1981).

45. I. Lindau and W.E. Spicer, in *Synchrotron Radiation Research*, p. 159, Plenum Press, New York (1980); L. Ley and M. Cardona, *Photoemission in Solids*, Topics in Applied Physics, Vol. 27, Springer-Verlag, New York (1979).
46. F.C. Brown, R.Z. Bachrach, and N. Lien, *Nucl. Instr. Methods* **152**, 73 (1978); F.C. Brown, R.Z. Bachrach, S.B. hagstrom, N. Lien, and C.H. Pruett, *Vacuum Ultraviolet Radiation Physics*, p. 785 (E.E. Koch, R. Haensel, and C. Kunz, eds.), Pergamon, Vieweg (1974).
47. V. Heine, *Phys. Rev.* **138**, A1689 (1965).
48. J. van Laar and A. Huijser, *J. Vac. Sci. Technol.* **13**, 769 (1975).
49. R.Z. Bachrach and B.S. Krusor, *J. Vac. Sci. Technol.* **18**, 756 (1981).
50. A. Amith and P. Mark, *J. Vac. Sci. Technol.* **15**, 1344 (1978).
51. J.M. Palau, E. Testemale, I. Ismail, and L. Lassabatere, *Solid State Electron.* **25**(4), 285 (1981).
52. C.B. Duke, *Crit. Rev. Solid State Mater. Sci.* **8**, 69 (1978).
53. D.J. Chadi, *J. Vac. Sci. Technol.* **17**, 989 (1980).
54. D.E. Eastman, *J. Vac. Sci. Technol.* **17**, (1979).
55. C.B. Duke, R.J. Meyer, and P. Mark, *J. Vac. Sci. Technol.* **17**, 971 (1980).
56. A.Y. Cho and P.D. Dernier, *J. Appl. Phys.* **49**, 3328 (1978).
57. D.J. Chadi, *J. Vac. Sci. Technol.* **16**, 1290 (1979).
58. R. Feder, W. Monch, and P.P. Auer, *J. Phys. C* **12**, 1179 (1979).
59. D. Haneman, *Adv. Phys.* **31**, 165 (1982).
60. K.C. Pandey, *Phys. Rev. Lett.* **47**, 1913 (1981).
61. D.J. Chadi, *Surf. Sci.* **99**, 1 (1980); *Phys. Rev. B* **26**, 4762 (1982).
62. M.J. Cardillo, *Phys. Rev. B* **23**, 4279 (1981).
63. D.J. Chadi, *Phys. Rev. B* **19**, 2074 (1979).
64. A. Huijser and J. van Laar, *Surf. Sci.* **52**, 202 (1975).
65. A.Y. Cho, *J. Appl. Phys.* **47**, 2841 (1976).
66. R.Z. Bachrach, *Prog. Crys. Growth Charact.* **2**, 115 (1979).
67. R.Z. Bachrach, R.S. Bauer, G.V. Hansson, and P. Chiaradia, *J. Vac. Sci. Technol.* **18**, 797 (1981).
68. D.J. Chadi, C. Tanner, and J. Ihm, *Surf. Sci.* **120**, 1425 (1982).
69. G.F. Prinz, J.M. Ferrari, and M. Goldenberg, *Appl. Phys. Lett.* **40**, 155 (1982).
70. D.J. Chadi and R.Z. Bachrach, *J. Vac. Sci. Technol.* **16**, 1159 (1979).
71. J. Ihm and J.D. Joannopoulos, *Phys. Rev. B* **26**, 4429 (1982).
72. J.J. Barton, C.A. Swarts, W.A. Goddard, and T.C. McGill, *J. Vac. Sci. Technol.* **17**, 164 (1980).
73. A. Zunger, *Phys. Rev. B* **24**, 4372 (1981). This work provides extensive references to other work on this subject.
74. L.J. Brillson, R.Z. Bachrach, R.S. Bauer, and J.C. McMenamin, *Phys. Rev. Lett.* **42**, 397 (1979).
75. R.Z. Bachrach and A. Bianconi, *J. Vac. Sci. Technol.* **15**, 525 (1978).
76. R.Z. Bachrach and R.S. Bauer, *J. Vac. Sci. Technol.* **16**, 525 (1979).
77. P.R. Skeath, I. Lindau, P. Pianetta, P.W. Chye, C.Y. Su, and W.E. Spicer, *J. Electron. Spectrosc.* **17**, 259 (1979).
78. A. Zunger, *J. Vac. Sci. Technol.* **19**, 690 (1981).
79. R. Ludeke and Landgren, *J. Vac. Sci. and Technol.* **19**, 667 (1981).
80. J.R. Chelikowski, S.G. Louie, and M.L. Cohen, *Solid State Commun.* **20**, 641 (1976).
81. H.I. Zhang and M. Schluter, *Phys. Rev. B* **18**, 1923 (1978).
82. K. Mednick and L. Kleinman, *Phys. Rev. B* **22**, 5768 (1980); E.B. Caruthers, L. Kleinman, and G.P. Alldredge, *Phys. Rev. B* **9**, 3330 (1974).

83. I. Batra and Herman, *J. Vac. Sci. Technol.* (1983).
84. K.N. Tu and J.W. Mayer, Silicide formation, in *Thin Films—Interdiffusion and Reactions*, Chap. 10 (J.M. Poate, K.N. Tu, and J.W. Mayer, eds.), John Wiley and Sons, New York (1978).
85. A. Christou and H.M. Day, *J. Appl. Phys.* **47**, 4217 (1976).
86. R.Z. Bachrach, R.S. Bauer, J.C. McMenamin, and A. Bianconi, *AIP Conf. Ser.* **43**, 1073 (1979).
87. G.P. Schwartz and A.Y. Cho, *J. Vac Sci. Technol.* **19**, 607 (1981).
88. X.F. Zeng and D.D.L. Chung, *Thin Solid Films* **93**, 207 (1982); *J. Vac. Sci. Technol.* **21**, 611 (1982).
89. J.M. Vandenberg and R.A. Hamm, *J. Vac. Sci. Technol.* **19**, 84 (1981).
90. L.J. Brillson, R.S. Bauer, R.Z. Bachrach, and G.V. Hansson, *Phys. Rev. B* **23**, 6204 (1981).
91. L.J. Brillson, C.F. Brucker, N.G. Stoffel, A.D. Katnani, and G. Margaritondo, *Phys. Rev. Lett.* **46**, 838 (1981).
92. P.M. Petroff, L.C. Feldman, A.Y. Cho, and R.S. Williams, *J. Appl. Phys.* **52**, 7317 (1981).
93. R.S. Nowicki and M.A. Nicolet, *Thin Solid Films* **96** 317 (1982).
94. L.J. Brillson, *Phys. Rev. Lett.* **38**, 735 (1977).
95. J.C. Phillips, *J. Vac. Sci. Technol.* **11**, 947 (1974).
96. J.M. Palau, A. Ismail, E. Testemale, and L. Lassabatere, Proceedings 16th International Conference on the Physics of Semiconductors (1982).
97. J.M. Rowe, S.B. Christman, and H. Margaritondo, *Phys. Rev. Lett.* **35**, 1471 (1975).
98. D.E. Eastman and W.D. Grobman, *Phys. Rev. Lett.* **28**, 1378 (1978); L.F. Wagner and W.E. Spicer, *Phys. Rev. Lett.* **28**, 1381 (1972).
99. R.H. Williams, R.R. Varma, and A. McKinley, *J. Phys. C: Solid State Phys.* **10**, 4545 (1977).
100. A.J. Bennett and C.B. Duke, *Phys. Rev.* **160**, 541 (1967); **162**, 578 (1967).
101. B. Pellegrini, *Phys. Rev. B* **7**, 5299 (1973).
102. J.C. Phillips, *J. Vac. Sci. Technol.* **11**, 947 (1974).
103. M.S. Daw and D.L. Smith, *Phys. Rev. B* **20**, 5150 (1979); *J. Vac. Sci. Technol.* **17**, 1028 (1980).
104. R.E. Allen and J.D. Dow, *Phys. Rev. B* **25**, 1423 (1982).
105. I. Lindau and W.E. Spicer, *Electron Spectroscopy: Theory, Techniques, and Applications*, Vol. 4, Chap. 4 (C.R. Brundle and A.D. Baker, eds.), Academic Press, New York (1981).
106. W.E. Spicer, P.W. Chye, P.R. Skeath, C.Y. Su, and I. Lindau, *J. Vac. Sci. Technol.* **16**, 1422 (1979).
107. D. Bolmont, P. Chen, V. Mercier, and C.A. Sebenne, Proceedings 16th International Conference on the Physics of Semiconductors (1982).
108. M.L. Cohen, *Adv. Electron. Electron Phys.* **51**, 1 (1980).
109. J.J. Lander and J. Morrison, *J. Appl. Phys.* **36**, 1706 (1965).
110. G.V. Hansson, R.Z. Bachrach, R.S. Bauer, and P. Chiaradia, *Phys. Rev. Lett.* **46**, 1033 (1981).
111. G.V. Hansson, R.Z. Bachrach, R.S. Bauer, and P. Chiaradia, *J. Vac. Sci. Technol.* **18**, 550 (1981).
112. S.S. Lau and W.F. van der Weg, Solid phase epitaxy, in *Thin Films—Interdiffusion and Reactions*, Chap. 12 (J.M. Poate, K.N. Tu, and J.W. Mayer, eds.), John Wiley and Sons, New York (1978).
113. D.J. Chadi, R.S. Bauer, R.H. Williams, G.V. Hansson, R.Z. Bachrach, J.C. Mikkelsen, Jr., F. Houzay, G.M. Guichar, R. Pinchaux, and Y. Petroff, *Phys. Rev. Lett.* **44**, 799 (1980).
114. N.W. Cheung, P.J. Grunthaner, F.J. Grunthaner, J.W. Mayer, and B.M. Ullrich, *J. Vac. Sci. Technol.* **18**, 917 (1981).
115. S.P. Muraka, *J. Vac. Sci. Technol.* **17**, 775 (1980).

116. J.L. Freeouf, *Solid State Commun.* **33**, 1059 (1980).
117. P.E. Schmid, P.S. Ho, H. Foll, and G.W. Rubloff, *J. Vac. Sci. Technol.* **18**, 937 (1981); P.E. Schmid, P.S. Ho, H. Foll, and T.Y. Tan, *Phys. Rev. B* **28**, 4593 (1983).
118. P.R. Skeath, I. Lindau, P.W. Chye, C.Y. Su, and W.E. Spicer, *J. Vac. Sci. Technol.* **16**, 1143 (1979).
119. J.M. Woodall, C. Lanza, and J.L. Freeouf, *J. Vac. Sci. Technol.* **15**, 1436 (1978); J.M. Woodall and J.L. Freeouf, *J. Vac. Sci. Technol.* **19**, 794 (1981).
120. G.W. Gobelli and F.G. Allen, *Semiconductors and Semimetals*, Vol. 2, Chap. 11 (R.K. Willardson and Albert C. Beer, eds.), Academic Press, New York (1966).
121. R.H. Williams, *J. Vac. Sci. Technol.* **18**, 929 (1981).
122. R.Z. Bachrach, R.S. Bauer, P. Chiaradia, and G.V. Hansson, *J. Vac. Sci. Technol.* **19**, 335 (1981).
123. R.S. Bauer, R.Z. Bachrach, G.V. Hansson, and P. Chiaradia, *J. Vac. Sci. Technol.* **19**, 674 (1981).
124. C.B. Duke, A. Paton, R.J. Meyer, L.J. Brillson, A. Kahn, D. Kanani, J. Carelli, J.L. Yeh, G. Margaritondo, and A.D. Katnani, *Phys. Rev. Lett.* **46**, 440 (1981).
125. P. Skeath, I. Lindau, C.Y. Su, P.W. Chye, and W.E. Spicer, *J. Vac. Sci. Technol.* **17**, 869 (1980).
126. R.R. Daniels, A.D. Katnani, Te-Xiu Zhao, G. Margaritondo, and A. Zunger, *Phys. Rev. Lett.* **49**, 899 (1982).
127. R. Ludeke and G. Landgren, *J. Vac. Sci. Technol.* **19**, 667 (1981).
128. J. Massies, J. Chaplart, and N.T. Linh, *Solid State Commun.* **32**, 707 (1979).
129. J. Massies and N.T. Linh, *Surf. Sci.* **114**, 147 (1981).
130. J.S. Best, *Appl. Phys. Lett.* **34**, 522 (1979).
131. P.W. Chye, I. Lindau, P. Pianetta, C.M. Garner, C.Y. Su, and W.E. Spicer, *Phys. Rev. B* **18**, 5545 (1978).
132. H.H. Wieder, *J. Vac. Sci. Technol.* **15**, 1498 (1978).
133. R.H. Williams, V. Montgomery, R.R. Varma, and A. McKinley, *J. Phys. C* **11**, 1989 (1978).
134. L.J. Brillson, C.F. Brucker, A.D. Katnani, and N.G. Stoffel, *Appl. Phys. Lett.* **37**, 917 (1980).
135. C.A. Mead, *Appl. Phys. Lett.* **6**, 103 (1965).
136. C.F. Brucker and L.J. Brillson, *J. Vac. Sci. Technol.* **18**, 787 (1981).
137. L.J. Brillson, *J. Vac. Sci. Technol.* **16**, 1137 (1979).

3

Fabrication and Characterization of Metal–Semiconductor Schottky Barrier Junctions

B.L. Sharma

1. INTRODUCTION

Metal–semiconductor contacts showing rectifying properties are finding more and more applications in modern semiconductor devices technology.(1) Apart from the fact that they are comparatively easy to fabricate and incorporate into integrated circuits, the main reason for their wide usage is that they do not exhibit minority carrier effects (e.g., long reverse recovery time, diffusion capacitance, etc.) similar to those observed in *p–n* junction devices. Various metal–semiconductor systems, investigated during the last two decades, are tabulated in Table 1. It can be seen from this table that the investigations are mainly centered round Schottky contacts based on Si and GaAs. Amongst the two, Si-based Schottky contacts are at present being used in a wide variety of devices and integrated circuits. Recent advances in GaAs material and processing techniques have, however, shown that not only can all of the semiconductor device structures realized in Si be fabricated in GaAs, but optical and very high speed integrated circuits can also be realized in the not-too-distant future. It is because of this as well as metallurgical and reliability considerations that the activity pertaining to GaAs-based Schottky contacts has increased appreciably during the last few years.

Although the fabrication and characterization aspects of metal–semiconductor Schottky contacts have been briefly reviewed from time to time

---

B.L. Sharma • Solid State Physics Laboratory, Lucknow Road, Delhi 110007, India.

TABLE 1. Metal Systems Used for Schottky Barrier Contacts to Semiconductors

Semiconductor	Metal systems
(*n*)Ge	Al(2–4), Ag(4–6), Au(3, 4, 6, 7), Cu(4, 6, 8), Fe(8), Mg(3), Ni(3, 4, 6, 8), Pb(6), Pd(3,8), Pt(3, 4, 8), W(9)
(*n*)Si	Al(10–13), Ag(14–16), Au(15, 17–19), Ca(20), Cd(21), Cr(22, 23), Cs(24), Cu(16, 25), Fe(8), In(21), Ir(26), K(24), Mg(27), Mo(23, 28, 29), Na(24), Ni(8, 29, 30), Os(31), Pb(25), Pd(8, 32–34), Pt(8, 28, 35), Re(31), Rh(18, 36), Sn(16), Ta(37), Ti(38, 39), V(40), W(37), Zn(41), Ag–Cu(42), Au–Cu(42), $CoSi_2$(43), $CrSi_2$(44), $DySi_2$(45), $ErSi_2$(45), $GdSi_2$(45), HfSi(44), $HoSi_2$(45), In–Cd(21), IrSi(26), Mo–Au(29), $MoSi_2$(43), Ni–Cr(29), NiSi(46, 47), $NiSi_2$(44), PdSi, Pd_2Si(48–50), PtSi, Pt_2Si(48, 49, 51), RhSi(52), $TaSi_2$(44), $TiSi_2$(44), WSi, WSi_2(44, 53), YSi_2(45), $ZrSi_2$(54)
(*p*)Si	Al(11, 55, 56), Ag(56), Au(56), Cr(57), Cs(24), Cu(56), Er(58), Hf(59, 60), In(16, 18), Mn(61), Ni(56), Pb(56), Pd(29), Sb(16), Tb(58), Ti(39), Y(58), Yb(58, 62), $DySi_2$(45), $ErSi_2$(45), $GdSi_2$(45), NiSi(47), RhSi(63)
(*n*)AlAs	Au(64), Pt(64)
(*n*)AlSb	Au(64)
(*n*)GaAs	Al(65–69), Ag(70, 71), Au(68, 69, 72, 73), Be(74), Bi(69), Co(75), Cr(76, 77), Cu(78), Hf(79), Hg(80), In(69, 78), Mg(69), Mo(81, 82), Ni(77, 78, 83–86), Pb(78, 87), Pd(88–90), Pt(82, 91–94), Rh(95), Sb(69), Sn(70, 77), Ta(82, 96), Ti(74, 82, 97), W(82, 98–100), Au–Cr(76), Au:Ga(101), Au:Ge(102), Au–Mo(103), Au–Ni(83, 104), Au–Sn(105), Au–W(106, 107), Cr–Ni(108), Pd–Ge(109), Pt–Cr(110, 111), Pt–Ni(112), Pt–W(107, 111), Ti–Ag(113), Mo–Pt–Au(114), Ti–Mo–Au(115), Ti–Pd–Au(116, 117), Ti–Pt–Au(114, 116, 117)
(*p*)GaAs	Al(74), Au(74), Nb(118), Pd(89), Sn(74)
(*n*)GaP	Al(119, 120), Ag(119, 121), Au(119, 122–124), Cr(119), Cu(119, 120), Mg(120), Mo(119), Ni(119, 124–127), Pt(119, 121, 128)
(*p*)GaP	Al(129), Au(130, 131)
(*n*)GaSb	Ag(132), Au(132–134)
(*p*)InAs	Au(135)
(*n*)InP	Al(136–140), Ag(136, 139, 141, 142), Au(138–145), Cr(142), Cu(139, 142), Hg(146), In(142), Ni(142), Pd(89), Pt(142), Sn(142), Ag–Ga(147), Au–Ti(140, 148), Au–Zn(149), Ti–Pd–Au(143)
(*p*)InP	Au(64), Pd(89)
(*n*)InSb	Al(150), Ag(64, 150), Au(64, 150)
(*p*)InSb	Ag(151), Au(151), Cu(151), In(151)
(*n*)CdS	Al(152), Ag(64, 153), Au(152–156), Cu(64, 153, 157), Ni(64), Pd(153), Pt(64, 153), Au–Ti(158)
(*n*)CdSe	Al(159), Ag(160, 161), Au(160, 161), Cu(161), Pt(160, 161)
(*n*)CdTe	Al(64, 162), Ag(64, 163), Au(64, 162, 164, 165), Cu(162, 164), In(166), Ni(162, 164), Pt(64, 162), Sn(164)
(*n*)ZnS	Al(167, 168), Ag(71, 168, 169), Au(167–173), Cu(168, 169), Cr(71), In(168, 169), Mg(168), Mn(169), Pd(168), Pt(168, 169), Ti(168)
(*n*)ZnSe	Al(167, 170, 171), Ag(170, 171), Au(167, 170–173), Bi(171), Cd(170, 171), Cu(174), Fe(171), Hf(173), In(171), Mg(171), Pb(171), Pt(174), Sb(171), Ti(174)

(continued)

TABLE 1. (cont.)

Semiconductor	Metal systems
(p)ZnTe	Al(175), Au(175), Ni(175), Ta(175)
(p)GaS	Al(176), Au(176), Cu(176), Pd(176), Sn(176)
(p)GaSe	Al(176, 177), Ag(176), Au(176), Ca(176), Cs(176), Cu(176), In(176), Li(176), Mg(176), Pd(176), Pt(176), Sn(176)
(p)GaTe	Al(176), Ag(176), Au(176), Mg(176)
(p)PbS	Pb(178)
(p)PbSe	Pb(179)
(n)PbTe	Au(180), Pt(180)
(p)PbTe	Au(180), Cu(181), In(180, 181), Pb(135, 180–182), Sn(135, 180), Zn(180, 181)
(n)WSe_2	Au(183)
(n)$Al_xGa_{1-x}As$	Au(184–187), In(187)
(n)$Al_xGa_{1-x}Sb$	Au(188), Ni(189), Pd(189)
(n)$GaAs_{1-x}P_x$	Au(190–193), Mo(194)
(n)$Ga_xIn_{1-x}Sb$	Au(195)
(n)$In_xGa_{1-x}As$	Au(138, 196), Au(138, 197, 198), Pd(199), Pt(200), Au–Ni(201)
(p)$Cd_xHg_{1-x}Te$	Al(202), Au(203), Cr(202), In(203), Zn(203)
(p)PbS_xSe_{1-x}	In(178), Pb(178)
(p)$Pb_{1-x}Sn_xSe$	Pb(204)
(p)$Pb_{1-x}Sn_xTe$	Pb(180, 205, 206)
(n)$AgGaSe_2$	Al(207), Ni(207)
(n)$CdIn_2S_4$	Au(208)
(n)$CdSnAs_3$	Al(209), Au(209)
(p)$CuGaSe_2$	Au(207), Cu(210), Ni(207)
(n)$CuInSe_2$	Au(211)
(p)$CuInSe_2$	Al(207), Ni(207)

in the literature,[1,52,212–215] an attempt in this chapter is made to consolidate the available information on these aspects in one place. This will not only act as a ready reference source but will also provide the readers with an insight into the areas which need further probing. In particular, design considerations, fabrication processes, and characterization aspects of metal–silicon and metal–gallium arsenide junctions are discussed in this chapter.

2. SELECTION OF SEMICONDUCTOR MATERIALS

Metal–semiconductor junction properties of nearly all important semiconductors in contact with a variety of metals have been investigated in one form or another. Although the selection of a semiconductor and a Schottky contact metal system depend on the particular application of Schottky barrier junction, certain general conclusions can be drawn by comparing the selected properties of the four important semiconductors (viz., Ge, Si, InP, and GaAs) whose present-day technologies are well advanced (refer to Table 2). It is clear

TABLE 2. Comparison of the Selected Properties of Ge, Si, InP, and GaAs

Properties	Ge	Si	InP	GaAs	Comments
Melting point (°C)	936	1412	1058	1238	The higher the melting point the better is the processing maneuverability.
Energy gap (eV)	0.67	1.11	1.35	1.43	The higher the energy gap the higher is the power handling capacity, the lower is the saturation current, and the better is the operating capability at higher operating temperatures.
Type of energy gap	Indirect	Indirect	Direct	Direct	Direct energy gap requirement essential for some special optoelectronic and microwave components.
Maximum mobility of electrons ($cm^2/V\,sec$)	3950	1900	4600	8000	The higher the mobility the smaller are the transit time and parasitic resistance, and the higher is the frequency operations.
Maximum mobility of holes ($cm^2/V\,sec$)	1900	450	150	400	
Effective electron mass (m_0)	0.55	1.1	0.077	0.067	The lower the effective mass the smaller are the transit time and parasitic resistance, and the higher is the frequency operations.
Effective hole mass (m_0)	0.37	0.59	0.64	0.48	
Intrinsic carrier concentration (cm^{-3})	1.7×10^{13}	1×10^{10}	1.1×10^{7}	1.8×10^{6}	The lower is the intrinsic carrier concentration the better is the doping maneuverability.
Semiinsulating material	No	No	Yes	Yes	Necessary for planar processing of components and ICs for high-frequency applications.
(i) doping	—	—	Fe	Cr	
(ii) resistivity ($\Omega\,cm$)	—	—	$\sim 10^8$	$> 10^8$	The higher the resistivity the better is the mutual insulation of elements in ICs and suppression of parasitic capacitances.

Electron affinity (eV)	4.13	4.01	4.4	4.07	Important parameter required in the selection of materials for heterojunctions and Schottky barrier junctions.
Thermal expansion coefficient (10^{-6}/°C)	5.5	2.44	4.5	6	Thermal expansion coefficient helps in determining the suitability of the depositing materials as their differences may possibly create destructive stresses on cooling.
Thermal conductivity (W/cm °C)	0.59	1.4	0.68	0.54	The higher the thermal conductivity the better is the power handling capability.
Dielectric constant (relative)	16	11.9	12.4	13.1	The lower the dielectric constant the lower are the junction capacitance and carrier storage effects, and the better is the high-frequency performance.
Lattice-matched heterostructure possible	Yes	No	Yes	Yes	Such heterostructures can be used to improve the device performance or ohmic contactibility.
Vapor pressure at melting point (Torr)	8.4×10^{-7}	5.6×10^{-4}	1.5×10^{4}	740	The lower the vapor pressure the better is the high-temperature process compatibility.
Maximum useful temperature of operation (°C)	100	150	250	400	This is the temperature at which thermally generated carrier concentration approaches uncompensated doping density.

from Table 2 that GaAs, by virtue of its electronic properties, has an edge over other semiconductors. It, however, has not been provided with the inherent advantages of a stable, passivating native oxide or convenient high temperature processing technology. In fact, it is these advantages which are responsible for widespread use of Si. In the case of InP the theoretically predicted advantages have not been realized to date because of relatively newer technology and low barrier heights of metal–InP structures. On the basis of the above conclusions and available information,[117,216–218] it can be indicated that the use of GaAs for some time to come will be confined to millimeter-wave and high-frequency microwave devices and monolithic integration of circuit elements. Another advantage of using GaAs is that the substrate material can either be a highly doped *n*-type or a semi-insulating wafer. With the trend towards integration of circuit elements for higher frequency and speed and virtually no package parasitics, the semi-insulating GaAs substrates are finding more and more use. Apart from these considerations, GaAs is inherently more tolerant to radiation damage than Si because of relatively short minority carrier lifetimes in it.

The requirements of processing, dimensions, electrical properties (conductivity and mobility), and the quality of the semiconductor can only be met by having at least a homoepitaxial layer (referred as an active layer) on top of the substrate. The active layer has to be nearly as thin as required for the Schottky diode, since excessive etching prior to metallization yields poor ideality factor values and "soft" reverse characteristics. This is probably due to the build-up of defects during extended chemical activity. Some of the basic structures of the Schottky diodes for high-frequency applications are schematically shown in Fig. 1. Although buffer layers between active layer and substrate are not shown in these structures, in practice they are invariably used in epitaxially grown structures. One or more techniques—namely, chemical vapor deposition, molecular beam epitaxy, ion implantation, and liquid phase epitaxy—have been used to produce active layers on different semiconductor substrates. Hammond[219] has surveyed the technology of Si epitaxy and has shown that the demand for thinner epitaxial layers of high crystalline perfection, control of autodoping, and buried layer pattern shift require low-temperature and/or low-pressure techniques. Between the two chemical vapor deposition (viz., $SiCl_4$ and SiH_4) techniques, the trend is towards using SiH_4 because of lower growth temperature. Another technique that combines the advantages of both low pressure and low temperature is molecular beam epitaxy. This technique, still in its infancy in terms of epitaxial quality and doping uniformity, may become an effective technique in the near future. However, in the case of Si, ion implantation to produce an active layer is now an established technique with its advantages of independent control of the doping level and the thickness of the active layer, good uniformity and

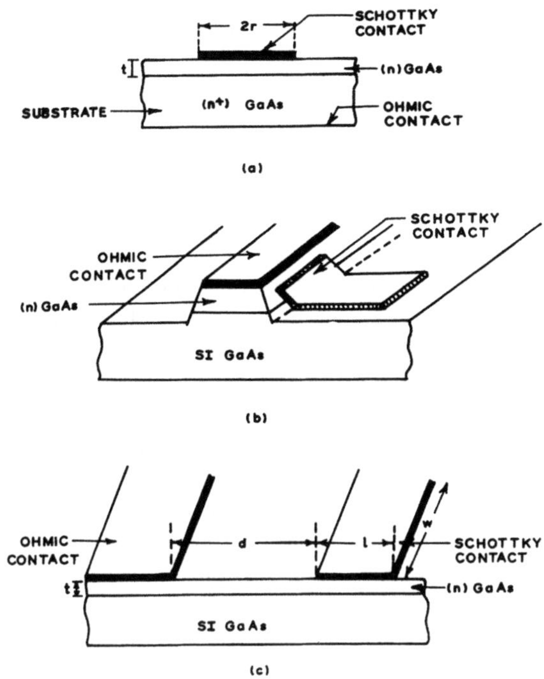

Figure 1. Some basic structures of the Schottky barrier diodes for high-frequency applications.

reproducibility of doping, and its compatibility with integrated circuits well recognized by the technologists. One of the main drawbacks of ion implantation (or for that matter of any low-temperature technique) is that it requires postimplantation (or postdeposition) high-temperature annealing to remove the crystalline defects induced during the process. With the availability of cw laser/electron beam annealing techniques it is possible to anneal out these defects in Si without raising the temperature of the whole wafer. On the other hand, DiLorenzo[(220)] has reviewed the status of the various techniques mentioned above to produce active layers for GaAs microwave and high-speed devices and has concluded that for discrete microwave devices chemical vapor deposition is at present the best technique to grow active layers, but for high-speed devices ion implantation happens to be an essential technique. It must, however, be noted that n/n^+Si structure for diodes can not be fabricated by ion implantation. For planar GaAs integrated circuits, the optimization of different devices has to be accomplished by using multiple localized ion implants into a high-resistivity epitaxial buffer layer or directly into the semi-insulating GaAs substrate.[(216,221)] Molecular beam epitaxy and organometallic chemical vapor deposition techniques will, however, be required in future to form heterostructures on the substrates (e.g., $In_{0.53}Ga_{0.47}As$ on lattice matched InP) to improve performances of the devices.

3. METAL–SEMICONDUCTOR SYSTEMS

Apart from the selection of a semiconductor, barrier height considerations also require a proper choice of the Schottky contact metal system. The important factors on which this choice depends include the mode and ease of deposition, convenient temperature of deposition, good adherence, no (or controlled) interface reaction, good electrical and thermal behavior, no surface tension effects, inertness to ambient, ease of difining contact geometry, and adaptability to integrated circuit fabrication technology. Since the reproducibility and reliability of Schottky barrier junctions are very much process dependent and to a very great extent depend on the interdiffusion and/or reactivity between semiconductor and contact metal, various metal–semiconductor systems with Si and GaAs as semiconductor are discussed below.

3.1. Metal–Silicon Systems

Formation of Schottky barrier junctions by evaporating the metal on an etched Si surface in a vacuum chamber having a pressure of about 10^{-6} Torr or by evaporating the metal on *in situ* cleaved Si surface in an ultrahigh vacuum ($< 10^{-8}$ Torr) have been discussed in detail by Rhoderick.[52] In the former case, the Si surface is usually covered with an ultrathin (10–20 Å) native oxide. This leads to pinning of the Fermi level at the interface and, consequently, to the formation of Schottky barrier junctions which are to a great extent independent of the metals and which age with time. In the latter case, the junctions can be considered chemically ideal as there is no likelihood of any chemical contamination and/or formation of native oxide. The complications may, however, arise in such contacts by some physical damage caused to the Si surface during cleaving and/or due to chemical reactions at the interface under ultrahigh-vacuum conditions. In fact, it has been reported that, in spite of elaborate experimental setup in the latter case, there appears to be no marked improvement in the electrical characteristics of the junctions.

Although it has been known for quite some time that Si forms compounds with most elements in the Periodic Table,[222] the use of metal silicides to obtain an interface with Si that is stable and free from native oxide has aroused considerable interest during the past few years. Tu and Mayer[223] and recently Mohammadi[44] have discussed the general behavior of various

TABLE 3. Interfacial Reactions in Metal–Silicon Systems

Metal	Phases	Formation temperature (°C)	Growth rate	Lowest eutectic temperature (°C)	Melting point (°C)	Dominant diffusing species	Comments
Al				577		Si	No compound forms; solid solubility of Si in Al increases rapidly from 0.008 w/o at 250°C to 1.3 w/o at 550°C[44]
							For Schottky barrier junction Al deposition followed by thermal treatment at 400–500°C; long thermal annealing leads to *p*-layer formation[10,11,224]
							In Au/Al/Si long-time annealing leads to formation of $AuSiAl_4$ at Si interface and presence of Au–Al phases[225]
Au				370		Si	No compound forms; at temperatures substantially below eutectic temperature (100–300°C) migration of Si through Au film occurs[226,227]
Co	Co_2Si	350–500	$t^{1/2}$	1195	1332	Co	Co_2Si formed as a layer between metal and CoSi; $CoSi_2$ grows epitaxially on Si substrate; silicides also form with SiO_2[223]
	CoSi	425–500	$t^{1/2}$		1460		
	$CoSi_2$	550			1326		
Cr	Cr_3Si				1730		Reaction not only confined to Si but extends several microns laterally over oxide protected region; $CrSi_2$ forms predominant phase at temperatures 450–1000°C[228]
	CrSi				1600		
	$CrSi_2$	450	t	1320	1550		
							$CrSi_2$ is observed to have semiconducting properties with $E_g \sim 0.3\,eV$[44]
Hf	HfSi	550–675	$t^{1/2}$		2200	Si	Final phases of silicides are $HfSi_2$ and HfSi; soluable in HF[44]; reacts with SiO_2[223]
	$HfSi_2$	750		~1300	1950	Si	
Ir	IrSi	400		1470			Partial interface reaction starts converting Ir contact area to IrSi at 300°C; annealing at 500°C produces a mixtures of IrSi and Ir_2Si_3[229]
	Ir_2Si_3	600					
	$IrSi_3$	960					
Mg	Mg_2Si	≥200		637	1102	Mg	Mg_2Si forms at a temperature as low as 200°C; decomposes slowly by water and rapidly by acids[44,230]

(continued overleaf)

TABLE 3. (cont.)

Metal	Phases	Formation temperature (°C)	Growth rate	Lowest eutectic temperature (°C)	Melting point (°C)	Dominant diffusing species	Comments
Mo	$MoSi_2$ Mo_5Si_3	525	t	~1410	2050 2120	Si	$MoSi_2$ and Mo_3Si and Mo_5Si_3 phases exist in thermal annealing; Mo–silicides resistant to acids and alkalis and also oxidation resistant[44] In Au/Mo/Si structure Au diffuses through Mo and intermixes with Si; nitrited Mo instead of Mo stops interdiffusion at this temperature[231]
Nb	$NbSi_2$	650		1295	1930		$NbSi_2$ and Nb_5Si_3 are final phases; dissolve in HF[44]
Ni	Ni_2Si NiSi $NiSi_2$	200–350 350–750 750	$t^{1/2}$ t	 964 966	1318 992 993	Ni	Ni films start reacting with Si at temperatures as low as 200°C[232] Ni_2Si and NiSi phases appear first at Si surface and move progressively towards the outer surface of Ni; above 750°C $NiSi_2$ phase appears[52]
Pd	Pd_2Si PdSi	200–700 ≥700	$t^{1/2}$	760	1330 ~1100	Pd, Si	Pd_2Si forms at temperatures as low as 200°C with both Si and Pd as moving species (Si faster than Pd)[89,234] Al overlayer to be deposited at temperature lower than 50°C to avoid interaction between Al and silicide[49] Cr/Pd/Si thermal annealing at 450°C leads to Cr/$CrSi_2$/Pd_2Si/Si structure[235]
Pt	Pt_2Si PtSi	200–500 ≥300	$t^{1/2}$ $t^{1/2}$	830	1100 1229	Pt, Si	Thermal annealing of thin Pt film on Si leads to formation of Pt_2Si. After all Pt is consumed, PtSi starts forming until all Pt_2Si converts into PtSi[223]

							PtSi/Si structure forms by thermal annealing at 400°C; Al overlayer deposited at temperature lower than 50°C to avoid interaction between Al and silicide[49]
							Cr/Pt/Si thermal annealing at 450°C leads to Cr/$CrSi_2$/Pt_2Si/Si structure[236]
							In Al/W/Ti/Pt/Si structure PtSi/Si forms Schottky barrier; Ti/W (99 w/o W) behaves as inter-diffusion barrier and Ti in it improves adhesion to SiO_2 and is corrosion resistant[237]
Rh	RhSi	377				Si	RhSi starts forming at about 377°C[52]
Ta	$TaSi_2$	650		1385	2200		$TaSi_2$ and Ta_5Si_3 phases exists; slowly dissolve in HF[44]
	Ta_5Si_3				2500		
Ti	Ti_5Si_3				2120		When Ti reacts with Si first TiSi is formed and finally $TiSi_2$ and Ti_5Si_3 phases are formed; they are soluable in HF[44,238]
	TiSi	500			1760		
	$TiSi_2$	600	$t^{1/2}$	1330	1540	Si	
							Pd/Ti/Si is an effective contact metallization; electropolated Ag or Au used as overlayer[239]
							In Pt/Ti/Si substantial interdiffusion between Ti and Pt forming PtTi; $PtTi_3$ at Pt/Ti interface; Ti/TiN/Pt an effective diffusion barrier with Au overlayer[240]
V	V_3Si				2070		V reacts with Si forming VSi_2 above 500°C with Si as the predominant moving specie[44,230]
	V_5Si_3				2150		
	VSi_2	500	t and $t^{1/2}$	1400	1750	Si	
W	W_5Si_3						WSi_2 and W_5Si_3 silicides form with WSi_2 at the interface; initial rate of formation depends on Si surface preparation; WSi_2 resistant to acids and alkalis[44,241]
	WSi_2	650	t and $t^{1/2}$	1400	2160	Si	
Z	ZrSi				2150		Initial nucleation phase is ZrSi but the final phases includes $ZrSi_2$, Zr_2Si and Zr_5Si_3 along with ZrSi[44]
	$ZrSi_2$	700		~1360	1700		

silicides and their properties. The solid phase interdiffusion and interfacial reactions during conventional furnace annealing of various metal–Si systems, from the point of view of Schottky barrier contacts, are summarized in Table 3. It can be seen from this table that Al and Au do not form any intermetallic compound with Si.

Among the metals which react with Si to form intermetallic compounds, the low resistivity and process compatibility make some transition-metal silicides suitable for making Schottky contacts with Si. By and large transition-metal silicides are commonly formed by solid phase interdiffusion between Si and a transition metal layer at temperatures $T < T_e/2$, where T_e is the lowest eutectic temperature in the appropriate binary phase diagram (refer to Fig. 2). The silicides formed by this process can broadly be divided into three classes, i.e., metal-rich silicides, monosilicides, and disilicides with their temperature of formations lying in the range 200–400°C, 300–500°C, and 500–700°C, respectively. The initial silicide nucleated at the transition-metal/Si interface, however, has the composition of the silicide adjacent to the lowest eutectic temperature in the binary system.[242]

Although furnace annealing has been the most widely used process to date, its limitations (refer to Table 4) and the need for fabricating finer and finer device structures have recently led to the use of energetic beams for forming metal silicides on a Si surface. Both laser and electron beam and ion beam mixing have been used for this purpose and each of these has led to

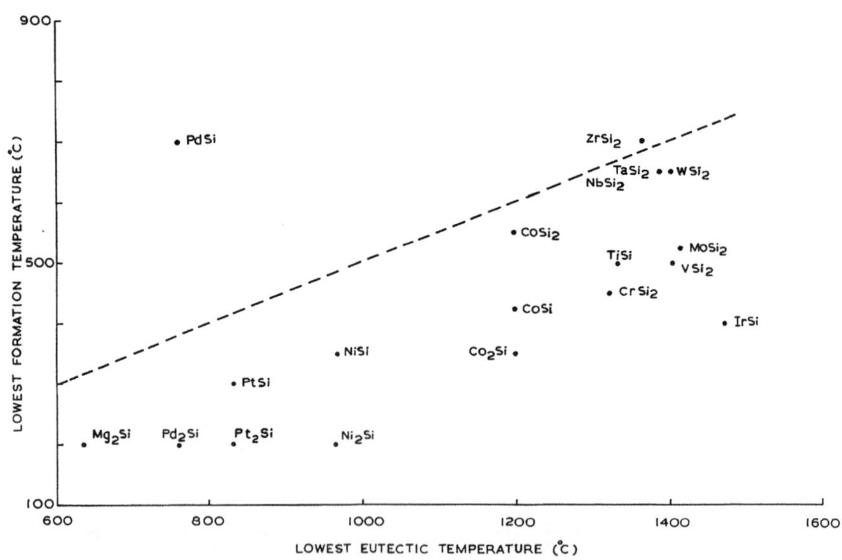

Figure 2. Plot of lowest temperature of formation of silicides against lowest eutectic temperature obtained from respective binary phase diagrams.

TABLE 4. Silicide Formation Induced by Various Processes

Formation process	Main features
Furnace annealing[223]	(i) Duration for silicide growth 10–60 min (ii) Formation via solid phase interdiffusion and reaction (iii) Layers uniform but controlled thicknesses not always possible (iv) Depending on annealing temperature and time single phase silicide formation possible (e.g., $Pt \rightarrow Pt_2Si \rightarrow PtSi$) (v) Localized heat treatment not possible (vi) Compatibility not always possible with other device fabrication processes (vii) Not possible to avoid interaction with device defining oxide (e.g., SiO_2)
CW laser beam annealing[243,244]	(i) Duration 10^{-3}–10^{-2} sec (ii) Formation via solid phase reaction (iii) Uniform layers of controlled thicknesses possible (iv) Localized heat treatment and scanning possible (v) Depending on beam power level single phase silicide formed (e.g., $Pt \rightarrow Pt_2Si$ or PtSi) (vi) To avoid reflectance a thin amorphous Si coating can be used (vii) Interaction with device defining oxide can be avoided (viii) Interference effects damage the perimeter of the device (ix) Compatible with other device fabrication processes (x) Reaction with thicker metal films possible by multiple laser scans
CW electron beam annealing[243,245]	CW electron beam annealing produces results very similar to those obtained by cw laser beam annealing except that interference effects are not present; however, in this case space charge effects appear
Pulsed laser annealing[244,246,247]	(i) Laser pulse duration $\sim 10^{-7}$ sec (ii) Formation via melting, interdiffusion and rapid solidification (iii) Mixed phase silicide layers (e.g., $Pt \rightarrow Pt_4Si$, $Pt_3Si, \ldots,$ PtSi) and in some cases amorphous silicides formed (iv) Laterally nonuniform cellular structure layers formed (v) Localized heat treatment possible (vi) Reaction with thicker metal films not possible (vii) Diffusion of surface impurities cannot be ignored (viii) Annealing through oxide layer possible (ix) Damage at the perimeter due to interference effects
Pulsed electron beam annealing[248]	Pulsed electron beam annealing produces results very similar to those obtained by pulsed laser beam annealing; possible to form thermally stable compounds whose reaction temperatures exceed melt temperature of Si
Ion-beam mixing[244,249]	(i) Ions: Xe^+, Kr^+, or Si^+ (ii) No significant reaction if ion beam does not reach interface (iii) When ion beam penetrates the interface, the initial ion-induced reactions are similar to those observed in furnace annealing (e.g., $Pt \rightarrow Pt_2Si$) (iv) Uniform silicide layers form with thickness $\propto$ (dose)$^{1/2}$ for metal-rich silicides and $\propto$ (dose) for disilicides (v) Higher doses and when complete metal is consumed formation of more Si-rich phase or amorphous mixture is observed

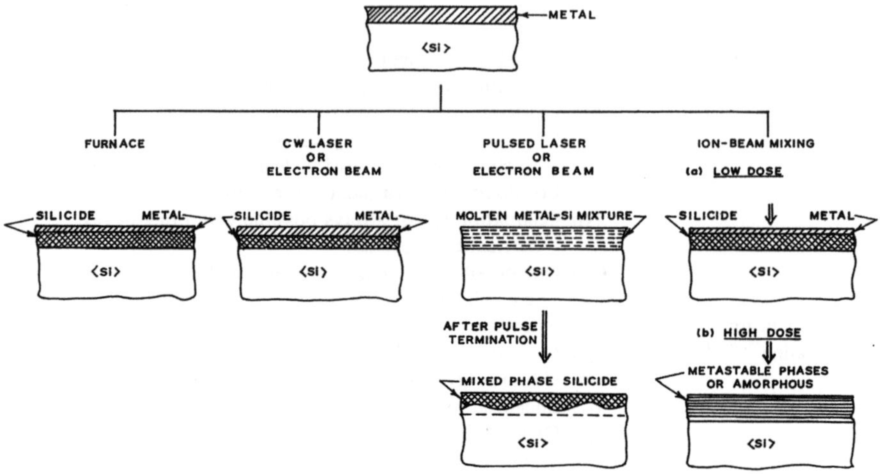

Figure 3. Schematic representation of silicide formation by various processes.

several unique results. The main features of various processes are listed in Table 4 and the schematic diagrams of the silicide formation by them are shown in Fig. 3. Energetic beam annealing unlike furnace annealing is, in principle, a localized rapid heating–cooling process. This makes precise control of silicide thickness both laterally and depthwise basically possible. Among the processes available, both scanned cw laser and electron beam annealing[244,250] provide uniform, single-phase $MoSi_2$, WSi_2 or $NbSi_2$ and essentially single-phase Pd_2Si layers on a Si surface. In the case of a Pt/Si system, however, single-phase PtSi and mixed-phase silicides with Pt_2Si_3 as dominant phase are formed by cw electron and laser beam annealings, respectively.[243] In the case of pulsed beam annealing, not only the layers are laterally nonuniform (refer to Fig. 3) but invariably are of mixed phase silicide consisting of several compounds including Si precipitation.[251,252] It is, however, possible by this process to induce metal/Si reactions which are otherwise not possible by other processes.

Of metals (Al, Ag, Au) which do not form intermetallic compounds with Si, only Al has been widely used for making Schottky as well as ohmic contacts. Reasonably high eutectic temperature (refer to Table 3), good adherence with both Si and SiO_2, and capability of forming Schottky barrier junction with lightly doped *n*-type Si and ohmic contacts with heavily doped *n*-type and *p*-type Si are some of the reasons for its extensive use in Si technology. The electrical properties of Al/(*n*)Si Schottky barrier junctions are, however, very sensitive to the method of preparation. Amongst the various steps involved in the fabrication, the postmetallization heat treatment is found to play the most important role in changing the electrical properties of Al/(*n*)Si Schottky barrier contacts. It removes the native oxide layer invariably present on the chemically cleaned Si surface and also forms (p^+) Al doped surface layer at the Al/Si interface. The (p^+) surface layer formation at the interface by heat treatment around 500°C has been explained on the basis of Si dissolution in Al during heating followed by recrystallization of dissolved Si on cooling.[10] It has also been suggested that the lower the cooling rate[10,11] and/or the larger the amount of Al available for reaction[11] the thicker is the recrystallized layer and the higher the barrier height. Recently Dascalu *et al.*[13] have observed that, when contacts are heat treated at 550°C, electrical properties, especially the barrier height, are not influenced by the above-mentioned reasons in spite of large nonuniformities in the contact geometry. At lower-temperature heat treatment the variation in barrier height may, however, be due to nonsaturation of Al film with Si and/or due to persistence of an interfacial oxide layer. This observed barrier insensitivity can only be explained by assuming that the process-dependent sensitivity is covered under the effect of excessive interface damage in this case. In fact, controlled experiments using cw energy beam annealing of an Al–Si system are necessary to understand the exact nature of the process dependency and to find a suitable process for making reliable Al/(*n*)Si Schottky barrier junctions. Thin Au films evaporated onto a freshly etched *n*-type Si surface form Schottky contact, but the migration of Si through Au at a relatively low temperature of about 100°C makes it a highly undesirable metal system for device applications. This migration/interdiffusion can, however, be prevented by growing or depositing a thin oxide film onto the Si surface prior to Au metallization.

3.2. Metal–GaAs Systems

Among the various metal systems used with *n*-type GaAs (refer to Table 1), Al has received considerable attention for Schottky contact metallization because of its good thermal stability, very good adherence with SiO_2 especially during bonding, and its potential for use in MESFETs. Al deposited on a chemically cleaned GaAs surface exhibited Schottky barrier heights in the range 0.73–0.89 eV[66,68,72,120,253] depending on the carrier concentration, surface preparation, and deposition temperature, while on atomically clean MBE GaAs surfaces its value was reported to be about 0.67 eV.[66,67] The increased barrier height in the case of chemically cleaned GaAs has been attributed due to the presence of natural oxide and/or formation of Al_2O_3 during Al deposition at the Al/GaAs interface. The experiments of Sakaki *et al.*[66] have clearly shown the presence of an interfacial layer. They have observed that, in the case of heat-cleaned GaAs surfaces when cleaning duration was kept constant, the barrier height decreased systematically from 0.89 to 0.74 eV by increasing the heat-cleaning temperature from 20°C to 580°C. According to them this decrease was brought about by gradual removal and compositional variation of interfacial oxide layers with increase in temperature (e.g., the residual oxide layer changed its composition at temperature around 350°C when volatile components left the layer and the oxide layer itself was removed around 560°C). While growing epitaxial Al layers on GaAs by MBE, Landgren and Ludeka[254] observed that at elevated temperatures (400°C) Al reacted with As to form AlAs and the free Ga at the interface outdiffused through Al layer. Although it is difficult to correlate the observations of various reported investigations due to varying experimental conditions, many of them[65,254–257] point to significant interdiffusion at Al/GaAs interface at temperatures $\geq$ 250°C.

Because of technological reasons other metal–GaAs systems which have been extensively investigated have Au as metal either for Schottky contact or for an overlayer (refer to Table 1). In the case of the Au–GaAs system, it has been observed that there is significant migration of Ga into Au at temperatures $\geq$ 300°C. This, however, is limited by solid solubility below the Au/GaAs eutectic temperature of 450°C. At this temperature and above out-diffusion of Ga and As to the Au surface and in-diffusion of Au in GaAs become predominant. It has also been reported that these interdiffusions lead to considerable degradation in the near ideal properties of the Schottky contact. One way of avoiding such degradation on heat treatment is to deposit an Au:Ga eutectic alloy layer on GaAs[101] (as it will prevent taking up of any more Ga from GaAs below the eutectic temperature), while another is to introduce a diffusion barrier metal layer between Au and GaAs.

In addition to the Al and Au systems discussed above, the brief comments on the reported interdiffusion and degradation studies of various metal–GaAs systems are tabulated in Table 5. Although safe operating temperatures of various metal–GaAs Schottky barriers can be estimated from these types of investigations, they are inadequate to predict the exact nature of the interface in the operating temperature regions. However, there are now available a few experimental results of atomic scale interdiffusion at the metal–GaAs interface[261,267] at room temperature which demonstrate representatively that Schottky barrier formation is governed by interdiffusion even at room temperature. This interdiffusion may be via crystal defects created at the surface by the heat of condensation of the metal.

With increasing utilization of GaAs-based heterostructures[268] for various device fabrication, the metal systems mentioned in this section have to be reexamined with respect to the ternary and quaternary III–V alloys lattice matched to GaAs.

TABLE 5. Interfacial Reaction in Some Metal–GaAs Systems

Metal system[a]	Comments
Al	Significant interdiffusion at Al/GaAs interface at temperatures $\geqslant 250°C$[65,254–257]
	Depending on GaAs surface preparation and condition for Al deposition natural oxide and/or Al_2O_3[66,72] or AlAs[254,257] or $Al_xGa_{1-x}As$[65] interfacial layer may be formed
Au	Thin Au film/GaAs structures annealed at temperatures $\geqslant 250°C$ in different ambients show both nonstoichiometric out-diffusion of Ga and As to the Au surface and in-diffusion of Au into GaAs[258–260]
	Interdiffusion on an atomic scale at Au/GaAs interface even at room temperature[261]
Cr	Short period heating of Cr/GaAs structure at 200°C to 300°C does not significantly change the Schottky barrier parameters: higher temperature heating results in considerable deterioration[76]
Ni	Ni film reacts with GaAs above 200°C in the solid–solid phase to form a metastable reaction product having Ni_2GaAs composition. Above 400°C it decomposes into NiAs and β NiGa[262]
Pd	Pd reacts with GaAs to produce $PdAs_2$ during annealing at 300°C and PdGa at temperatures $\geqslant 400°C$; interdiffusion takes place prior to alloy formation[263]
Pt	Out-diffusion of Ga into Pt and also formation of $PtAs_2$ at the Pt/GaAs interface starts at temperatures $\geqslant 300°C$. As reaction progresses Pt–Ga compounds (Pt_3Ga and PtGa) start forming[264]
Ti	No interdiffusion occurs at 350°C. Annealing at 500°C TiAs is formed. The Schottky barrier parameters, however, remain the same when annealed between 350°C and 500°C. No reaction detected up to 450°C[82]
Cr–Au	Au/Cr/GaAs degrades during short annealings at temperatures $\geqslant 200°C$ due to Au diffusion through Cr layer[76]
Cr–Pt	Heat treatment above 300°C increases the barrier height due to diffusion of Pt through Cr and reaction of Pt with GaAs[110]
Ga:Au	Presence of Ga in the Ga:Au eutectic alloy layer prevents taking up of any more Ga from GaAs and thus avoids degradation on annealing[101]
Ge–Pd	Pd/Ge/GaAs structure, sintered at temperatures below 350°C, shows interdiffusion of Pd and Ge to form PdGe and Pd_2Ge_3 while at about 500°C PdGe, $PdAs_2$, and PdGa[265]
Ni–Au	AuGa and NiAs intermediate phases form during annealing at 350°C[104]
Pt–W	W/(PtGa, $PtAs_2$)/GaAs with an over layer at Au forms an ideal Schottky structure[97,107]; W film forms an excellent barrier for Pt reacting with Au at temperatures up to 500°C[107]
Ti–Au	Au/Ti/GaAs shows interdiffusion involving grain boundary diffusion at temperatures as low as 200°C and growth of an AuGa intermediate phase[266]
Ti–Pt	Pt/Ti/GaAs behaves like Ti/GaAs up to 300°C; TiAs formed at Ti/GaAs interface at 500°C which is ineffective to prevent formation of Pt_3Ga/TiAs/$PtAs_2$/GaAs[97]
W–Au	Thin W films forms an excellent barrier to diffusion of Au up to 500°C[107]
	Au/W/GaAs shows diffusion of Au through W at 475°C probably through pin holes and formation of AuGa at the interface[106]

[a]Metal deposited first on the GaAs surface is put first.

3.3. Multilayer Metallization Systems

In practice, multilayer metallization is often required to achieve high reliability and thermal stability of Schottky barrier contacts. Two metal layers (i.e., one for providing "intimate" contact with the semiconductor and another on top of it to provide a low resistance path for connection) are normally required for this purpose. Invariably, however, it is necessary to introduce a third layer to minimize interdiffusion or reaction between the two layers. In Si Schottky diodes, silicide-forming metal is generally used as contact metal (see Section 3.1) while Al is used as top layer because of its compatibility with established bonding and packaging technologies.(225,269) Very often, with such metallization systems, the reaction of Al with silicide-forming metal and/or silicide comes in the way of forming good, thermally stable Schottky barrier junctions. A number of investigations (e.g., Al/PtSi/Si,(224,270) $Al/Pd_2Si/Si$,(270–272) Al/NiSi/Si,(46) $Al/CoSi_2/Si$,(43) $Al/MoSi_2/Si$(43)) to study such reactions in the temperature range 200–500°C and their effects on Schottky barrier junction properties have been reported in the literature. In most cases, it has been observed that reaction between Al and silicides starts with dissolution of silicide at the Al/silicide interface and then diffusion of Al through the silicide to the silicide/Si interface and silicide metal into Al to form binary or ternary intermetallic compound and Si precipitation takes place. Van Gurp *et al.*(273) have reported that $PtAl_2$(275°C), $NiAl_3$(275°C), Co_2Al_9(400°C), $MoAl_{12}$(535°C) intermetallic binary compounds start forming in respective Al/silicide/Si systems at temperatures shown in parentheses. Ho *et al.*(272) have found formation of a ternary compounds in Al/Pd_2Si reaction. The difficulty in forming thermally Al/silicide/Si contacts due to formation of the above-mentioned compounds can be overcome by introducing a layer of W(43) or Ti:W(237) pseudoalloy between Al and silicide. This layer has proved to be an effective barrier at temperature up to 500°C. In fact, Al/W:Ti/PtSi metallization has been used for Schottky diodes in integrated circuits with excellent results.(274) Another structure, formed by sequential evaporation of Ni, W, and Al on Si and subsequent thermal annealing to form NiSi/(n)Si Schottky barrier junction, shows thermal stability up to 450°C.(275)

In the case of metal–GaAs systems, the need for a multilayer metallization system is all the more important both because out-diffusion of Ga and/or As to the metal surface often leads to considerable degradation in the electrical properties of Schottky contact, and because subsequent formation of Ga and As oxides at the metal surface is undesirable for making contact. In order to avoid these, various two-component systems having Au as an overlayer (see Table 5) have been investigated by various workers. It has been observed that, in the case of the Au/W/GaAs system, there is no appreciable interdiffusion at and in the vicinity of the W/GaAs interface even after annealing at 500°C for two hours in vacuum and that thin film of W can act as an excellent diffusion barrier between GaAs and Au. In fact, Sinha[107] has observed that not only the electrical properties of Au/W/(*n*)GaAs but Pt/W/(*n*)GaAs Schottky diodes are found to remain very stable with thermal aging at up to 500°C. Other thermosteady multilayer systems reported in literature[97,276,277] are Au/W/Pt/GaAs and Au/W:Ti/Pt/GaAs which have Pt as Schottky contact metal. In a report on GaAs reliability[116] Au/Pt/Ti and Au/Pd/Ti systems appear superior to Al as Schottky contact used in GaAs MESFETs provided temperatures are kept below 150°C. Lundgren and Ladd[114] have observed that Au/Pt/Ti/GaAs and Au/Pt/Mo/GaAs have no degradation up to 350°C while Au/Pt/Cr has significant degradation at 350°C. Another metallization system using Ti as Schottky contact metal (Au/Mo/Ti) has been successfully used for low barrier height GaAs mixer diodes.[115] Although Cr has certain advantages over Ti (e.g., less susceptibility to oxidation), Au/Cr/(*n*)GaAs Schottky contacts[76] have been found to degrade even during short annealings at temperatures ≥ 200°C due to Au diffusion through Cr layer. Although it can be thought that introduction of Pt or W between Cr and Au will improve thermal stability of such contacts, electrical properties of Pt/Cr/(*n*)GaAs contacts have also been found to degrade above 300°C (see Table 5).

A number of other metallization system have been tried, but dislocation due to differential thermal expansion and interactions at comparatively lower temperatures hold up their use. It is clear from the reported investigations that most metallization system currently in use for Si and GaAs consist of a Schottky contact metal (e.g., Pt, Ni for Si and Pt, Ti for GaAs), a diffusion barrier metal (e.g., Ti, W, Ti:W for Si and W, Ti:W for GaAs), and a bonding or contact metal (e.g., Al, Au, Ag, for Si and Au for GaAs).

4. DESIGN CONSIDERATION

The two important properties of a Schottky barrier junction are its current–voltage ($I-V$) and capacitance–voltage ($C-V$) characteristics. While these properties are discussed in the chapter dealing with the physics of Schottky barrier junctions, their relevant features from the point of view of designing Schottky diodes are discussed in this section.

The forward $I-V$ characteristics of an actual Schottky barrier diode can be expressed as[212]

$$\begin{aligned} I &= A^*ST^2\exp(-\phi_B/kT)\exp(qV/\eta kT)[1-\exp(-qV/kT)] \\ &= I_0\exp(qV/\eta kT)[1-\exp(-qV/kT)] \end{aligned} \tag{1}$$

where A^* is the effective Richardson constant, S the junction area, T the absolute temperature, ϕ_B the zero bias barrier height, k the Boltzmann constant, η the ideality factor, q the electronic charge, and V the applied bias. It is clear from this expression that current I depends on the values of the two basic parameters, namely, ϕ_B and η. In spite of wide usage of Schottky barrier junctions in a variety of devices and integrated ciruits, their effectiveness is often compromised by discrete set of barrier heights available with those metal–semiconductor systems which also satisfy a great many other technological requirements. For example, it is not always possible by conventional fabrication techniques to incorporate Schottky barrier diodes of specified barrier height and compatible metallization scheme in integrated circuits.

Ion implantation has been found to be an effective tool to modify the barrier height of Schottky barrier diodes. Shannon[278,279] has shown that the effective barrier height of a Schottky barrier diode can be adjusted over a wide range by using shallow ion-implanted highly doped layers. The barrier height reduction, by fabricating metal–(n^+–n) structure, has been reported in the literature.[51,115,280–282] In these studies the (n^+) surface layers have been obtained by low-energy ion implantation. Wu[283] has shown theoretically that these thin highly doped surface layers control the electric field at the contact surface and, thereby, contribute to lowering the effective barrier height. The barrier height enhancement, on the other hand, has also been carried out by implanting low-energy ions of opposite conductivity type into the semiconductor surface prior to contact metallization. The metal–(n–p) or metal–(p–n) Schottky barrier structures, formed by this process, have been studied[279,284–286] and theoretical models have been developed[287–289] to characterize the expected performance of such modified barriers as a function of surface layer doping and thickness and substrate doping level. In the case of Si, it appears from the reported studies that PtSi/(n)Si system is a viable one for achieving a system with continuously variable Schottky barrier height.[51,283]

As mentioned earlier in Section 2, ion implanted surface required high-temperature annealing to remove implantation damages and to activate impurity dopants. In the case of compound semiconductors such as GaAs, it is necessary to critically examine the implications of this process while considering ion implantation as a tool for obtaining tailor-cut barrier heights. The temperatures (800–900°C) usually employed for postimplantation annealing of GaAs are such that, unless preventive measures are taken, significant loss of As takes place during this process.[290] The most commonly used method to prevent As loss is to encapsulate or cap the implanted surface with a material such as Si_3N_4 or SiO_2.[221] Since the implantation doping results are strongly affected by the properties of this cap layer,[290,291] the transient annealing of an ion-implanted GaAs surface with laser or electron beam irradiation has been tried without encapsulation. In this case local heating raises selective

GaAs surfaces to high temperatures for short times without significantly altering the bulk temperature and thus prevents As loss. Another approach is to use a proximity cap. This involves placing a second polished GaAs wafer in physical contact with the wafer to be annealed in the hydrogen atmosphere and may or may not have an overpressure of As.[292] These proximity cap or capless annealing techniques and for that matter creation of ion-implanted thin (p^+) or (n^+) layer on n- or p-type GaAs are still in their infancy, and investigations currently being carried out will only indicate their comparative superiority.

Another approach for modifying the barrier height of Schottky contacts can be to provide a thin epitaxial layer at the metal–semiconductor interface. In fact White and Brookbanks[143] have achieved reduction of effective barrier height of metal–(n)InP Schottky contacts from 0.4 to 0.24 eV by introducing a thin (n^+) epitaxial layer at the interface. It has also been proposed[293] that metal–GaAs Schottky contacts with an interfacial layer of $Al_xGa_{1-x}As$ may provide a technique to have tailor-cut barrier height. It can be seen from the

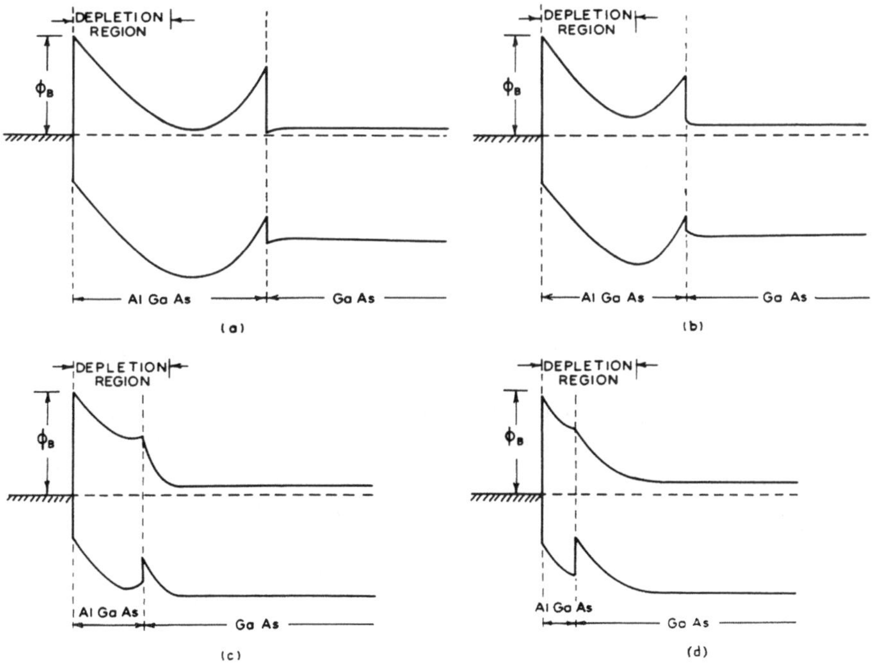

Figure 4. Equilibrium energy band diagrams of metal/(n)AlGaAs/(n)GaAs structures.

equilibrium energy band diagrams of metal/(*n*)AlGaAs/(n)GaAs heterostructures, shown in Figs. 4a–4d, that for efficient functioning the thickness of AlGaAs has to be such that it is fully depleted of mobile carriers at zero bias. A few metal–heterostructure systems (e.g., Au/AlGaAs/GaAs,[186] Au–Pt–Ti/AlGaAs/GaAs,[294] Al/AlInAs/GaInAs,[295] Au–Mo–Ti–Pt/Ge/GaAs[296]) to enhance or reduce the barrier height and improve the device performance have been reported in literature. The capability of growing very thin heteroepitaxial films by molecular beam epitaxy[297] is currently being assessed by various workers and may, in future, be utilized to make Schottky barrier junctions of specified barrier heights.

In addition to the above-mentioned approaches, a thin insulating oxide layer at the metal–semiconductor interface can also be used to increase the effective barrier height. Although excellent RF performance of the devices utilizing pseudo-Schottky barriers with an ultrathin film of SiO_2 between metal/(*n*)InP has been reported,[138] persistently high surface state densities and instability of such structures based on III–V compound semiconductors have limited their usage for devices.

For high-frequency performance evaluation of Schottky barrier diodes, the cutoff frequency is used as a simple figure of merit. The expression for cutoff frequency, f_c, is given by

$$f_c = 1/(2\pi R_s C_j) \tag{2}$$

where R_s is the series resistance and C_j is the junction capacitance. Considering a metal–(n–n^+)-type structure (shown in Fig. 1a) and taking R_s (in its simplest form) as $\rho_{ep} t/\pi r^2$, the above expression can be rewritten as

$$f_c = \frac{\mu}{\pi t}\left(\frac{qN_d V_D}{2\epsilon_s}\right)^{1/2} \tag{3}$$

where μ is the electron mobility, t the epitaxial layer thickness, N_d the donor density, ε_s the permittivity of the semiconductor, V_D the diffusion potential, and ρ_{ep} $(= 1/qN_d\mu)$ is the resistivity of the epitaxial layer. It can be seen from this simple expression that for good high-frequency performance it is necessary to reduce the epitaxial layer thickness t and increase the product $\mu N^{1/2d}$. Since the election mobility μ decreases with increasing donor density N_d, a suitable combination of μ and N_d has to be arrived at to maximize the product $\mu N_d^{1/2}$. For example, in the case of GaAs the maximum of this product is obtained at electron concentration of about $10^{17}/cm^3$. On the other hand, the limit to the reduction of t is governed by the depletion layer width. In an ideal case t should be only slightly larger than that required to accommodate the depletion layer, but it may not always be possible to achieve this in practice.

In spite of the fact that the conclusions drawn above may be applicable qualitatively to various Schottky barrier diode structures, the complex dependence of R_s on various parameters for smaller geometries and planar structures makes it difficult to arrive at a suitable optimization from the cutoff frequency point of view without numerical analysis. For example, one of the approximate expressions for R_s in the axial structure is given by[298]

$$R_s \simeq \frac{\rho_{ep}}{2\pi}\left[\frac{t}{r^2 + t^2} + \frac{1}{r}\tan^{-1}\left(\frac{t}{r}\right)\right] + R_{su} \tag{4}$$

where ρ_{ep} and t are the resistivity and thickness of the epitaxial layer, r is the radius of the Schottky contact, and R_{su} is the spreading resistance of the substrate (see Fig. 1a), while for planar structure on a semi-insulating substrate (shown in Fig. 1c) it is given by[299]

$$R_s \simeq \frac{1}{w}\left[(\rho_c \rho_{ep}/t)^{1/2} + \frac{1}{t}\rho_{ep} d + \frac{1}{3} R_{\square} l\right] \tag{5}$$

where ρ_{ep} and t are the resistivity and thickness of the epitaxial layer, ρ_c is the specific contact resistance, $R_{\square}$ is the diode resistance per unit area, d is the distance between ohmic and Schottky contact, and l and w are the length and width of the Schottky contact. From the geometrical considerations alone, it can be easily shown with the help of the above expressions that the planar diodes with a single n-type layer on semi-insulating GaAs have lower cutoff frequencies than axial diodes with a single n-type layer on n^+-type GaAs substrate. In fact, it has been observed that diodes on (n^+) substrates easily reach cutoff frequencies of 3000GHz[300] while the highest value reported so far for a diode with a single n-type layer on semi-insulating GaAs has been 200 GHz.[301] Cutoff frequencies above 500GHz have, however, been achieved with double or triple layers on semi-insulating GaAs.[302–305]

5. FABRICATION TECHNOLOGY

The present-day technology of fabricating discrete Schottky barrier diodes involves surface processing, dielectric film deposition, ohmic contact formation, contact pattern generation, Schottky contact metallization, metal pattern generation, bonding, and packaging. Their sequential order, however, depends on the basic structure of the diode under consideration. Since processing of Si Schottky barrier diodes and their incorporation in integrated circuits is now a well-established technology and is readily available in the literature, the main discussion in this section is confined to GaAs processing techniques.

5.1. Surface Processing

The electrical characteristics of Schottky barrier diodes are very sensitive to surface processing at various stages of fabrication. Amongst these, the surface preparation prior to Schottky contact metallization is considered to be the most critical. Since a monolayer of SiO_2 is formed in air on a clean Si surface in about 1 msec,[213] or native oxide in the 20–30 Å thickness region is normally present on the surface of any GaAs sample kept under standard laboratory conditions[306] and a $3H_2SO_4$:H_2O_2:H_2O solution produces an ultrathin oxide of about 10 Å on GaAs surfaces,[194] a clean semiconductor surface from the point of view of Schottky contact has to be atomically clean. The traditional methods of producing atomically clean surfaces involve either cleaving in ultrahigh vacuum or heating in vacuum. *In situ* cleaving may be an ideal method from the chemical point of view as there is no oxide layer at the surface, but it is known to produce physical damages and, therefore, cannot be physically perfect. In addition to this, the semiconductor surfaces of device interest are not the natural cleavage planes. On the other hand, vacuum heating of GaAs or InP may lead to removal of surface impurities and decomposition of surface. Even for Si, it may lead to a redistribution of impurities near the surface, which in turn may vary the depth of the Fermi level position. McKinley *et al.*[307] have described a method of producing atomically clean Si and InP surfaces by laser irradiation. Although its applicability is not yet established, if successful, it may provide an easy *in situ* method to clean semiconductor surfaces prior to metal evaporation. At present, however, atomically clean surfaces of III–V compounds without resorting to cleavage can be produced by molecular beam epitaxy with its related surface cleaning methods.[308] In fact, Sakaki *et al.*[66] have reported fabrication of nearly ideal ($\eta \simeq 1$) Al/(n)GaAs Schottky barrier diodes by *in situ* deposition of Al at room temperature on oxygen-free MBE surfaces of GaAs and have also observed that vacuum heating (under As pressure) at temperatures $\geq 580°C$ were required to remove the surface oxide. Although various other dry etching techniques [309,310] have been tried to remove interfacial oxide or contamination layers prior to Schottky contact metallization, fast wet chemical etching followed by rinsing in a suitable solvent and drying in an inert atmosphere is still the most prevalent cleaning technique for surface processing. A set of chemical etchants along with their respective etching times, used for cleaning Si, GaAs, and InP surfaces prior to loading in a vacuum chamber for Schottky contact metallization, are listed by Hokelek and Robinson.[89]

Apart from surface cleaning, high-frequency considerations often require accurate thinning of the active layer. This is also carried out by dry etching techniques, but a more promising technique in the case of GaAs appears to be anodic oxidation followed by oxide etching. This technique is either based on anodic oxidation in dark[311] or under light illumination.[312] Of these two, the former has been used to improve the uniformity of the breakdown voltage in the thick epitaxial layers while the latter has been utilized to optimize the thickness of the active layer with respect to the depletion width irrespective of local variations in doping and thickness of the starting material.

5.2. Dielectric Film Deposition

One of the important fabrication steps consists of depositing dielectric films[313,314] on active layers. These films are used not only for defining contact geometries and edge protection, but are sometimes used for tailoring electrical properties of Schottky diodes. For these purposes, the most widely used films are of SiO_2, which can be deposited on Si either thermally by reacting Si with oxygen at about 1200°C or by the pyrolysis of tetraethoxysilane in the temperature range at 650–750°C or at a much lower temperature range of 400–450°C by reacting silane with oxygen. Although the quality of the low-temperature deposited films is not quite as good as films thermally grown on Si surfaces, nevertheless, these are the films normally used in device fabrication. The SiO_2 films, having thicknesses in the range 5000–8000 Å, are found suitable for high-frequency Schottky diodes.

Considerable efforts have also been made to find a suitable technique for the formation of insulating layers on the surfaces of III–V compounds. Even though native oxide layers on GaAs[315–318] and on InP[319,321] have been grown by thermal, anodic, and plasma oxidation, SiO_2 is still the most widely deposited dielectric film on these compounds. It is normally grown at about 400–500°C either by the pyrolysis of tetraethoxysilane in oxygen atmosphere[322] or by reacting silane with oxygen.[115] Setzer and Mattauch[323] have reported that boron-doped SiO_2 (i.e., B_2O_3/SiO_2) films are better suited for this purpose than SiO_2 because addition of B_2O_3 reduces the interfacial strain characteristically associated with thick SiO_2 films deposited on GaAs. A reliable Ta–GaAs Schottky junction and native oxide passivation in a quasiplanar configuration has also been reported in the literature.[324]

Silicon nitride films have also been used for this purpose especially in the fabrication of planar GaAs integrated circuits.[216] In addition to having a higher dielectric constant, the Si_3N_4 films are much more impervious to all kinds of diffusants than SiO_2 and are resistant to oxidation and to corrosive environments. Despite these properties, Si_3N_4 films are still not able to compete with SiO_2 because of the problem of poor adherence of sputtered films and high deposition temperature requirements of CVD films.

5.3. Ohmic Contact Formation

Since the frequency characteristics, noise characteristics and power capability of a Schottky diode can depend substantially on the quality of ohmic contact, a uniform reproducible low-resistance ohmic contact is an important consideration in its fabrication. In the case of ohmic contacts to n-type GaAs, two groups of metallization systems, namely, Au-base eutectic systems (e.g., Au:Ge) with an overlay of Au, Ni or Pt and noneutectic mixtures (e.g., Ge/Pd) have been tried with different thermal cycles.(325,326) The most widely used technique for making large area ohmic contact to n^+-type GaAs substrate consists of sequentially evaporating Au:Ge eutectic alloy, Ni and/or Au and furnace alloying at a temperature of about 450°C for 3 min in forming gas. Typical specific contact resistance ρ_c value of about $9 \times 10^{-5} \Omega cm^2$, obtained by this technique, has been improved to $1 \times 10^{-6} \Omega cm^2$ by rapidly raising the temperature from room temperature to 450°C in less than 10 sec.(327) It must, however, be noted that, in spite of wide usage, furnace alloying often leads to nonreproducible contacts as it produces surface roughness, poor edge definition, uneven penetration into underlying semiconductor, and considerable depth damage. Many of these problems can either be avoided or greatly diminished if laser or electron beam irradiation is used as a heat source(328) or ohmic contact formation is carried out at a temperature lower than the eutectic temperature.(329,330)

Eckhardt(328) has reported the results of the investigation in which the annealing of the multilayer metallization systems of the type Au–X–Au:Ge/(n)GaAs and X–Au:Ge/(n)GaAs, where X stands for the metals Ni, Ag, Pt, Ti, and In, were carried out by using pulsed CO_2, Nd:YAG, pulsed ruby, cw Ar-ion lasers, and pulsed electron beam. Amongst these, the best results (i.e., $\rho_c \sim 1.3 \times 10^{-6} \Omega cm^2$, excellent adhesion between contact and GaAs and highest reproducibility) were obtained with cw Ar-ion laser and In–Au:Ge/(n)GaAs structure. A lowest ever value of specific contact resistance ($\sim 4 \times 10^{-7} \Omega cm^2$) has been obtained by Tondon *et al.*(331) *for* ohmic contacts formed by pulsed electron beam alloying of Pt–Au:Ge/(n)GaAs/SIGaAs structure. This value of ρ_c is still higher than one normally obtained with silicide contacts on Si ($\sim 10^{-7} \Omega cm^2$).(332)

Generally for good quality ohmic contact formation at temperatures lower than the alloying temperatures, a thin buffer (n^+) layer between contact metal system and n-type GaAs has to be formed prior to metal deposition. Several techniques, namely, growing doped layer by molecular beam epitaxy,(333) generating layers by dual implantation (e.g., Se and Ga) followed by thermal annealing,(334) producing layers by high-dose ($\geq 10^{15} cm^{-2}$) donor ion implantation followed by subsequent annealing by pulsed laser(335) or electron beam,(330) and growing doped layers by chemical vapor deposition(336) have been reported in the literature. With shrinkage of contact dimensions and intracontact spacing in planar devices, the technological importance of nonalloyed ohmic contacts has increased greatly. The best specific contace resistance value of $6 \times 10^{-7} \Omega\,cm^2$ for nonalloyed type of ohmic contacts have been obtained by Mozzi *et al.*[330] for contacts formed by pulsed electron beam alloying of ion implanted GaAs followed by deposition of Ti/Pt/Au. This result is quite encouraging as a single metallization step of the Ti/Pt/Au system can be used to form both Schottky and ohmic contacts in planar fabrication.

In addition to the above-mentioned techniques, high density of crystal defects created near the semiconductor surface by sputtering or ion bombardment may act as efficient recombination centers and may significantly decrease the contact resistance when a contact metal system is deposited on it. In fact, a few studies[91,337–339] have indicated that such damages are helpful in decreasing the contact resistance, but more systematic investigations are needed to prove its technological usefulness in improving the contactibility.

5.4. Metal Deposition

Amongst the various techniques,[340,341] thermal evaporation by resistance and electron-beam heating in vacuum are the most widely used ones for depositing thin metal films on semiconductor surfaces. The resistance heating is applicable only to metals having relatively low vaporization temperature ($< 1500°C$), while electron-beam heating is used for evaporating high vaporization refractory metals. Archibald and Parent[342] have discussed the limitations, advantages, and deposition characteristics of various source–evaporant systems available for thermal evaporation. Since the evaporation ratios, in the above two techniques, are controlled by the vapor pressure of the evaporant at the temperature and background chamber pressure, it is difficult to deposit films of alloys and mixtures having the same composition as that of the starting material. This problem, however, is overcome by using the flash evaporation technique. In this case a fine-sized powder of the alloy is dropped at a controlled rate onto a hot "boat," maintained well above the minimum vaporization temperature of the individual elements of the alloy, so that the arriving powder grains instantly flash off without fractionation.

Sputtering[343–345] is an alternative to thermal evaporation for refractory metals and is often used for making refractory metal silicide Schottky barrier junctions on Si. In comparison to conventional dc or rf sputtering, ion beam sputtering offers the advantages of better vacuum and film adherence and of reduced substrate heating due to the absence of electron bombardment. In the ion beam sputtering technique, the source metal (cathode or target) is bombarded by energetic gaseous ions in a high electric field. This dislodges the atoms from the surface of the source metal as the kinetic energy of the gaseous ions exceeds the bonding energy of the metal atoms. In this energy-momentum transfer process most of the sputtered atoms acquire negative charges and are accelerated towards the semiconductor (anode) and condense on it. This technique is useful for depositing alloys of desired stoichiometry and multilayer films. The latter is carried out by varying or exchanging the targets during the deposition. In spite of the above-mentioned advantages, the sputtering techniques, in general, are as far as possible avoided for depositing a contact metal system on III–V compound semiconductors because of low sputtering rates, greater degree of surface damage, and difficulty in accurate monitoring of thicknesses.

Chemical techniques[346–349] for depositing metallic films on semiconductors have received comparatively less attention for many obvious reasons. For example, one of the main reasons against the use of electrolytic and electroless plating techniques is that the fresh semiconductor surface in contact with chemical solution gets contaminated and invariably leads to formation of an undesirable interface. Anyway amongst the available chemical techniques, electro- and electroless plating are, in principle, more convenient than evaporation for selective area deposition (i.e., for forming Schottky contact at very small windows opened in SiO_2). Although a few dc electroplating studies[21,77,151,350] have been reported for depositing metals (e.g., Au, Cd, Cu, In, Ni, etc.) on Si and III–V compound semiconductors, the technique[351] which utilized high-field pulses of extremely short duration for deposition of Pt on GaAs may in future prove useful for the formation of submicron Schottky contacts. Electroless plating, on the other hand, has some basic advantages, namely, better adhesion and no external connections in comparison to electroplating, and has also been used to deposit metals on III–V compound semiconductors.[75,124,352]

Another approach for depositing strongly adherent films of Ag, Au, Pd, and Pt from commercial plating solution onto Si and GaAs[91,337] consists of first damaging the selected area either mechanically or by bombardment with charge particles and then immersing the slice in a plating solution and illuminating it by a light whose quantum energy exceeds the bandgap. In practice, the ion bombardment is carried out either by ac glow-discharge[337] or by dc discharge produced by electrode system[91] and illumination by an ordinary tungsten-filament bulb. Both rectifying and ohmic contacts are formed by this technique depending on the particular semiconductor, the nature of the bombarding particles, the contact metal, and subsequent annealing treatment. Postdeposition annealing is necessary if good rectifying characteristics are to be obtained. It must, however, be noted that even though the Schottky contacts obtained this way are far from ideal, they seem suitable for many applications. Ehrlich *et al.*[353] have demonstrated that laser photolysis of gas phase metal alkyl molecules such as $Al(CH_3)_3$ can be used to produce metal deposits with micron dimensions.

5.5. Other Steps

In addition to the fabrication steps discussed above, there are others which are briefly mentioned here. Schottky contact pattern generation involves defining and etching a matrix of windows of required junction diameter in an oxide deposited on the semiconductor. This is normally carried out with the help of photolithography,[(354)] which consists of photoresist coating, prebaking, mask aligning, exposure, development, postbaking, chemical etching, and stripping of the photoresist. When very small contact areas are required, this technique may lead to appreciable undercutting of the masked area and chemical contamination of the contact surface. For example, the best photolithographically prepared circular windows have a diameter variation of $\pm 0.1\ \mu m$, which is quite appreciable for windows of diameters $\lesssim 1.5\ \mu m$ required for very high-frequency (150 GHz or above) applications.[(117)] In such cases, electron beam lithography[(355)] not only improves the reproducibility but provides close control of diode parasitics. Along with electron beam lithography, dry etching techniques like sputter etching, plasma etching, and ion-beam etching are being tried to achieve desired small device geometries and/or submicron dimensions in planar Schottky junction devices. The contact sensitivity to residual defects formed during etching is, however, one of the severest problem coming in the way of their popularization. Goken and Esho[(356)] have reported a simple technique for obtaining fine gap geometries down to $0.3\ \mu m$ using conventional photolithography and ion-beam etching.

Direct wet chemical etching to remove unwanted multilayer contact metal from the oxide surface may not be a suitable technique. This technique results in coarse metal edge topography, and it is also sometimes difficult to remove all traces of organic residue by stripping and cleaning sequence without damaging the contact metal layers. An effective alternative to this is to use the lift-off tecnique. This technique essentially consists of coating the oxide surface with positive photoresist prior to metallization, defining the pattern such that the areas in which metal is desired are not covered with photoresist, and stripping the photoresist after metal system deposition in acetone with ultrasonic agitation. In this way unwanted metal comes off with the photoresist having the metal system in the desired area. Because of excellent resolution available with positive photoresist and incompatability of many metal etchants with GaAs, this technique is invariably used in the fabrication of GaAs-based devices. It must, however, be noted that wafer temperature during metal deposition cannot be raised above baking temperature in this case.

Another important step, specifically in the fabrication of reverse bias-operated Schottky diodes (e.g., avalanche photodetector, avalanche transit-time diode, etc.), is edge protection. In conventional Schottky diode structure (see Fig. 1a), the electric field peaking along the periphery of the Schottky contact gives rise to edge breakdown under reverse bias. Various modified diode structure, namely, MOS guard ring,[213] single and double diffused guard ring,[23,357,358] metal guard ring,[213] locally oxidized guard ring,[359,360] metal-overlap laterally diffused,[12] moat-etched,[23] moat-etched air isolated,[213] and air isolated mesa[213] have either been tried or suggested to prevent edge leakage currents in Si Schottky diodes. Some of these structures can also be used for III–V compound semiconductor based Schottky diodes.

The final steps in the fabrication of nonplanar conventional Schottky barrier diode consist of dicing into chips by scribe-and-break technique and packaging the individual chips. Although packaging depends on the geometry and end use of the diode, it often requires die bonding and wire bonding by thermocompression and/or ultrasonic bonding techniques. The wire bonding requirements (e.g., 15-μm-diam Au wire to be bonded directly to Schottky contact of 10-μm-diam in GaAs diodes for 12-GHz operation) often leads to low yield. In order to overcome such difficulties, metallized beam-lead structures have been reported in the literature.[361–363]

6. CHARACTERIZATION

As mentioned earlier and as can be seen from expression (1), the two most important parameters from the standpoint of devices are ϕ_B and η. According to the simple Schottky model, ϕ_B is equal to $(\phi_m - \chi_s)$ and $(E_g + \chi_s - \phi_m)$ for metal contact with n-type and p-type semiconductors, respectively. In this case E_g and χ_s are the band gap and electron affinity of the semiconductor and ϕ_m is the work function of the contact metal. It is, however, observed that such simple relationships do not exist in most practical metal–semiconductor contacts. These departures have been explained by various workers as either due to the interface states originating from surface states[364,365] or from metal-induced gap states[366,367] or due to narrowing of the band gap rather than the surface states[368] and/or due to interfacial chemical reactions.[369–373] In the absence of enough microscopic information, under controlled conditions, about metal–semiconductor interfaces it is difficult to ascertain the validity of various explanations at this juncture, but the experimental data (e.g., see Figs. 5 and 6) do confirm an empirical relationship of the type $\phi_B = C_1\phi_m + C_2$, where C_1 and C_2 are constant characteristics of the semiconductors. From such an empirical relationship, both the ideal Schottky barrier and Bardeen barrier can be visualized by assuming $C_1 = 1$ and $C_1 = 0$, respectively. In fact, these parameters have been used in the past

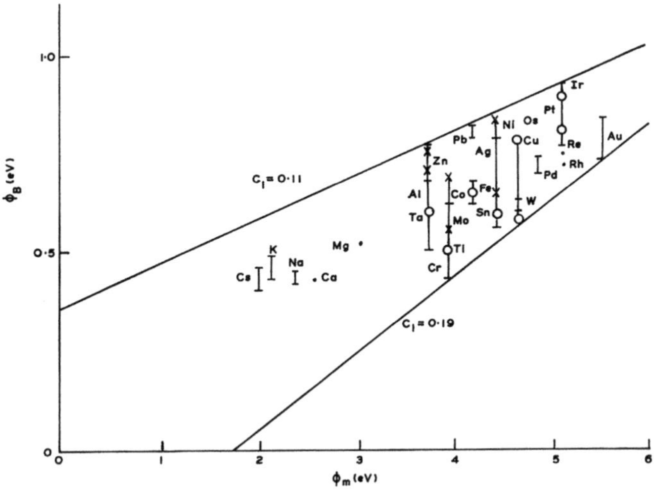

Figure 5. Experimental ϕ_B values for various metal/(*n*)Si Schottky junctions plotted against metal work function.

by some workers to describe the Fermi level stabilization or pinning and to estimate the interface state density. Sharma and Gupta[1] have obtained limiting values of C_1 and C_2 for various *n*-type semiconductors from the plots of experimentally available ϕ_B versus ϕ_m (see Figs. 5 and 6) and have mentioned that interface state density limits can thus be estimated by using an expression of the type given by Cowley and Sze.[365]

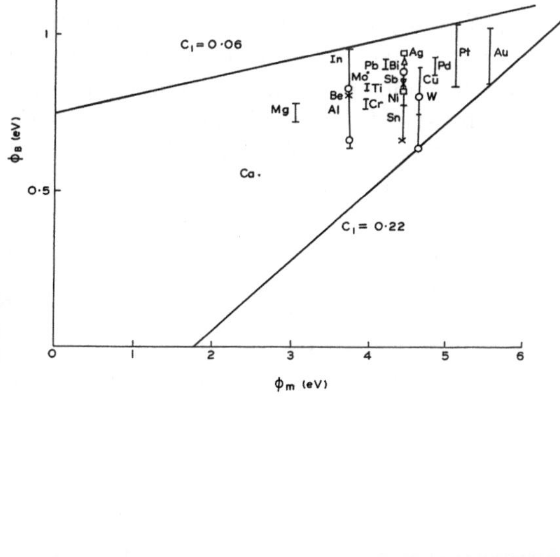

Figure 6. Experimental ϕ_B values for various metal/(*n*) GaAs Schottky junctions plotted against metal work function.

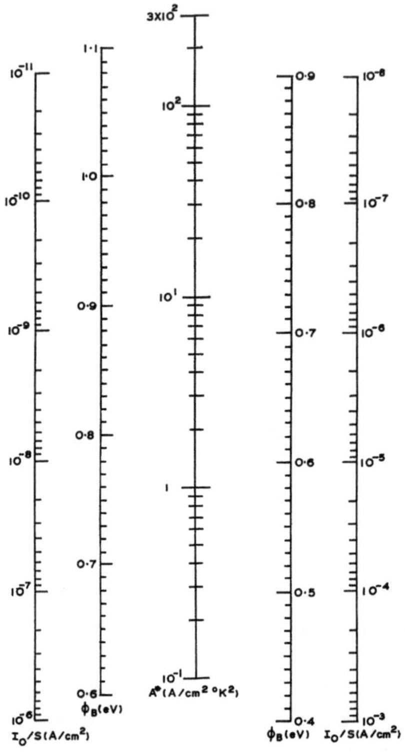

Figure 7. Nomograph for determining any one of the three parameters (i.e., A^*, I_0, and ϕ_B) at 300K for any Schottky barrier diode by knowing the other two.

Experimentally ϕ_B and η can be determined from the intercept and slope of the straight line plots of $\ln\{I/[1-\exp(-qV/kT)]\}$ versus V. The intercept of the straight line on the vertical axis gives the value of I_0 and a knowledge of A^* and S leads to the estimation of ϕ_B. In fact, the nomograph(1) shown in Fig. 7 can be used to evaluate any one of the three parameters, namely, A^*, I_0/S, and ϕ_B by knowing the other two. Although it is difficult to predict these parameters when a particular metal makes contact with a semiconductor, it is often possible to correlate the increasing or decreasing trend with the chemistry of the interface (e.g., with whether or not a chemical reaction takes place at the interface or native oxide is present at the interface). Boutrit *et al.*(374) have described a computer technique to determine series resistance, shunt conductance, saturation current, and ideality factor and hence ϕ_B of metal–semiconductor devices from forward I–V characteristics using a least-squares fitting of experimental results with the theoretical expression (1).

According to the thermionic-emission theory, the reverse current of an ideal Schottky diode should saturate at a value $I = A^*ST^2\exp(-\phi_B/kT)$, but vary often localization of a strong electric field in the defect regions of the interface leads to the appearance of an excess tunneling-recombination current. This "softening" of reverse $I-V$ characteristics and reduction in the breakdown voltage depend, therefore, to a very great extent on the structural perfection of the interface. In fact, Konakova *et al.*[375] have shown that, since structural perfection of the GaAs epitaxial structure without buffer layer is considerably less than with buffer layer, the "softening" is dominant in diodes made on wafers having epitaxial layer without a buffer layer.

The value of junction capacitance C_j at zero-bias and its variation with reverse bias are also required for the electrical characterization of a metal–semiconductor Schottky barrier junction. In an ideal case, the plot of C_j^{-2} versus V is a straight line and its intercept on voltage axis can also be used to estimate ϕ_B. It has, however, been observed that inhomogeneities in the epitaxial layers and presence of an interfacial layer between metal and epitaxial layer formed during surface preparation prior to metal deposition and/or prior to contact formation often lead to nonlinear variation. Vasudev *et al.*[376] have suggested a simple graphical method for determination of a quantity called the "excess capacitance" C_0 which playes the role of a lumped parameter incorporating most of the effects pertaining to nonlinear variation. The contribution to C_0 may arise from surface states and/or deep traps and its value can be used as an approximate quality index for the epitaxial layer and interface. Using this value of C_0, the nonlinear C_j^{-2} versus V plots showing downwards concave curvature can be transformed into linear $(C_j - C_0)^{-2}$ versus V plots.

In addition to the electrical characterization of Schottky contacts discussed above, their structural characterization is equally important, not only because it provides insight into the metal–semiconductor interfaces but also because of the ever-increasing demand for improved performance, reliability, product yield, and aging of various Schottky barrier devices. Some of the methods developed for thin film and surface analysis have been used for structural characterization of metal–semiconductor contacts and are listed in Table 6. Since many methods given in this table are complementary to each other, it is desirable to use more than one method to ascertain the metal–semiconductor contact morphology and metallurgy. This is all the more necessary as multicomponent metal systems are generally used for metallization.

Amongst the various methods (see Table 6), Auger electron spectroscopy combined with ion sputtering is the most popular one while secondary ion mass spectrometry in conjunction with sputtering can be considered as the most suitable one for obtaining depth profiles of the constituent elements in the vicinity of the metal–semiconductor interface. This latter method not only covers the whole mass range but has a very high detection sensitivity and very fine depth and lateral resolutions. Since the sensitivity is sufficient to measure

TABLE 6. Various Methods Used for Structural Characterization of Metal–Semiconductor Contacts

Method	Comments
Optical microscopy	(i) Examination of microscopic and structural defects
	(ii) Study of structural damages associated with transient processing of metal or oxide covered semiconductor surfaces[377]
	(iii) Examination of cross-sectional view of metal–semiconductor interface[378]
Scanning electron microscopy (SEM)	(i) Surface morphology and local defects
	(ii) Examination of various fabrication steps and degradation of contacts with the help of sectional photomicrographs[13,117,274,363]
	(iii) Observation of semiconductor surfaces after annealing and removal of metal and/or oxide films[327,379,380]
Transmission electron microscopy (TEM)	(i) Examination of microstructural changes at the interface caused by annealing[244,246,381]
	(ii) Residual damage study of implanted layers[382–384]
X-ray diffraction analysis	Identification of compound formation in metal–semiconductor interaction and of crystallographic phases present in an intermixed layer[48,385–388]
Low-energy electron diffraction (LEED)	Study of surface structure of As-deposited and annealed metal layers and the reconstruction of clean surfaces by transient annealing[164,389–391]
Reflection high-energy electron diffraction (RHEED)	Identification of reaction products formed during annealing[262,263]
Auger electron spectroscopy (AES)	(i) Depth profiling of the constituent elements in metal–semiconductor systems before and after annealing[46,106,262,392–395]
	(ii) Chemical composition of upper 10–20-Å layer of the surface[396]
	(iii) Determining presence of foreign elements on surface[396]
Secondary ion mass spectrometry (SIMS)	(i) Investigation of metallurgical changes in the vicinity of metal–semiconductor interface due to heat treatment[263,379,385,397]
	(ii) Compositional characterization of monolayers near the surface
	(iii) Investigation of diffusion in semiconductor from thin metal layers and implanted layers[398,399]
Ion beam backscattering analysis	(i) Low-energy ion backscattering used for surface analysis[400]
	(ii) Low-energy backscattering in conjunction with sputter etching used for depth profiles
	(iii) MeV ion backscattering for depth-sensitive depth microscopy[36,244,394,401–403]
	(iv) MeV ion backscattering in conjunction with angle lapped samples for deep profiles[404]
X-ray photoelectron spectroscopy (XPS)	Adsorption phenomena at surfaces, reactivity and interface chemistry[164,405–408]
UV photoelectron spectroscopy (UPS)	Surface electronic structure determination[164,405,409]
Proton-induced x-ray analysis	(i) Composition of outermost layers of a surface
	(ii) Detection of trace amount of elements[410]
Electron-probe microanalysis (EPMA and EDXA)	(i) Elemental profiles in the vicinity of interface by exposing beveled or cleaved surfaces[328,411,412]
	(ii) Elemental analysis and lateral distribution[327,413]

trace concentration, it can also be used to obtain diffusion profiles of very slowly diffusing elements in semiconductors.[398,399] There are several problems associated with the interpretation of the depth profiles developed by ion sputtering. The first among these is the homogeneity of the sputtered area due to the spread in the ion beam current density. This, however, can be partially compensated for by rastering. Another problem stems from the uncertainty of the true sputtering efficiencies. A very interesting example, though an extreme case, of this is the sputtering rate of molybdenum off various substrates.[414] It is for these reasons that proper calibration and care must be taken in interpreting the sputtering profiles. The difficulties encountered in converting the observed Auger electron signals as functions of ion dose to atomic concentrations as functions of depth are discussed in detail by Hofmann.[415] It has also been proposed[416] that mechanical angle lapping of the surface prior to analysis may be advantageous in terms of depth resolution in certain cases. In fact, Auger electron spectroscopy, in the form of scanning Auger microprobe, can be used to measure lateral distribution of such surfaces with resolution of 3–5 μm. In addition to these two, methods based on high-energy backscattering and energy-dispersive x-ray analysis (EDXA) have also been used to determine depth profiles. The high-energy (Mev He^+) ion backscattering method is capable of providing quantitative information on composition without recourse to sample erosion by sputtering and has been used successfully by various workers. Because of a number of limitations (e.g., poor lateral and depth resolutions) of this method in its conventional form, it is often difficult to interpret the experimental backscattering spectra for depth distributions. The focused ($\sim$10-μm spot) ion beam backscattering method[417] in conjunction with angle-lapped surface overcomes some of the limitations of conventional method. Similarly methods involving x-ray generation, though not suitable for determining depth profiles, when used with angle-lapped samples may provide similar useful information about compositional depth distribution.

In order to have a complete understanding of interfaces, the identification of intermetallic compounds and crystallographic phases present in the intermixed layers are also very important. A number of methods (see Table 6) have been used for this purpose. In fact, glancing angle x-ray diffraction in a Read x-ray camera has been one of the most widely used method for the identification of metal silicides in metal–Si interactions. The presence of intermetallic compounds within alloyed metal–semiconductor contacts can, sometimes, be a potential source of stress. Its value can be estimated by using the x-ray diffraction lattice curvature measurement method.[418] The electron diffraction methods, on the other hand, have been used for identifying the surface structures of metal as well as clean reconstructed semiconductor surfaces before and after transient processing. Transmission electron microscopy combined with electron diffraction, a versatile method to study structural defects in this crystals, can be used to investigate microstructural changes caused by annealing at the interface.

Since surface processing also plays a vital role in the fabrication of Schottky barrier junctions (see Section 5.1), examination of surfaces, both from the point of view of defects and of contamination at various stages of fabrication, forms an important part of structural characterization. As can be seen from Table 6, a number of complementary methods are available for this purpose, and their use depend on the ease, availability, and nature of the study.

Finally, it can be said that the task of perfecting various metallizations as well as incorporating them within very small vertical and lateral dimensions (especially in the case of III–V compound based contacts) will require not only, in the near future, a clear understanding of contact morphology and metallurgy, but newer nondestructive methods for in-process characterization and evaluation.

REFERENCES

1. B.L. Sharma and S.C. Gupta, *Solid State Technol.* **23**(5), 97 (1980); **23**(6), 90 (1980).
2. P.A. Chen, *Solid State Electron.* **22**, 277 (1979).
3. H.L. Malm, *IEEE Trans. Nucl. Sci.* **NS-22**(1), 40 (1975).
4. A. Thanailakis and D.C. Northrop, *Solid State Electron.* **16**, 1383 (1973).
5. J.C. Manifacier and J.P. Fillard, *Solid State Electron.* **19**, 289 (1976); *Jpn. J. Appl. Phys.* **15**, 457 (1976).
6. E.Y. Chan and H.C. Card, *IEEE Trans. Electron. Devices* **ED-27**, 78 (1980).
7. T. Nishino, F. Yamano, and Y. Hamakawa, *Jpn. J. Appl. Phys.* **14**, 1835 (1975).
8. H.J. Jager and W. Kosak, *Solid State Electron* **12**, 511 (1969).
9. D. Tsang and S.E. Schwarz, *Appl. Phys. Lett.* **30**, 263 (1977).
10. J. Basterfield, J.M. Shannon, and A. Gill, *Solid State Electron.* **18**, 209 (1975).
11. H.C. Card, *IEEE Trans. Electron Devices* **ED-23**, 538 (1976).
12. A. Rusu, C. Bulucea, and C. Postolache, *Solid State Electron.* **20**, 499 (1977).
13. D. Dascalu, G.H. Brezeanu, P.A. Dan, and C. Dima, *Solid State Electron.* **24**, 897 (1981).
14. I. Derrcen, G. Lelay, and F. Salvan, *J. Phys. (Paris) Lett.* **39**, 287 (1978).
15. R.R. Varma, A. McKinley, R.H. Williams, and I.J. Higgenbotham, *J. Phys. D: Appl. Phys.* **10**, L171 (1977).
16. M. Hirose, N. Altaf, and T. Arizumi, *Jpn. J. Appl. Phys.* **9**, 260 (1970).
17. R.H. Williams, V. Montgomery, R.R. Verma, and A. McKinley, *J. Appl. Phys.* **10**, 1253 (1977).
18. U.A. Shakirov and M.S. Yunusov, *Phys. Status Solidi A* **37**, 681 (1976).
19. P. Panayotatos and H.C. Card, *Solid State Electron.* **23**, 41 (1980).
20. C.A. Crowell, H.B. Shore, and E.E. Labate, *J. Appl. Phys.* **36**, 3843 (1965).
21. R.M. Mitra, S.B. Roy, K. Ghosh, and A.N. Daw, *Solid State Electron.* **23**, 793 (1980).
22. A. Martinez, D. Esteve, A. Guivarch, P. Auvray, P. Henoc, and G. Pelous, *Solid State Electron.* **23**, 55 (1980).
23. C. Rhee, J. Saltich, and R. Zaerman, *Solid State Electron.* **15**, 1181 (1972).
24. N. Szydlo and R. Poirier, *J. Appl. Phys.* **44**, 1386 (1973).

25. M.J. Turner and E.H. Rhoderick, *Solid State Electron.* **11**, 291 (1968).
26. J. de'sousa Pieres, P. Ali, B. Crowder, F. d'Herule, S. Peterson, L. Scholt, and P.A. Tove, *Appl. Phys. Lett.* **35**, 202 (1979).
27. J.M. Wilkinson, J.D. Wilcock, and M.E. Brinson, *Solid State Electron.* **20**, 45 (1977).
28. E. Calleja, J. Garrido, J. Piqueras, and A. Martinez, *Solid State Electron.* **23**, 591 (1980).
29. Y. Anand, *IEEE Trans. Electron Devices* **ED-24**, 1330 (1977).
30. S.G. Askerov and R.K. Mamedov, *Sov. Phys. Semicond.* **12**, 1236 (1978).
31. A. Smirnov, P.A. Tove, J. de Sousa Pires, and H. Norde, *Appl. Phys. Lett.* **36**, 313 (1980).
32. L.L. Tongson, B.E. Knox, T.E. Sullivan, and S.J. Fonash, *J. Appl. Phys.* **50**, 1535 (1979).
33. C.R. Wronski, D.E. Carlson, and R.E. Daniel, *Appl. Phys. Lett.* **29**, 602 (1976).
34. A.K. Datta, K. Ghosh, R.N. Mitra, and A.N. Daw, *Solid State Electron.* **23**, 99 (1980).
35. P.H. Gerzon, J.W. Barnes, D.W. Waite, and D.C. Northrop, *Solid State Electron.* **18**, 343 (1975).
36. D.J. Coe, E.H. Rhoderick, P.H. Gerzon, and A.W. Tinsley, in *Metal–Semiconductor Contacts* (M. Pepper, ed.), p. 74, Conf. Ser. No. 22, Institute of Physics, London (1974).
37. F.J. Landkammer, *Solid State Commun.* **5**, 247 (1967).
38. J.M. Wilkinson, in *Metal–Semiconductor Contacts* (M. Pepper, ed.), p. 27, Conf. Ser. No. 22, Institute of Physics, London (1974).
39. A.M. Cowley, *Solid State Electron.* **13**, 403 (1970).
40. K.J. Miller, S.M. Sze, and M.J. Greece, Abstract Vol. 14, No. 81, p. 1, Electrochem. Soc. Spring Meeting (1965).
41. C.Y. Chang, P.L. Chiu, and C.H. Ma, *Solid State Electron.* **16**, 646 (1973).
42. T. Arizumi, M. Hirose, and N. Altaf, *Jpn. J. Appl. Phys.* **8**, 1310 (1969).
43. G.J. van Gurp and W.M. Reukers, *J. Appl. Phys.* **50**, 6923 (1979).
44. F. Mohammadi, *Solid State Technol.* **24**(1), 65 (1981).
45. K.N. Tu, R.D. Thompson, and B.Y. Tsaur, *Appl. Phys. Lett.* **38**, 626 (1981).
46. E. Hokelek and G.Y. Robinson, *Thin Solid Films* **53**, 135 (1978).
47. J.M. Andrews and F.B. Koch, *Solid State Electron.* **14**, 901 (1971).
48. M. Eizenberg, H. Foell, and K.N. Tu, *J. Appl. Phys.* **52**, 861 (1981).
49. C. Canali, F. Catellani, S. Mantovani, and M. Prudenziati, *J. Phys. D: Appl. Phys.* **10**, 2481 (1977).
50. D.J. Fertig and G.Y. Robinson, *Solid State Electron.* **19**, 407 (1976).
51. J.B. Bindall, W.M. Moller, and E.F. Labuda, *IEEE Trans. Electron Devices* **ED-27**, 420 (1980).
52. E.H. Rhoderick, *Metal–Semiconductor Contacts*, Clarendon Press, Oxford (1978).
53. J.M. Andrews, *J. Vac. Sci. Technol.* **11**, 1972 (1974).
54. J.M. Andrews and J.C. Phillips, *Phys. Rev. Lett.* **35**, 56 (1975).
55. D.E. Loannou, C.A. Dimitriadis, and S.M. Davidson, in *Development of Electromicroscopy and Analysis*, Glasgow 12–14 Sept. 1977, p. 255, Conf. Ser. No. 36, Institute of Physics, London (1977).
56. B.L. Smith and E.H. Rhoderick, *Solid State Electron.* **14**, 71 (1971).
57. W.A. Anderson, A.E. Delahoy, and R.A. Milano, *J. Appl. Phys.* **45**, 3913 (1974).
58. H. Norde, J. de Sousa Pires, F. d'Heurle, F. Pesavento, S. Petersson, and P.A. Tove, *Appl. Phys. Lett.* **38**, 865 (1981).
59. A.N. Saxena, M. Nage-Ali, P. Siffert, and V.I. Mitchell, in *Metal–Semiconductor Contacts* (M. Pepper, ed.), p. 160, Conf. Ser. No. 22, Institute of Physics, London (1974).
60. M. Beguwala and C.T. Crowell, *J. Appl. Phys.* **45**, 2792 (1974).
61. M.P. Ali, P.A. Tove, and M. Ibrahim, *J. Appl. Phys.* **50**, 7250 (1979).
62. M. Leppihalme and T. Tuomi, *Phys. Status Solidi* (*a*) **33**, 125 (1976).
63. J.M. Andrews and M.P. Lepselter, *Solid State Electron.* **13**, 1011 (1970).

64. C.A. Mead and W.G. Spitzer, *Phys. Rev. A* **134**, 713 (1964).
65. A.K. Srivastava and B.M. Arora, *Solid State Electron.* **24**, 1049 (1981).
66. H. Sakaki, Y. Sekiguchi, D.C. Sun, M. Taniguchi, H. Ohno, and A. Tanaka, *Jpn. J. Appl. Phys. Lett.* **20**, L107 (1981).
67. A.Y. Cho and P.D. Dernier, *J. Appl. Phys.* **49**, 3328 (1978).
68. J.M. Borrego, R.J. Gutmann, and S. Ashok, *Appl. Phys. Lett.* **30**, 169 (1977).
69. M.S. Tyagi, *Jpn. J. Appl. Phys.* **16**, Suppl. 1, 333 (1977).
70. G.B. Seiranyan and Y.A. Tkhorik, *Phys. Status Solidi (a)* **13**, K115 (1972).
71. R.D. Baertsch and J.R. Richardson, *J. Appl. Phys.* **40**, 229 (1969); *Solid State Electron.* **12**, 393 (1969).
72. A.K. Srivastava, B.M. Arora, and S. Guha, *Solid State Electron.* **22**, 185 (1981).
73. S. Ashok, J.M. Borrego, and R.J. Gutmann, *Solid State Electron.* **22**, 621 (1979).
74. W.G. Spitzer and C.A. Mead, *J. Appl. Phys.* **34**, 3061 (1963).
75. G.E. Eimers and E.H. Stevens, *Solid State Electron.* **17**, 721 (1972).
76. O.Y. Borkovskaya, N.L. Dimitruk, R.V. Konakova, and M.Y. Filatov, *Electron. Lett.* **14**, 700 (1978).
77. Y. Sato, M. Uchida, K. Shimada, M. Ida, and T. Imai, *Rev. Elec. Commun. Lab.* **18**, 638 (1970).
78. B.L. Smith, *Electron. Lett.* **4**, 332 (1968); Thesis, Manchester University (1969).
79. K. Kajiyama, S. Sakata, and O. Ochi, *J. Appl. Phys.* **46**, 3321 (1975).
80. P.H. Ladbrooke, *Solid State Electron.* **16**, 743 (1973).
81. P.M. Batev, M.D. Ivanovich, E.I. Kafedjiiska, and S.S. Simenov, *Int. J. Electron.* **48**, 511 (1980).
82. J.G. Smith and L.D. Clough, Proc. 7th Biennial Cornell Electrical Engineering Conf. p. 93, Cornell University, Ithaca, NY (1979).
83. B.K. Sehgal, S. Mohan, S.N. Mukerjee, B.L. Sharma, D.B. Agarwal, B.P. Jain, and K. Chand *J. Inst. Electron. Telecom. Eng.* **28**, 610 (1982).
84. A.M. Patwari and H.L. Hartnagel, *Phys. Status Solidi (a)* **26**, 469 (1974).
85. P.D. Taylor and D.V. Morgan, *Solid State Electron.* **19**, 473, 481, 935 (1976).
86. R. Heckmann and P. Harrop, *IEEE Trans. Electron Devices* **ED-19**, 1231 (1972).
87. P. Guetin and G. Schreder, *Phys. Rev. B* **5**, 3979 (1972).
88. A.P. Vyaktin, A.V. Dubinin, N.K. Maksimova, and N.G. Filonov, *Sov. Phys. Semicond.* **15**, 276 (1981).
89. E. Hokelek and G.Y. Robinson, *Solid State Electron.* **24**, 99 (1981).
90. J.A. Calviello, J.W. Wallace, and R.B. Paul, *IEEE Trans. Electron Devices* **ED-21**, 624 (1974).
91. A.A. England and E.H. Rhoderick, *Solid State Electron.* **24**, 337 (1981).
92. H. Gautier, C. Maerfeld, and P. Tournois, *Appl. Phys. Lett.* **32**, 517 (1978).
93. C. Barret and A. Vapaille, *Solid State Electron.* **21**,1209 (1978).
94. S.P. Murarka, *Solid State Electron.* **17**, 869 (1974).
95. A. Moser, *Appl. Phys. Lett.* **20**, 244 (1972).
96. J.A. Calviello, J.W. Wallce, and P.R. Bie, *Electron. Lett.* **15**, 509 (1979).
97. A.K. Sinha, T.E. Smith, M.H. Reed, and J.M. Poate, *Solid State Electron.* **19**, 489 (1976).
98. P.M. Batev, M.D. Ivanovich, E.I. Kafeduska, and S.S. Simenov, *Phys. Status Solidi (a)* **45**, 671 (1978).
99. K.J. Linden, *Solid State Electron.* **19**, 843 (1976).
100. A.K. Sinha and J.M. Poate, *Appl. Phys. Lett.* **23**, 666 (1973).
101. S. Guha, B.M. Arora, and V.P. Salvi, *Solid State Electron.* **20**, 431 (1977).
102. W.J. Moroney and Y. Anand, *Proc. 3rd. Int. Symp. GaAs*, p. 259, Institute of Physics, London (1971).
103. W.J. Devlin, R. Stall, C.E.C. Wood, and L.F. Eastman, *Proc. 7th Biennial Cornell Electrical Engineering Conf.*, p. 189, Cornell University, Ithaca, New York (1979).
104. I.H. Scobey, C.A. Wallace, and R.C.C. Ward, *J. Appl. Crystallogr.* **6**, 425 (1973).

105. L. Buene, T. Finstad, K. Rimstad, O. Lonsjo, and T. Olsen, *Thin Solid Films* **34**, 149 (1976).
106. H. Morkoc, A.Y. Cho, C.M. Stanchak, and T.J. Drummond, *Thin Solid Films* **69**, 295 (1980).
107. A.K. Sinha, *Appl Phys. Lett.* **26**, 171 (1975).
108. T. Arakai, *IEEE Trans. Electron Devices* **ED-25**, 1091 (1978).
109. H.R. Grinolds and G.Y. Robinson, *Appl. Phys. Lett.* **34**, 575 (1979).
110. M. Hagio, H. Takagi, A. Nagashima, and G. Kano, *Solid State Electron.* **22**, 347 (1979).
111. P.S. Ho and J.O. Olowolafe, *IBM Tech. Disclosure Bull.* **21**(4), 1752 (1978).
112. M. Ogawa, D. Shinoda, N. Kawamura, T. Nozaki, and S. Asanabe, *Proc. 3rd. Int. Symp. GaAs*, p. 268, Institute of Physics, London (1971).
113. J.C. Irvin and N.C. Vanderwal, in *Microwave Semiconductor Devices and Their Circuit Applications* (H.A. Watson, ed.), pp. 340–369, McGraw-Hill, New York (1969).
114. R.E. Lundgren and G.O. Ladd, *16th. Annual Proceedings on Reliability Physics*, San Diego, California, p. 255, IEEE, New York (1978).
115. Y. Anand, A. Christou, and H. Dietrich, *Proc. 6th Biennial Cornell Electrical Engineering Conf.*, p. 161, Cornell University, Ithaca, New York (1977).
116. Y.A. Brodzinsky and J.E. Davey, GaAs reliability programs in the United States, NATO Symp. on Microwave Components for the Frequency Range above 6 GHz, Brussels (1978).
117. N.J. Keen, *Proc. Inst. Electr. Eng.* 1, **127**, 188 (1980).
118. Y. Sugiyama, M. Tacano, S. Sakai, and S. Kataksa, *IEEE Electron Devices Lett.* **EDL-1**, 236 (1980).
119. T.F. Lei, C.L. Lee, and C.Y. Chang, *Solid State Electron.* **22**, 1035 (1979).
120. S.M. Sze, *Physics of Semiconductor Devices*, Chap. 8, John Wiley and Sons, New York (1969).
121. C.R. Wronski, *J. Appl. Phys.* **41**, 3805 (1970).
122. M. Somogyi, *Acta. Phys. Acad. Sci. Hungary* **44**(1), 49 (1978).
123. G. Margaritondo, P. Migliorato, and P. Perfetti, *Solid State Electron.* **16**, 523 (1973).
124. Y.A. Goldberg, E.A. Posse, and B.V. Tsarenkov, *Electron. Lett.* **7**, 601 (1971).
125. A.I. Ivashchenko, B.E. Samorukov, S.V. Slobodchikov, and A.I. Solomonov, *Sov. Phys. Semicond.* **13**, 451 (1979).
126. B.V. Tsarenkov, Y.A. Goldberg, A.P. Izergin, E.A. Posse, V.N. Ravich, T.Y. Rafiev, and N.F. Silvestrova, *Sov. Phys. Semicond.* **6**, 610 (1971).
127. Y. Nannichi and G.L. Pearson, *Solid State Electron.* **12**, 341 (1969).
128. A.M. Cowley, *J. Appl. Phys.* **37**, 3024 (1966).
129. A.G. Golovko, *Sov. Phys. Semicond.* **12**, 1418 (1978).
130. C.A. Mead, *Solid State Electron.* **9**, 1023 (1966).
131. H.C. White and R.A. Logan, *J. Appl. Phys.* **34**, 1990 (1963).
132. L.P. Krukovskaya, L.S. Berman, A.Y. Vul, and A.Y. Shik, *Sov. Phys. Semicond.* **11**, 1109 (1977).
133. Y. Nagao, T. Hariu, and Y. Shibata, *IEEE Trans. Electron Devices*, **ED-28**, 407 (1981).
134. P.W. Chye, I. Lindau, P. Pianetta, C.M. Garner, and W.E. Spicer, *Phys. Rev. B* **17**, 2682 (1978).
135. J.N. Walpole and K.W. Nill, *J. Appl. Phys.* **42**, 5609 (1971).
136. D.K. Skinner, *J. Electron. Mat.* **9**, 67 (1980).
137. A. Christou and W.T. Anderson, Jr., *Solid State Electron.* **22**, 857 (1979).
138. H. Morkoc, T.J. Drummond, and C.M. Stanchak, *IEEE Trans. Electron Devices* **ED-28**, 1 (1981).
139. R.H. Williams, R.R. Varma, and A. McKinley, *J. Phys. C: Solid State Phys.* **10**, 4545 (1977).
140. H.B. Kim, A.F. Loves, G.G. Sweeney, and T.M.S. Heng, *GaAs and Related Compounds 1976* (L.F. (Eastman, ed.), Conf. Ser. No. 33B, p. 145, Institute of Physics, London (1977).
141. R. Williams, Surface and Interface Properties of InP Single Crystals, Annual Report RU 39-3 (Applied Physics), Department of Physics, New University of Ulster, Coleraine, Northern Ireland (1976).

142. G.S. Korotchenko and I.P. Molodyan, *Sov. Phys. Semicond.* **12**, 141 (1978).
143. P.M. White and D.M. Brookbanks, *Appl. Phys. Lett.* **30**, 348 (1977).
144. N. Szydlo and J. Oliver, *J. Appl. Phys.* **50**(3), Pt. 1, 1445 (1979).
145. O. Wada and A. Majorfeld, *Electron Lett.* **14**, 125 (1978).
146. B. Tuck and B.P. Hayes-Gill, *Phys. Status Solidi* (*a*) **60**, 215 (1980).
147. H.D. Rees, in *Metal–Semiconductor Contacts* (M. Pepper, ed.), Conf. Ser. No. 22, p. 105, Institute of Physics, London (1974).
148. G.G. Roberts and K.P. Pande, *J. Phys. D: Appl. Phys.* **10**, 1323 (1977).
149. B. Tuck, K.T. Ip, and L.F. Eastman, *Thin Solid Films* **55**, 41 (1975).
150. M.L. Korwin-Pawlowski and E.L. Heasell, *Solid State Electron.* **18**, 849 (1975).
151. A.S. Volkov, Y.A. Goldberg, D.N. Masledov, N.G. Neromova, and Z.A. Saimkulov, *Sov. Phys. Semicond.* **6**, 1987 (1973).
152. C.K. Campbell and C.H. Morgan, *Thin Solid Films* **26**, 213 (1975).
153. A.M. Goodman, *J. Appl. Phys.* **35**, 573 (1964).
154. E. Hufnagl, D. Schreyer, A. Chatelain, and U.T. Hochili, *Helv. Phys. Acta.* **52**, 354 (1980).
155. B. Lepley and S. Ravelet, *Phys. Status Solidi* (*a*) **33**, 517 (1976).
156. D.J. Wheeler and D. Haneman, *Solid State Electron.* **16**, 875 (1973).
157. S.J. McCarthy and S.S. Yee, *Solid State Electron.* **16**, 115 (1973).
158. K.P. Pande, *Phys. Status Solidi* (*a*) **42**, 615 (1977).
159. H. Tsuge and Y. Onuma, *Jpn. J. Appl. Phys.* **16**, 1973 (1977).
160. R.L. Consigny and J.R. Madigan, *Solid State Commun.* **7**, 187 (1969).
161. C.A. Mead, *Appl. Phys. Lett.* **6**, 103 (1965).
162. T. Takeba, J. Saraie, and T. Tanaka, *Phys. Status Solidi* (*a*) **47**, 123 (1978).
163. T. Touskova and R. Kuzel, *Phys. Status Solidi* (*a*) **40**, 309 (1977).
164. M.H. Patterson and R.H. Williams, *Vacuum* **31**, 639 (1981).
165. T. Touskova and R. Kuzel, *Phys. Status Solidi* (*a*) **36**, 747 (1977).
166. J.M. Pawlikowski and J. Zylinski, *Acta Phys. Polonica A* **54**, 155 (1978).
167. N.T. Gordon, *IEEE Electron Dev.* **ED-28**, 434 (1981).
168. M. Aven and C.A. Mead, *Appl. Phys. Lett.* **7**, 8 (1965).
169. N.B. Lukyanchikova, G.S. Pekar, N.N. Tekachenko, H.M. Shin, and M.K. Sheinkman, *Phys. Status Solidi* (*a*) **41**, 299 (1977).
170. D.D. Nedeoglo, D.H. Lam, and A.V. Simaskevich, *Phys. Status Solidi* (*a*) **44**, 83 (1977).
171. M.S. Tyagi and S.M. Arora, *Phys. Status Solidi* (*a*) **32**, 165 (1975).
172. M.E. Ozsan and J. Woods, *Solid State Electron.* **18**, 519 (1975).
173. R. Mach, W. Gericke, H. Treptow, and W. Ludwog, *Phys. Status Solidi* (*a*) **49**, 667 (1978).
174. C.A. Mead, *Phys. Lett.* **18**, 218 (1965).
175. W.D. Baker and A.G. Milnes, *J. Appl. Phys.* **43**, 5152 (1972).
176. S. Kurtin and C.A. Mead, *J. Phys. Chem. Solids* **30**, 2007 (1969).
177. A.G. Kyazymzada, A.O. Guliev, and V.I. Tagirov, *Sov. Phys. Semicond.* **15**, 102 (1981).
178. R.B. Schoolar, J.D. Jensen, and G.M. Black, *Appl. Phys. Lett.* **31**, 620 (1977).
179. D.K. Hohnke and H. Holloway, *Appl. Phys. Lett.* **24**, 633 (1974).
180. K.W. Nill, J.N. Walpole, A.R. Calawa, and T.C. Harman, in *The Physics of Semimetals and Narrow-Gap Semiconductors*, (D.L. Carter and R.T. Bate, eds.) p. 383, Pergamon Press, New York (1970).
181. J. Baars, D. Bassett, and M. Schulz, *Phys. Status Solidi* (*a*) **49**, 483 (1978).
182. E.M. Logothetis, H. Holloway, A.J. Varga, and E. Wilkes, *Appl. Phys. Lett.* **19**, 319 (1971).
183. C. Clemen, A. Moller, P. Muny, J. Honigschmid, and E. Bucher, in *Proc. Photovoltaic Solar Energy Conf.*, Luxembourg, Sept. 1977, p. 638, D. Reidel Publishing Co. (1978).
184. A. Diligenti, B. Pellegrini, G. Salardi, P.E. Bagnoli, and C. Flores, *Solid State Electron.* **23**, 799 (1980).

185. J.S. Best, *Appl. Phys. Lett.* **34**, 522 (1979).
186. Y.D. Shen and G.L. Pearson, *Solar Energy Mat.* **2**, 31 (1979).
187. Y.A. Goldberg, T.Y. Rafiev, B.V. Tsarenkov, and Y.P. Yakovlev, *Sov. Phys. Semicond.* **6**, 398 (1972).
188. R. Chin, R.A. Milano, and H.D. Law, *Electron. Lett.* **16**, 626 (1980).
189. G.K. Arbuzova and M.K. Maksimova, *Sov. Electrochem.* **15**, 911 (1979).
190. D.V. Morgan, J. Frey, and W.J. Devlin, *J. Electrochem. Soc.* **127**, 1202 (1980).
191. V.I. Osinskii, E.N. Vigdorovich, G.I. Bolot'ko, S.A. Malyshev, and L.F. Plavich, *J. Appl. Spectrosc.* **31**, 917 (1979).
192. G.P. Donzelli, G. Guarini, and F. Vidimari, *Thin Solid Films* **55**, 25 (1978).
193. T. Nishino, H. Nishizawa, H. Takakura, and Y. Homakawa, *Jpn. J. Appl. Phys.* **15**, 807 (1976).
194. E. Calleja and J. Piqueras, *Electron. Lett.* **17**, 37 (1981).
195. W.J. Keeler, A.P. Roth, and E. Fortin, *Can. J. Phys.* **58**, 63 (1980).
196. H. Markoc, J.D. Oliver, Jr. and L.F. Eastman, Prof. 7th Biennial Cornell Electrical Engineering Conf., p. 71, Cornell University, Ithaca, New York (1979).
197. D.V. Morgan and J. Frey, *Electron. Lett.* **14**, 738 (1978).
198. K. Kajiyama, Y. Mizushima, and S. Sakata, *Appl. Phys. Lett.* **23**, 458 (1973).
199. T.P. Lee, C.A. Burrus, M.A. Pollock, and R.E. Nahory, Dev. Res. Conf., Ottawa, Canada, 24–26 June (1975).
200. G.E. Stillman, C.M. Wolfe, A.G. Foyt, and W.T. Lindley, *Appl. Phys. Lett.* **24**, 8 (1974).
201. K. Kajiyama, M. Ida, S. Sakata, and Y. Mizushima, *IEEE Trans. Electron Devices* **ED-26**, 244 (1979).
202. D.L. Polla and A.K. Sood, *J. Appl. Phys.* **51**, 4908 (1980).
203. J.M. Pawlikowski, P. Becla, K. Lubowski, and K. Rosziewioz, *Acta Phys. Polonica* **A49**, 563 (1976).
204. D.K. Hohnke, H. Holloway, K.F. Yeung, and M. Harley, *Appl. Phys. Lett.* **29**, 98 (1976).
205. S. Buchner, T.S. Sun, W.A. Beck, N.E. Byer, and J.M. Chen, *J. Vac. Sci. Technol.* **16**, 1171 (1979).
206. S.G. Parker, *J. Electrochem. Soc.* **123**, 920 (1976).
207. P. Robinson and J.I.B. Wilson, in *Proc. 3rd. Int. Conf. on Ternary Compounds* (G.D. Holah, ed.), p. 229, Conf. Ser. No. 35, Institute of Physics, London (1977).
208. S.V. Bulyarsku and G.A. Kudintseva, *Sov. Phys. Semicond.* **11**, 1184 (1977).
209. R.P. Parkhomenko, T.T. Kokhanenko, and O.V. Voevodina, *Sov. Phys. J.* **22**, 656 (1979).
210. J. Stankiewicz and W. Giriat, *Appl. Phys. Lett.* **35**, 70 (1979).
211. J. Parkes, R.D. Tomlinson, and M.J. Hampshire, *Solid State Electron.* **16**, 773 (1973).
212. E.H. Rhoderick, *Proc. Inst. Electr. Eng.* **129**, 1 (1982).
213. V.L. Rideout, *Thin Solid Films* **48**, 261 (1978).
214. V.I. Strikha, E.V. Buzanera, and I.A. Rodzievskii, *Semiconductor Devices with Schottky Barriers* (Russian), Sovetskoc Radio, Moscow (1974).
215. A.G. Milnes and D.L. Feucht, *Heterojunctions and Metal–Semiconductor Junctions*, Academic Press, New York (1972).
216. B.M. Welch, *Solid State Technol.* **23**(2), 95 (1980).
217. I. Deyhimy, R.J. Anderson, R.C. Eden, and J.S. Harris, Jr., *Proc. Inst. Electr. Eng. Pt.* 1, **127** 278 (1980).
218. W. Kellner, N. Enders, D. Ristow, and H. Kniepkamp, *Solid State Electron.* **23**, 9 (1980).
219. M.L. Hammond, *Microelectron. J.* **10**(2), 4 (1979).
220. J.V. DiLorenzo, *Proc. 7th Biennial Cornell Electrical Engineering Conf.*, p. 131, Cornell University, Ithaca, New York (1979).
221. D.V. Morgan, F.H. Eisen, and A. Ezis, *Proc. Inst. Electr. Eng. Pt. I*, **128**, 109 (1981).

222. G.V. Samsonov, *Handbook of the Physico-Chemical Properties of Elements*, Plenum Press, New York (1968).
223. K.N. Tu and J.W. Mayer, in *Thin Films—Interdiffusion and Interactions* (J.M. Poate, K.N. Tu, and J.W. Mayer, eds.), p. 359, John Wiley and Sons, New York (1978).
224. R. Rosenberg, M.J. Sullivan, and J.K. Howard, in *Thin Films—Interdiffusion and Interactions* (J.M. Poate, K.N. Tu, and J.W. Mayer, eds.), p. 13, John Wiley and Sons, New York (1978).
225. E. Philofsky, *Solid State Electron.* **13**, 1391 (1970).
226. H.L. Gaigher and N.G. van der Berg, *Thin Solid Films* **68**, 373 (1980).
227. T. Narusawa, S. Komiya, and A. Hirake, *Appl. Phys. Lett.* **20**, 272 (1972).
228. R.W. Bower and J.W. Mayer, *Appl. Phys. Lett.* **20**, 359 (1972).
229. I. Ohdomari, T.S. Kuan, and K.N. Tu, *J. Appl. Phys.* **50**, 7020 (1979).
230. W.K. Chu, S.S. Lau, J.W. Mayer, H. Muller, and K.N. Tu, *Thin Solid Films* **25**, 393 (1975).
231. R.S. Nowicki, *Solid State Technol.* **23**(12), 83 (1980).
232. K.N. Tu, W.K. Chu, and J.W. Mayer, *Thin Solid Films* **25**, 403 (1975).
233. T.G. Finstad, *Thin Solid Films* **51**, 411 (1978).
234. R.W. Bower, D. Sigurd, and R.E. Scott, *Solid State Electron.* **16**, 1461 (1973).
235. J.W. Olowolafe, M.A. Nicolet, and J.W. Mayer, *J. Appl. Phys.* **47**, 5182 (1976).
236. R. Pretorius, J.O. Olowolafe, and J.W. Mayer, *Phil. Mag.* **A-37**, 327 (1978).
237. P.B. Ghate, J.C. Blair, C.R. Fuller, and G.E. McGuire, *Thin Solid Films* **53**, 117 (1978).
238. W.K. Chu, H. Krautle, J.W. Mayer, H. Muller, M.A. Nicolet, and K.N. Tu, *Appl. Phys. Lett.* **25**, 454 (1974).
239. R.B. Campbell and A. Rohatgi, *J. Electrochem. Soc.* **127**, 2702 (1980).
240. W.J. Garceau, P.R. Fournier, and G.K. Herb, *Thin Solid Films* **60**, 237 (1979).
241. L.D. Locker and C.D. Capio, *J. Appl. Phys.* **44**, 4366 (1973).
242. R.M. Walser and R.W. Bene, *Appl. Phys. Lett.* **28**, 629 (1976).
243. T. Shibata, T.W. Sigmon, J.L. Regolini, and J.F. Gibbons, *J. Electrochem. Soc.* **128**, 637 (1981).
244. S.S. Lau, J.W. Mayer, B.Y. Tsaur, and M. von Allmen, in *Laser and Electron Beam Processing of Materials* (C.W. White and P.S. Peercy, eds.), p. 511, Academic Press, New York (1980).
245. T.W. Sigmon, in *Laser and Electron Beam Solid Interactions and Material Processing*, Vol. I (J.F. Gibbons, L.D. Hess, and T.W. Sigmon, eds.), p. 511, North-Holland, New York (1981).
246. M. von Allmen, S.S. Lau, T.T. Sheng, and M. Wittmer, In *Laser and Electron Beam Processing of Materials* (C.W. White and P.S. Peercy, eds.), p. 524, Academic Press, New York (1980).
247. C.J. Doherty, C.A. Crider, H.J. Leamy, and G.K. Celler, *J. Electron. Mat.* **9**, 453 (1980).
248. P. Younger, A. Melas, J. Minnucci, A. Kirkpatrick, and A. Greenwald, in *Laser and Electron Beam Processing of Materials* (C.W. White and P.S. Peercy, eds.), p. 524, Academic Press, New York (1980).
249. B.Y. Tsaur, Z.L. Liau, and J.W. Mayer, *Appl. Phys. Lett.* **34**, 168 (1979).
250. T.W. Sigmon, J.L. Regolini, J.F. Gibbons, S.S. Lau, and J.W. Mayer, in *Laser and Electron Beam Processing of Electronic Materials* (C.L. Anderson, G.K. Celler, and G.A. Rozgonyi, eds.), p. 531, Electrochemistry Society, Princeton, New Jersey (1980).
251. J.M. Poate, H.J. Leamy, T.T. Sheng, and G.K. Celler, *Appl. Phys. Lett.* **33**, 918 (1978).
252. M. von Allmen and M. Wittmer, *Appl. Phys. Lett.* **34**, 68 (1979).
253. Authors Group (unpublished).
254. G. Landgren and R. Lukeka, *Solid State Commun.* **37**, 127 (1981).
255. A. Christou, J.E. Davey, H.M. Day, and A.C. Macpherson, *J. Appl. Phys.* **50**, 1139 (1979).

256. K. Sleger and A. Christou, *Solid State Electron.* **21**, 677 (1978).
257. N.M. Johnson, T.J. Magee, and J. Peng, *J. Vac. Sci. Technol.* **13**, 838 (1976).
258. C.A. Chang and N.J. Chou, *J. Vac. Sci. Technol.* **17**, 1358 (1980).
259. A.K. Sinha and J.M. Poate, *Thin Films—Interdiffusion and Reactions* (J.M. Poate, K.N. Tu, and J.W. Mayer, eds.), Chap. II, Wiley-Interscience, New York (1978).
260. C.J. Madams, D.V. Morgan, and M.J. Howes, *Electron. Lett.* **11**, 574 (1975).
261. L.J. Brillson, R.S. Bauer, R.Z. Bachrach, and G. Hansson, *Appl. Phys. Lett.* **36**, 326 (1980).
262. M. Ogawa, *Thin Solid Films* **70**, 181 (1980).
263. A. Oustry, M. Caumont, A. Escaut, A. Martinez, and B. Toprasertpong, *Thin Solid Films* **79**, 251 (1981).
264. V. Kumar, *J. Phys. Chem. Solids* **36**, 535 (1975).
265. A.K. Sinha, T.E. Smith, and H.J. Levinstein, *IEEE Trans. Electron Devices* **ED-22**, 218 (1978).
266. J.D. Speight and K. Cooper, *Thin Solid Films* **25**, 531 (1975).
267. L.J. Brillson, R.Z. Bachrach, R.S. Baur, and J.C. McMenamin, *Phys. Rev. Lett.* **42**, 397 (1979).
268. B.L. Sharma, in *Physics of Semiconductor Devices* (S.C. Jain and S. Radhakrishna, eds.), p. 96, Wiley Eastern, Delhi (1981).
269. J. Lymann, *Electronics* **48**(26), 61 (1974).
270. P.C. Parekh, R.C. Sirrine, and P. Lemieux, *Solid State Electron.* **19**, 493 (1976).
271. H. Grinolds and G.Y. Robinson, *J. Vac. Sci. Technol.* **14**, 75 (1975).
272. P.S. Ho, U. Koster, J.E. Lewis, and S. Libertini, in *Thin Film Phenomena—Interfaces and Interactions* (J.E.E. Baglin and J.M. Poate, eds.), p. 66, Electrochemistry Society, Princeton, New Jersey (1978).
273. G.J. van Gurp, J.L.C. Daams, A. van Oostrom, L.J.M. Augustus, and Y. Yamminaga, *J. Appl. Phys.* **50**, 6015 (1979).
274. C. Canali, F. Fantini, S. Gavinaghi, and A. Senin, *Microelectron. Reliab.* **21**, 637 (1981).
275. M. Bartur and M.A. Nicolet, *Appl. Phys. Lett.* **39**, 822 (1981).
276. A.K. Sinha and J.M. Poate, in *Proc. International Vacuum Congress*, Kyoto, Japan, p. 841 (1974).
277. T.L. Hierl and D.M. Collins, in *Proc. 7th Biennial Cornell Electrical Engineering Conf.*, p. 369, Cornell University, Ithaca, New York (1979).
278. J.M. Shannon, *Appl. Phys. Lett.* **24**, 369 (1974).
279. J.M. Shannon, *Solid State Electron.* **19**, 537 (1976).
280. W.K. Chu, M.J. Sullivan, S.M. Ku, and M. Shatzkes, *Radiat. Eff.* **47**, 7 (1980).
281. H. Ishiwara and S. Furukawa, *Jpn. J. Appl. Phys.* **16**, 53 (1976).
282. J.M. Andrews, R.M. Ryder, and S.M. Sze, United States Patent, 3,964,084 (1976).
283. C.Y. Wu, *J. Appl. Phys.* **51**, 4919 (1980).
284. R.L. Thornton, *Electron. Lett.* **17**, 485 (1981).
285. S.S. Li, J.S. Kim, and K.L. Wang, *IEEE Trans. Electron Devices* **ED-27**, 1310 (1980).
286. Y.P. Pai, H.C. Lin, M. Peckerar, and R.L. Kocher, *Proc. IEDM* 470 (1976).
287. S.S. Li, *Solid State Electron.* **21**, 435 (1978).
288. Y.P. Pai and H.C. Lin, *Solid State Electron.* **24**, 929 (1981).
289. C.Y. Wu, *Solid State Electron.* **24**, 857 (1981).
290. F.H. Eisen and J.W. Mayer, in *Treatise of Solid State Chemistry*, Vol. 6B (N.B. Hannay, ed.), p. 125, Plenum Press, New York (1976).
291. D.V. Morgon, F.H. Eisen, and A. Ezis, *Proc. Inst. Electr. Eng. Pt. I* **128**, 109 (1981).
292. R.P. Mandal and W.R. Scoble, *Gallium Arsenide and Related Compounds*, 1978 (C.M. Wolfe, ed.), p. 462, Conf. Ser. No. 45, Institute of Physics, London (1979).
293. B.L. Sharma and R.K. Purohit, *Surf. Technol.* **11**, 411 (1980).
294. K. Hikosaka, T. Mimura, and K. Joshin, *Jpn. J. Appl. Phys. Lett.* **20**, L847 (1981).

295. D.V. Morgan, H. Ohno, C.E.C. Wood, W.J. Schaff, K. Board, and L.F. Eastman, *Proc. Inst. Electr. Eng. Pt. I* **128**, 141 (1981).
296. A. Christou, J.E. Davey, and Y. Anand, *Electron. Lett.* **15**, 324 (1979).
297. L. Esaki and L.L. Chang, *Thin Solid Films* **36**, 285 (1976).
298. M. Ida, Y. Sato, M. Uchida, and K. Shimoda, *Rev. Elec. Comm. Lab.* **21**, 800 (1973).
299. W. Kellner, N. Enders, D. Ristaw, and H. Kniepkamp, *Solid State Electron.* **23**, 9 (1980).
300. G.T. Wrixon, *IEEE Trans. Microwave Theory Technol.* **MTT-24**, 702 (1976).
301. A.K. Wortmann and E.E. Kohn, *Solid State Electron.* **18**, 1095 (1975).
302. Y. Sato, M. Uchida, Y. Ishibashi, and T. Araki, *Rev. Elec. Comm. Lab.* **23**, 535 (1975).
303. W.C. Ballamy and A.Y. Cho, *IEEE Trans. Electron Devices* **ED-23**, 481 (1976).
304. R.A. Murphy, C.O. Bozler, C.D. Parker, H.R. Fetterman, P.E. Tannenwald, B.J. Clifton, J.P. Donnelly, and W.T. Lindley, *IEEE Trans. Microwave Theory Tech.* **MTT-25**, 494 (1977).
305. D. Bocon–Gibod and P. Harrop, *Proc. 8th Europ. Microwave Conf.*, p. 696 (1978).
306. B. Schwartz, *CRC Crit. Rev. Solid State Sci.* **6**, 609 (1975).
307. A. McKinley, A.W. Parke, G.J. Hughes, J. Fryar, and R.H. Williams, *J. Phys. D: Appl. Phys.* **13**, L193 (1980).
308. R.Z. Bachrach, in *Molecular Beam Epitaxy* (B.R. Pamplin, ed.), p. 115, Pergamon Press, Oxford (1980).
309. R.L. Maddox, *Microelectron. J.* **11**, 4 (1980).
310. J. Massies, P. Devoldere, and N.T. Linh, *J. Vac. Sci. Technol.* **15**, 1353 (1978).
311. R.L. Rode, B. Schwartz, and J.V. DiLorenzo, *Solid State Electron.* **17**, 119 (1974).
312. A. Shimano, H. Takagi, and G. Kano, *IEEE Trans. Electron Devices* **ED-26**, 1690 (1979).
313. B.L. Sharma, *Solid State Technol.* **21**(2), 48 (1978); **21**(4), 122 (1978).
314. A.H. Agajanian, *Solid State Technol.* **20**(1), 36 (1977).
315. K. Watanabe, M. Hashiba, Y. Hirota, M. Nishino, and Y. Yamashina, *Thin Solid Films* **56**, 63 (1979).
316. R.P.H. Chang, *Thin Solid Films* **56**, 89 (1979).
317. N. Yokoyama, T. Mimura, K. Odani, and M. Fukuta, *Appl. Phys. Lett.* **32**, 58 (1978).
318. T. Sugano and Y. Mori, *J. Electrochem. Soc.* **121**, 113 (1974).
319. K. Kamazawa and H. Matsunami, *Jpn. J. Appl. Phys. Lett.* **20**, L211 (1981).
320. J.F. Wager and C.W. Wilmsen, *J. Appl. Phys.* **51**, 812 (1980).
321. K.P. Pande and G.G. Roberts, *J. Vac. Sci. Technol.* **16**, 1470 (1979).
322. B.L. Sharma, P.L. Bharti, S.N. Mukerjee, and S. Mohan, *J. Inst. Electron. and Telecom. Eng.* **23**, 727 (1977).
323. C.S. Setzer and R.J. Mattauch, *Electron. Lett.* **17**, 555 (1981).
324. G.S. Sundaram, *Int. Defence Rev.* **2**, 271 (1979).
325. B.L. Sharma, in *Semiconductors and Semimetals*, Vol. 15, (R.K. Willardson and A.C. Beer, eds.), pp. 1–38, Academic Press, New York (1981).
326. V.L. Rideout, *Solid State Electron.* **18**, 541 (1975).
327. N. Yokoyama, S. Ohkawa, and H. Ishikawa, *Jpn. J. Appl. Phys.* **14**, 1071 (1975).
328. G. Eckhardt, in *Laser and Electron Beam Processing of Materials* (C.W. White and P.S. Peercey, eds.), p. 467, Academic Press, New York (1980).
329. J.G. Werthen and D.R. Scifres, *J. Appl. Phys.* **52**, 1127 (1981).
330. R.L. Mozzi, W. Fabian, and F.J. Piekarski, *Appl. Phys. Lett.* **35**, 337 (1979).
331. J.L. Tondon, C.G. Kirkpatrick, B.M. Welch, and P. Fleming, in *Laser and Electron Beam and Processing of Materials* (C.W. White and P.S. Peercey, eds.), p. 487, Academic Press, New York (1980).
332. C-Y. Ting and C.Y. Chen, *Solid State Electron.* **14**, 433 (1971).
333. P.A. Barnes and A.Y. Cho, *Appl. Phys. Lett.* **33**, 651 (1978).
334. T. Inada, S. Kato, T. Hara, and N. Toyoda, *J. Appl. Phys.* **50**, 4466 (1979).

335. P.A. Barnes, H.J. Leamy, J.M. Poate, S.D. Ferris, J.S. Williams, and G.K. Celler, *Appl. Phys. Lett.* **33**, 965 (1978).
336. J.W. Thompson, *Proc. 7th Biennial Cornell Electrical Engineering Conf.*, p. 165, Cornell University, Ithaca, New York (1979).
337. A.B.J. Sullivan, *Electron. Lett.* **12**, 133 (1976).
338. K. Heime, U. Konig, E. Kohn, and A. Wortmann, *Solid State Electron.* **17**, 835 (1974).
339. A.Y.C. Hu, H.J. Gopen, and R.K. Watts, Tech. Rep. No. AFAL-TR-70-196, Air Force At. Lab., Air Force Syst. Command, Wright-Patterson Air Force Base, Ohio (1970).
340. J.E. Varga and W.A. Bailey, *Solid State Technol.* **16**(12), 79 (1973).
341. L.I. Maissel and R. Glang, *Handbook of Thin Film Technology*, McGraw-Hill, New York (1970).
342. P. Archibald and E. Parent, *Solid State Technol.* **19**(7), 32 (1976).
343. J.L. Vossen and J.J. O'Neill, Jr., *RCA Rev.* **29**, 149 (1968).
344. S.J. Fonash, S. Ashok, and R. Singh, *Appl. Phys. Lett.* **39**, 423 (1981).
345. G.R. Thompson, Jr., *Solid State Technol.* **21**(1), 73 (1978).
346. H.G. Schneider, in *Advances in Epitaxy and Endotaxy* (H.G. Schneider and V. Ruth, eds.), p. 44, Elsevier Scientific Publishing Co., Amsterdam (1976).
347. B.N. Chapman and J.C. Anderson, *Science and Technology of Surface Coatings*, Academic Press, New York (1974).
348. Y.A. Goldberg, D.N. Nasledov, and B.V. Tsarankov, *Instrum. Exper. Tech.* **3**, 899 (1971).
349. W.J. Hillegas and G.L. Schmable, *Electrochem. Tech.* **1**, 228 (1963).
350. F.H. Dorbeck, *Solid State Electron.* **9**, 1135 (1976).
351. M. McColl, A.B. Chase, and W.A. Garbor, *J. Appl. Phys.* **50**, 8254 (1979).
352. J.C. Marinace, cited in Ref. 322.
353. D.J. Ehrlich, T.F. Deutsch, and R.M. Osgood, Jr., *Laser and Electron Beam Annealing of Materials* (C.W. White and P.S. Peercey, eds.), p. 671, Academic Press, New York (1980).
354. R.A. Colclaser, *Microelectronics: Processing and Device Design*, p. 39, John Wiley, New York (1980).
355. G.T. Wrixon and R.F.W. Pease, *GaAs and Related Compounds*, 1974; Conf. Ser. No. 24, p. 55, Institute of Physics, London (1975).
356. H. Gokan and S. Esho, *J. Vac. Sci. Technol.* **19**(1), 32 (1981).
357. R.A. Zettler and A.M. Cowley, *IEEE Trans. Electron. Devices* **ED-16**, 58 (1969).
358. J.L. Saltich and L.E. Clark, *Solid State Electron.* **13**, 857 (1970).
359. N.G. Anantha and K.G. Ashar, *IBM J. Res. Develop.* **15**, 442 (1971).
360. J.A. Appels, E. Kooi, M.M. Paffen, J.J.H. Schatorje, and H.C.G. Verkulylen, *Philips Res. Rep.* **25**, 118 (1970).
361. Y. Harada and H. Fukuda, *IEEE Trans. Electron Devices* **ED-26**, 1799 (1979).
362. J.A. Calviello, *Microwave J.* **22**(9), 92 (1979).
363. J. Kung, P. Pusateri, and L. Casner, *Proc. 7th Biennial Cornell Electrical Engineering Conf.*, p. 321, Cornell University, Ithaca, New York (1979).
364. J. Bardeen, *Phys. Rev.* **71**, 717 (1947).
365. A.M. Cowley and S.M. Sze, *J. Appl. Phys.* **96**, 3212 (1965).
366. V. Heine, *Phys. Rev.* **138A**, 1689 (1965).
367. S.G. Louie, J.R. Chelikowsky, and M.L. Cohen, *Phys. Rev.* **15B**, 2154 (1977).
368. J.C. Inkson, *J. Phys.* **5**, 2599 (1972).
369. G. Ottaviani, K.N. Tu, and J.W. Mayer, *Phys. Rev. Lett.* **44**, 284 (1980).
370. L.J. Brillson, *Phys. Rev. Lett.* **40**, 260 (1978).
371. K. Zdansky and Z. Sroubek, *Physics of Semiconductors* 1978 (B.L.H. Wilson, ed.), Conf. Ser. No. 43, p. 761, Institute of Physics, London (1979).
372. R.H. Williams, V. Montgomery, and R.R. Varma, *J. Phys. C: Solid State Phys.* **11**, L735 (1978).

373. M. Schluter, *Phys. Rev.* **17B**, 5044 (1978).
374. C. Boutrit, J.C. Georges, and S. Ravelet, *Proc. Inst. Electr. Eng. Pt.* 1 **127**, 250 (1980).
375. R.V. Konakova, G.D. Melinkov, Yu.A. Tkhorik, and M.Yu. Filatov, *Phys. Status Solidi* (*a*) **55**, K131 (1979).
376. P.K. Vasudev, B.L. Maties, E. Pietras, and R.H. Bube, *Solid State Electron.* **19**, 557 (1976).
377. C.W. White and P.S. Peercey (eds.), *Laser and Electron Beam Processing of Materials*, Academic Press, New York (1980).
378. J. Gyulai, J.W. Mayer, V. Rodriguez, A.Y.C. Yu, and H.J. Gopen, *J. Appl. Phys.* **42**, 3578 (1971).
379. B.L. Sharma, P.L. Bharti, S.N. Mukerjee, and S. Mohan, *Ind. J. Pure Appl. Phys.* **16**, 727 (1978).
380. J.M. Leas, P.J. Smith, A. Nagarajan, and A. Leighton, in *Laser and Electron Beam Processing of Materials* (C.W. White and P.S. Peercey, eds.), p. 645, Academic Press, New York (1980).
381. H. Foell, P.S. Ho, and K.N. Tu, *J. Appl. Phys.* **52**, 250 (1981).
382. J. Narayan, C.W. White, and R.T. Young, *Radiat. Eff.* **47**, 167 (1980).
383. J.L. Regolini, N.M. Johnson, R. Sinclair, T.W. Sigmon, and J.F. Gibbons, in *Laser and Electron Beam Processing of Materials* (C.W. White and P.S. Peercy, eds.), p. 297, Academic Press, New York (1980).
384. M. Servidore and I. Vecchi, *Solid State Electron.* **24**, 329 (1981).
385. H.B. Kim, G.G. Sweeney, and T.M.S. Heng, in *GaAs and Related Compounds* (J. Bok, ed.), Conf. Ser. No. 24, p. 307, Institute of Physics, London (1975).
386. G. Majni, F. Nava, G. Ottaviani, A. Luches, V. Nassisi, and G. Celotti, *Vaccum* **32**, 11 (1982).
387. T. Shibata, T.W. Sigmon, and J.F. Gibbons, in *Laser and Electron Beam Processing of Materials* (C.W. White and P.S. Peercy, eds.), p. 530, Academic Press, New York (1980).
388. S.S. Lau, W.K. Chu, J.W. Mayer, and K.N. Tu, *Thin Solid Films* **23**, 205 (1974).
389. K. Oura, S. Okada, Y. Kishikawa, and T. Hanawa, *Appl. Phys. Lett.* **40**, 138 (1982).
390. D.M. Zehner, C.W. White, G.W. Ownky, and W.H. Christie, in *Laser and Electron Beam Processing of Materials* (C.W. White and P.S. Peercy, eds.), p. 201, Academic Press, New York (1980).
391. J.B. Pendry, *Low-Energy Electron Diffraction*, Academic Press, New York (1974).
392. J.E. Baker, R.J. Blattner, S. Nadel, C.A. Evans, Jr., and R.S. Nowicki, *Thin Solid Films* **69**, 53 (1980).
393. A. Christou and K. Sleger, in *GaAs and Related Compounds* (L.F. Eastman, ed.), Conf. Ser. No. 33b, p. 191, Institute of Physics, London (1977).
394. P.A. Barnes and R.S. Williams, *Solid State Electron.* **24**, 907 (1981).
395. C.P. Lee, B.M. Welch, and J.L. Tondon, *Appl. Phys. Lett.* **39**, 556 (1981).
396. I. Shiota, K. Motoya, T. Ohmi, M. Miyamoto, and J. Nishizawa, *J. Electrochem. Soc.* **124**, 155 (1977).
397. W.A. Brantley, B. Schwartz, V.G. Keramidas, G.W. Kommlott, and A.K. Sinha, *J. Electrochem. Soc.* **122**, 434 (1975).
398. H.G. Parks and K. Rose, in *Laser and Electron Beam Processing of Materials* (C.W. White and P.S. Peercy, eds.), p. 549, Academic Press, New York (1980).
399. P.D. Vyas and B.L. Sharma, *Thin Solid Films* **51**, L21 (1978).
400. J.A. van der Berg and D.G. Armour, *Vacuum* **31**, 259 (1981).
401. W.K. Chu, J.W. Mayer, and M-A. Nicolet, *Backscattering Spectrometry*, Academic Press, New York (1978).
402. D. Sigurd, in *Metal–Semiconductor Contacts* (M. Pepper, ed.), p. 141, Conf. Ser. No. 22, Institute of Physics, London (1974).
403. A.H. Oraby, K. Murakama, Y. Yuba, K. Gamo, S. Namba, and Y. Masuda, *Appl. Phys. Lett.* **38**, 562 (1981).
404. C.J. Palmstrom, D.V. Morgan, and M.J. Howes, *Electron. Lett.* **13**, 504 (1977).

405. A.J. McEvoy, A. Parkes, K. Solt, and R. Bichsel, *Thin Solid Films* **69**, L5 (1980).
406. S.P. Kowalczyk, J.R. Waldrop, and R.W. Grant, *Appl. Phys. Lett.* **38**, 167 (1981).
407. P.W. Chye, I. Lindau, P. Pianetta, C.M. Garner, and W.E. Spicer, *Phys. Rev.* **178**, 2682 (1978).
408. P.J. Grunthaner, F.J. Grunthaner, A. Madhukar, and J.W. Mayer, *J. Vac. Sci. Technol.* **19**, 649 (1981).
409. J.L. Freouf, *J. Vac. Sci. Technol.* **18**, 910 (1981).
410. T.J. Gray, R. Lear, R.J. Dexter, F.N. Schwettman, and K.C. Wimer, *Thin Solid Films* **19**, 103 (1973).
411. K. Chino and Y. Wada, *Jpn. J. Appl. Phys.* **16**, 1823 (1977).
412. H. Hartnagel, K. Tomozawa, L.H. Herron, and B.L. Weiss, *Thin Solid Films* **36**, 393 (1976).
413. M. Wittmer, R. Pretorious, J.W. Mayer, and M-A. Nicolet, *Solid State Electron.* **20**, 433 (1977).
414. M.L. Tarng and G.K. Wehner, *J. Appl. Phys.* **43**, 2268 (1972).
415. S. Hofmann, *Surf. Interface Anal.* **2**, 148 (1980).
416. C. Lea and P. Seah, *Thin Solid Films* **75**, 67 (1981).
417. J.A. Cookson and F.D. Pilling, *Thin Solid Films* **19**, 381 (1973).
418. W.A. Brantley, V.G. Keramidas, B. Schwartz, M.H. Reed, and P.M. Petroff, *J. Electrochem. Soc.* **123**, 1582 (1976).

4

Schottky-Barrier-Type Optoelectronic Structures

Stephen J. Fonash

1. INTRODUCTION

This chapter focuses on optoelectronic structures which employ Schottky-barrier-type junctions. In the chapter we shall use the term "Schottky-barrier-type junction" to refer to rectifying metal–semiconductor (MS) junctions, metal–interfacial layer–semiconductor (MIS) junctions, and semiconductor–interfacial layer–semiconductor (SIS) junctions. The optoelectronic structures to be considered are light-emitting diodes, photodiodes, and photovoltaic devices.

In general, Schottky-barrier-type junctions are conducting structures which have only one semiconductor material, of one doping type, supporting the barrier. The barrier is a high electric field region depleted of majority carriers which is formed at the surface of the semiconductor. In MS and MIS Schottky-barrier-type configurations the metal layer (in the MS structure) or the metal–interfacial layer (in the MIS structure) has the functions of barrier former and current collector.[1] In the SIS version the semiconductor of the semiconductor–interfacial layer part of the structure is a degenerate, wide energy gap (window) material; hence this semiconductor–interfacial layer part has the functions of barrier former and current collector also.[1] A generalized schematic of a Schottky-barrier-type structure is seen in Fig. 1.

As we shall see, Schottky-barrier-type junctions are relatively simple structures that offer a number of potential advantages when used in optoelectronic applications.[1] Among these advantages are the following: (1)

Stephen J. Fonash • Engineering Science Program, The Pennsylvania State University, University Park, Pennsylvania 16802.

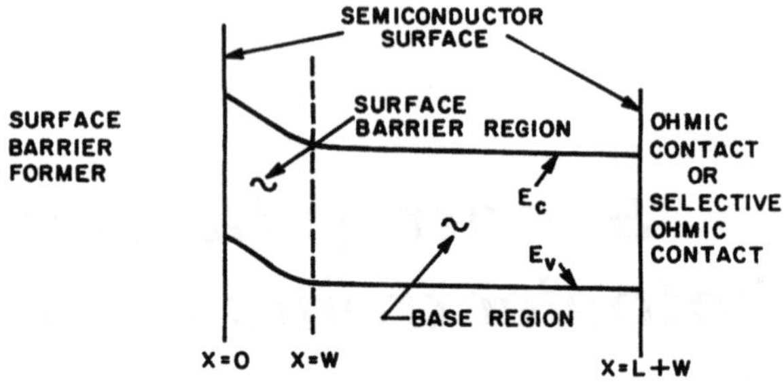

Figure 1. Generalized schematic of a Schottky-barrier-type junction.

The act of forming the barrier can be a low-temperature process that does not degrade material properties; (2) no diffusion or ion implantation is necessary to form the barrier; (3) in devices made on polycrystalline materials, diffusion down grain boundaries during junction formation is not a concern; (4) devices can be made using semiconductors available in just one doping type; and (5) short-wavelength spectral response for photodiodes and solar cells should be excellent, since the bare surface of the absorber (the semiconductor containing the barrier) can be passivated by the I layer (of MIS and SIS devices).

Schottky-barrier-type junctions can also have two distinct problem areas in optoelectronic applications.[1] One is that the metal layer of the MS and MIS configuration can reduce the amount of light coming from or entering into the semiconductor. This effect can be minimized by reducing the metal thickness, by reducing the extent of the metal coverage (using the metal in a grid pattern), or by using the transparent semiconductor–interfacial layer of the SIS configuration. The other problem area is that many MIS and SIS devices made to date rely on fixed charge in the interfacial layer for aiding barrier formation, and in some cases this charge may degrade under illumination.

In this chapter the chemistry and physics of barrier formation for Schottky-barrier-type junctions used in optoelectronic structures are outlined in Section 2. Transport in Schottky-barrier-type junctions is then

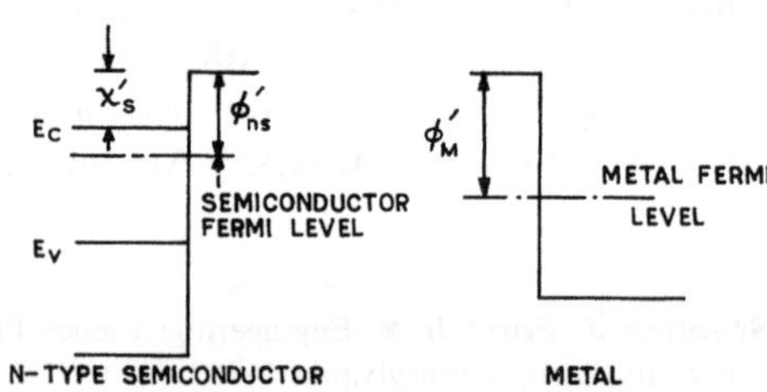

Figure 2. Metal and *n*-type semiconductor before contact. The vacuum level (escape energy) is the reference energy.

outlined in Section 3. These subjects are discussed briefly here since both are treated in greater detail elsewhere in this book. Section 4 surveys research and development work that has been done on Schottky-barrier-type optoelectronic structures. Section 4.1 treats Schottky-barrier-type light-emitting structures; Section 4.2 treats Schottky-barrier-type photodiodes; and Section 4.3 treats Schottky-barrier-type photovoltaic devices. The chapter closes with a summary in Section 5.

2. BARRIER FORMATION IN SCHOTTKY-BARRIER-TYPE JUNCTIONS

When two materials are placed into contact and allowed to come to thermodynamic equilibrium, there is just one temperature and one chemical potential for the system. We will henceforth refer to the chemical potential as the electrochemical potential since electric fields will be present in the junction region where the materials interface.

Figure 2 shows a metal and an n-type semiconductor before any contact has been made. We can see that the electrochemical potential of the metal is different from the electrochemical potential of the semiconductor. In fact by lining both energy bands up with respect to a common reference as is done in Fig. 2 it is seen that the electrochemical potential of the metal in this case lies below the electrochemical potential of the semiconductor. In terms of the notation of Fig. 2 these materials are such that $\phi'_M > \phi'_{ns}$.

If the metal and semiconductor of Fig. 2 are used to form an MS contact, there can be only one electrochemical potential once thermodynamic equilibrium is established (no bias, no light, and no temperature gradients). To force the electrochemical potentials to line up in thermodynamic equilibrium, a dipole charge layer (the junction) with its concomitant step in electrostatic potential energy develops at the MS interface. The total potential energy step ΔE_I set up across this dipole charge layer, once the junction is formed, must be

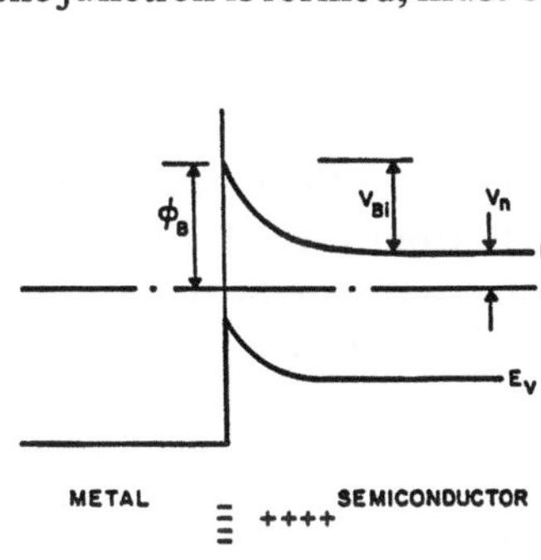

Figure 3. Ideal (Schottky–Mott model) metal–semiconductor junction for the case $\phi'_M > \phi'_{ns}$.

just enough to equate the electrochemical potentials; viz,

$$\Delta E_I = \phi'_M - \phi'_{ns} \tag{1}$$

The potential energy increment ΔE_I developed at the MS interface for $\phi'_M > \phi'_{ns}$ ideally appears as seen in Fig. 3. In this ideal situation ΔE_I is developed by a simple dipole layer (the Schottky–Mott model) consisting of a sheet negative charge at the metal surface and a positive space charge region in the semiconductor. The space charge region may exist as much as a micron or so into the semiconductor and it is seen to be depleted forming a Schottky barrier.†

In general, that part of the total potential energy step ΔE_I which is developed in the semiconductor space charge region, in thermodynamic equilibrium, is termed the built-in potential V_{bi}. In the ideal MS junction of Fig. 3 V_{bi}, the band bending in the positive space charge region of the semiconductor, is ΔE_I. Consequently, for the ideal MS structures

$$V_{bi} = \phi'_M - \phi'_{ns} \tag{2}$$

As we will see, the value of V_{bi} given by Eq. (2) is not valid for MIS (or SIS) configurations; it is only valid for the ideal Schottky–Mott MS junction.

For all Schottky-barrier-type junctions the Shottky barrier height ϕ_B is defined by

$$\phi_B = V_{bi} + V_n \tag{3}$$

In the case of ideal MS junctions this can be written as

$$\phi_B = \phi'_M - \phi'_{ns} + V_n \tag{4}$$

or as

$$\phi_B = \phi'_M - \chi'_s \tag{5}$$

where χ'_s is the semiconductor electron affinity (see Fig. 2). The $d\phi_B/d\phi'_B = 1$ behaviour predicted by Eq. (4) or Eq. (5) for MS structures fabricated on the same semiconductor is the characteristic feature of the Schottky–Mott model[(2)] for MS junctions.

† Had the electrochemical potentials been such that $\phi_M \leq \phi'_{ns}$, then a negative space charge region would have formed in the semiconductor in this simple dipole idealization. For a n-type semiconductor the result would be an accumulated layer giving an ohmic contact. We limit ourselves here to depleted regions in the semiconductor and, hence, to Schottky barriers.

Before we consider situations where there is, advertently or inadvertently, an interfacial layer between the metal and the semiconductor (i.e., MIS structures), we must face a fundamental problem that always occurs in assessing Schottky-barrier-type junction formation. This fundamental problem is the following: the ϕ_M' and ϕ_{ns}', which locate the metal and semiconductor electrochemical potentials, respectively, and which determine ΔE_I, should be *bulk material parameters*. Unfortunately, they are approximated with work function measurements and work functions simply are not a bulk parameter. In fact, work functions are surface sensitive; e.g., metal work function data are well known to vary with surface preparation and even (on clean surfaces) with crystallographic orientation.[3]

The error inherent in using work function data, which are surface sensitive, to approximate ϕ_M' and ϕ_{ns}' which should be surface-independent bulk quantities, must always be borne in mind. To try to avoid this problem one can turn to electronegativities[4] to assess the barrier-forming capabilities of a material. The larger the work function, the larger in the electronegativity, as seen in Fig. 4. Electronegativity data, such as the Pauling electronegativity scale of Fig. 4, have the advantage that they are characteristic of the material *only* and are *surface independent.*[5] As can be seen from Fig. 4 and Eqs. (2)–(5), to create a depleted region in a semiconductor with a metal–semiconductor contact (i.e., to form a Schottky barrier), a highly electronegative barrier former is needed for an n-type semiconductor and a highly electropositive (low electronegativity) barrier former is needed for a p-type semiconductor. These assessments are based on the idealization of the simple Schottky–Mott dipole at the junction; i.e., they are based on Eq. (2).

Turning now to the presence of interfacial layers, we note that Schottky barrier structures generally do not have the band diagram of Fig. 3. Often there is an interfacial layer (an I layer) present, either advertently or inadvertently, between the metal and the semiconductor. This I layer can be a purposefully deposited layer or a purposefully thermally grown layer, etc. The I layer could also inadvertently result from the act of junction formation. In the

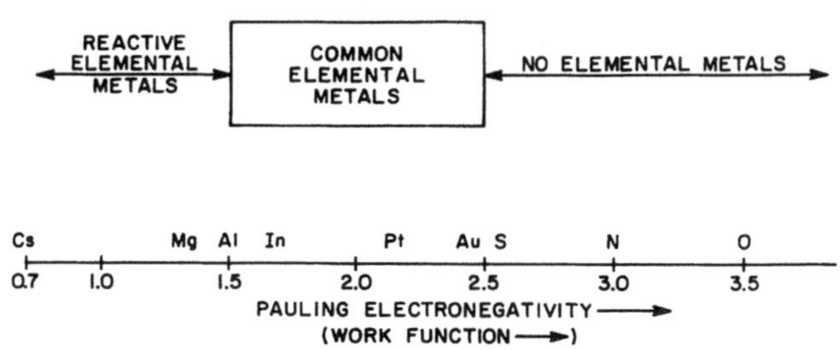

Figure 4. Pauling electronegativity of metals. Work function trend is also indicated. Electronegativity is not a surface sensitive quantity; work function, in reality, is surface sensitive.

latter case there can be an interfacial, or transition, layer extending into the semiconductor due to outdiffusion of constituents of the semiconductor,[6-8] in-diffusion of metal atoms,[9] chemical interactions[6-9] (perhaps non-stoichiometric), and semiconductor surface damage and rearrangement.[7,9]

The presence of an I layer can have two effects on barrier formation: one is a geometrical effect (it can simply further separate the charge in the metal from the charge in the semiconductor, giving a larger dipole length) and the other is a potentially significant modification of the dipole arrangement. If the interfacial layer contains no charge, then its only effect on barrier formation would be geometrical; i.e., the integral of the electric field ξ across the entire junction region must be ΔE_I and, therefore,

$$\left| \int_{\text{semiconductor}} \xi \, dx + \int_{\text{I layer}} \xi \, dx \right| = \Delta E_I \tag{6}$$

Equation (6) shows that the presence of a charge-free I layer inserted between a given metal and semiconductor, with ΔE_I predicted by Eq. (1), simply has the geometrical effect of decreasing the integral $\int_{\text{semiconductor}} \xi \, dx$ and, hence, of decreasing the built-in potential V_{bi} since it is always true that

$$V_{\text{bi}} = \left| \int_{\text{semiconductor}} \xi \, dx \right| \tag{7}$$

Note that Eq. (2) is no longer valid in MIS and SIS structures. However, Eq. (3) is kept, since it remains a measure of the barrier height in the semiconductor. Hence, the Schottky barrier height ϕ_B decreases in this case as follows from Eq. (3).

On the other hand the presence of an I layer *together with interface states* has more than a geometrical effect. It can strongly modify the dipole arrangement since these localized states can hold charge. Such interface states can be extrinsic arising from defects caused by cross diffusion, chemical interaction, and semiconductor surface damage and rearrangement. Interface states can also be intrinsic: they can be a basic feature of the semiconductor surface or they can arise from the extension of metal electron wave functions into the semiconductor gap.[10] Since interface states can store charge, they can modify the field in the I layer. Hence, through Eq. (6) it is seen that their presence, together with the I layer,† can modify V_{bi}

† Note that if these interface states were immediately adjacent to the metal (i.e., if there were localized states on the semiconductor surface but there were no I layer) they would have no effect since all their charge would reflect into the metal. Since they are distributed across the interfacial layer, some of their charge reflects into the semiconductor causing the electric field in the semiconductor to be modified and causing the shape of the variation of the band edges (the barrier) to be modified.

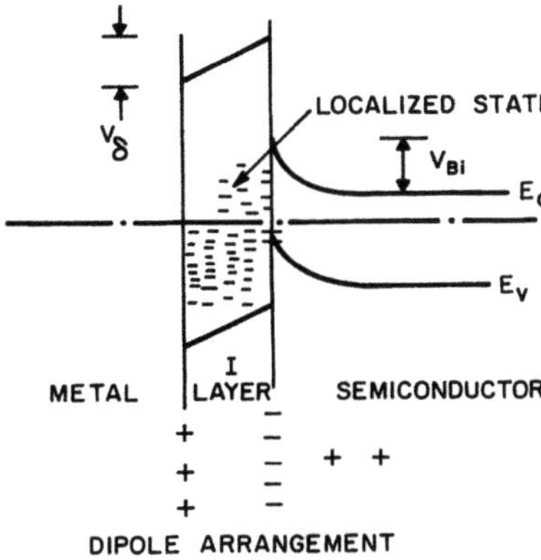

Figure 5. A possible MIS configuration. Here the I layer has a different band gap than the semiconductor. Interface states present in this particular MIS device are developing a negative charge, creating the complex dipole layer seen.

and ϕ_B. Unlike the simple geometrical case, it is now possible to increase V_{bi} and ϕ_B or decrease them, depending on the charge stored in the interface states.

Figure 5 is the energy band diagram for an MIS configuration. In this particular example the interfacial layer (I layer) is assumed to have a different energy band gap than the semiconductor and it is seen to contain interface states. In general, the I layer could have the same energy gap as the semiconductor, it could have a variable gap, or it could have the wide gap of an insulator as shown in the figure. The interface states present in Fig. 5 are seen to be developing negative charge, and the simple dipole layer of the Schottky–Mott model (Fig. 3) has now been modified to the more complex dipole seen in this figure, due to the presence of the I layer and interface states.

In the example of Fig. 5 there is so much negative charge held in the localized states that the sign of $\int_{\text{I layer}} \xi\, dx$ in Eq. (6) is now opposite to that of $\int_{\text{semiconductor}} \xi\, dx$. From Eqs. (6) and (7) it is seen that V_{bi} and ϕ_B for the MIS configuration are larger than V_{bi} and ϕ_B for the corresponding MS configuration in this example.

In general, we may rewrite Eq. (6) as

$$V_{bi} \pm V_\delta = \Delta E_I \tag{8}$$

for MIS and SIS structures,[1] where

$$V_\delta \equiv \left| \int_{\text{I layer}} \xi\, dx \right| \tag{9}$$

The plus sign is used in Eq. (8) when the electric field in the I layer is in the same direction as the electric field in the semiconductor. Note that, for a given semiconductor, the ΔE_I in Eq. (8) is dictated by the electronegativity of the barrier-forming metal (MIS) or barrier-forming semiconductor (SIS). Equation (8) demonstrates that, if an I layer has the role in barrier formation of simply further separating the components of the dipole (what we called the geometrical effect), then V_{bi} and ϕ_B are reduced. Equation (8) demonstrates that, if an I layer modifies the dipole configuration through the charge held in interface states, then V_{bi} and ϕ_B can be increased or reduced. To increase V_{bi} and ϕ_B in an n-type semiconductor the I layer should contain negative charge in localized states; to increase V_{bi} and ϕ_B in a p-type semiconductor the I layer should contain positive charge in localized states.[1]

In reality many supposedly intimate MS junctions are MIS structures where the inadvertent I layer has been created by out-diffusion of constituents of the semiconductor, in-diffusion of metal atoms, chemical interactions, semiconductor surface damage and rearrangement, etc. This inadvertent MIS structure must have a barrier region which obeys Eq. (1) and Eq. (8). The V_{bi} and ϕ_B of that inadvertent MIS structure may be strongly modified if a purposeful I layer is inserted. This purposeful I layer may block cross diffusion, etc. The resulting purposeful MIS structure will also satisfy Eq. (1) and Eq. (8) but the values of V_{bi} and ϕ_B that are now present may be substantially different from those of the inadvertent MIS structure.

3. TRANSPORT IN SCHOTTKY-BARRIER-TYPE STRUCTURES

3.1. MS and MIS Structures

The discussion here focuses on MIS structures; however, the simplification to MS devices is straightforward. Figure 6 shows an MIS structure in a forward-biasing configuration. For definitiveness an n-type semiconductor is depicted. Assuming a one-dimensional geometry, the current density J flowing for this bias V is a constant independent of position. We can develop a model

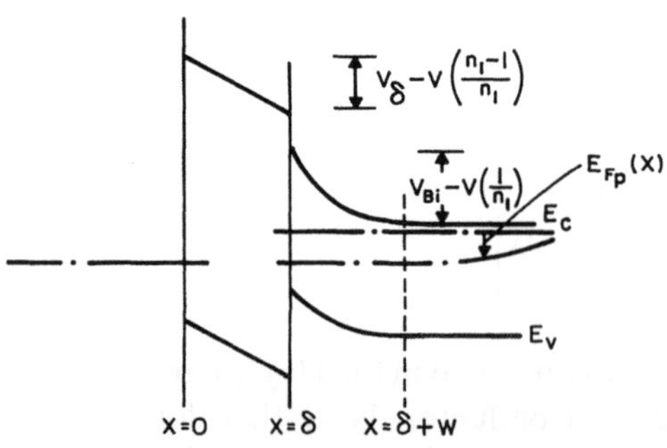

Figure 6. MIS device in forward bias. The total voltage V is seen to be split into $V[n_1 - 1)/n_1]$ and V/n_1.

for the $J-V$ characteristic in the dark by noting that J is made up of two components, J_n, the electron current, and J_p, the hole current, flowing at $x = \delta + W$ in the figure. The plane $x = \delta + W$ is chosen because it lets us immediately write[1]

$$J_p(\delta + W) = J_{\text{dif}}\{\exp[E_{F_p}(\delta + W)/kT] - 1\} \tag{10}$$

This is the hole current density flowing at the edge of the space charge region (i.e., $x = \delta + W$) in the dark. As can be seen from Eq. (10) and Fig. 6, it is a diffusion current driven by the hole quasi-Fermi level at $x = \delta + W$. In Eq. (10) E_{F_p} is measured positively down from the thermodynamic equilibrium position of the Fermi level in the gap. If the I layer and the space charge region of the semiconductor offer no impediment to the holes, then[1]

$$E_{F_p}(\delta + W) = V \tag{11}$$

If the I layer and the space charge region of the semiconductor offer an impediment to the holes, then in the dark[1]

$$E_{F_p}(\delta + W) < V \tag{12}$$

As we will see, in many MIS-type optoelectronic structures unimpeded flow of the minority carriers across the I layer and the semiconductor barrier region is a necessity. Hence, we will assume that the optimum situation [Eq. (11)] holds.

Since $J_p(\delta + W)$ is a diffusion current, the prefactor J_{dif} is given by[1]

$$J_{\text{dif}} = \left[\frac{eD_p p_{n0}}{L_p}\right]\left[\frac{\beta_7 \cosh\beta_5 + \sinh\beta_5}{\beta_7 \sinh\beta_5 + \cosh\beta_5}\right] \tag{13}$$

The β factors control the coupling of the back ohmic contact to the active junction; using the notation of Ref. 1 we have

$$\beta_5 \equiv L/L_p \tag{14}$$

and

$$\beta_7 \equiv L_p S_p / D_p \tag{15}$$

where L is the width of the quasineutral region in the base and S_p is the back surface, minority carrier recombination speed.

The component $J_n(\delta + W)$ flowing $x = \delta + W$ cannot be analyzed as being simply a diffusion current or simply a drift current. It is both, since it is a majority carrier current.[1] However, it can be seen that this electron current density flowing across the plane $x = \delta + W$ does so because of (1) the current density J_1 arising from electrons entering the I layer due to their being emitted over the semiconductor barrier, (2) the current density J_2 arising from electrons entering the I layer due to their tunneling through the semiconductor barrier, (3) the current density J_3 arising from electrons recombining in the semiconductor barrier region ($\delta < x \leq \delta + W$), and (4) the current density J_4 arising from electrons recombining at the semiconductor–interface layer boundary. Expressed mathematically, in the dark

$$J_n(\delta + W) = J_1 + J_2 + J_3 + J_4 \tag{16}$$

For the component J_1 to flow electrons must cross the barrier region in the semiconductor, they must be thermionically emitted over the Schottky barrier peak at $x = \delta$, and then they must cross the I layer. At least for thin I layers this current J_1 may be mathematically modeled by[11,12]

$$J_1 = \Xi_{\text{ob}} A^* T^2 \exp(-\phi_B/kT) \exp(V/n_1 kT) - \exp[-(n_1 - 1) V/n_1 kT] \tag{17}$$

where Ξ_{ob} is a transport factor ($\Xi_{\text{ob}} \leq 1$) characterizing the probability of an electron crossing the I layer,† A^* is Richardson's constant, and the n factor is defined by

$$n_1 \equiv V/V_s \tag{18}$$

This n factor accounts for the fact that not all of the bias V is developed across the semiconductor barrier region. Only a part of V, which we call V_s, is developed in the semiconductor in an MIS structure. It is the band bending in the semiconductor (i.e. $V_{\text{bi}} - V_s$) which controls emission over the barrier, as Eq. (17) indicates.[1]

In semiconductors with low mobilities, crossing the space charge region in the semiconductor and not the subsequent thermionic emission over the barrier may be rate-controlling step for J_1. In such materials $A^* T^2$ is replaced with[1]

$$e N_c \mu_n \xi_M$$

where N_c is the conduction band effective density of states and ξ_M is the maximum field in the semiconductor barrier region.

† A variety of transport modes are available to the electron for crossing the I layer including direct tunneling, multistep tunneling, etc.

The component J_2 arises due to electrons tunneling through the barrier in the semiconductor. These electrons then recombine at the semiconductor surface or cross the I layer. We assume tunneling through the semiconductor barrier region is the rate-determining step. For the semiconductor doping ranges of interest in optoelectronic configurations, the semiconductor space charge layer is too wide to permit direct tunneling. Hence the tunneling must be a multistep process involving an electron's tunneling from one localized state to another in the semiconductor barrier region.[1] A model for such a process, valid for $V - kT$, is[13]

$$J_2 = J_{02} e^{BT} e^{AV} \tag{19}$$

The two currents J_3 and J_4 both flow due to recombination. Usually they are modeled by [1]

$$J_3 = J_{03}(e^{V/n_3kT} - 1) \tag{20}$$

and

$$J_4 = J_{04}(e^{V/n_4kT} - 1) \tag{21}$$

The recombination n factors n_3 and n_4 are different from the voltage splitting n factor n_1 of Eq. (18). Contrary to the often-held view, n_3 and n_4 need not be 2.[1]

Combining Eqs. (10), (11), (16), (17), (19), (20), and (21) allows us to write

$$\begin{aligned} J = {} & J_{\text{dif}}(e^{V/kT} - 1) + \Xi_{\text{ob}} A^* T^2 e^{-\phi_B/kT} [e^{V/n_1kT} - e^{-(n_1-1)V/n_1kV}] \\ & + J_{02} e^{BT} e^{AV} + J_{03}(e^{V/n_3kT} - 1) + J_{04}(e^{V/n_4kT} - 1) \qquad (V > kT) \end{aligned} \tag{22}$$

This is the J–V characteristic of an MS of MIS structure in the dark.

If illumination is present, Eq. (22) is modified by subtracting off† the photogenerated current density J_L generated in the semiconductor, where[1]

$$\begin{aligned} J_L = e\Phi_0 \Bigg\{ & \left[\frac{\beta_6^2 e^{-\beta_4} e^{-\beta_6}}{\beta_6^2 - \beta_5^2}\right] \left[\frac{(\beta_7\beta_5/\beta_6) - 1}{\beta_7 \sinh\beta_5 + \cosh\beta_5}\right] \\ & + \left[\frac{\beta_6^2 e^{-\beta_4}}{\beta_6^2 - \beta_5^2}\right] \left[1 - \left(\frac{\beta_5}{\beta_6}\right)\left(\frac{\beta_7 \cosh\beta_5 + \sinh\beta_5}{\beta_7 \sinh\beta_5 + \cosh\beta_5}\right)\right] \Bigg\} + e\Phi_0(1 - e^{-\beta_4}) \end{aligned} \tag{23}$$

† Obtaining J–V under illumination by subtracting J_L from Eq. (22) is based on assuming superposition is valid. For some materials it may not be.[1]

Following Ref. 1, the new β factors introduced here are defined by

$$\beta_4 \equiv W\alpha$$

and

$$\beta_6 \equiv L\alpha$$

In these equations, Φ_0 is the number of photons with wavelengths in the range λ to $\lambda + \Delta\lambda$ per area per time impinging on the semiconductor at $x = \delta$. The quantity α is the absorption coefficient for these photons of wavelength λ. Equation (23) neglects any contribution to J_L from photons absorbed in the metallization.†

The first two terms in Eq. (23) arise from carriers photogenerated in the quasineutral region $x \geq \delta + W$. The mathematical modeling used to attain Eq. (23) assumes that this part of the photocurrent is set up by photogenerated minority carriers diffusing to the charge-separating electric field of the barrier. The last term in Eq. (23) arises from the direct sweeping out of photocarriers generated in the space charge region of the semiconductor by the electric field present there. This photocurrent of Eq. (23) is totally produced in the semiconductor; consequently it must cross $x = \delta$ as a hole current for the case of the n-type semiconductor of Fig. 6. Since we are assuming Eq. (11) is valid, we are assuming this hole current flows unimpeded across the I layer. Hence, I-layer properties do not enter into Eq. (23) because of Eq. (11).

Examination of Eq. (22) permits us to develop an overall picture of transport in MS and MIS diodes. We note that if the diode is an intimate MS contact, then Eq. (22) applies with suitable modification. If the diode is an MIS contact, then it can be seen that (1) the presence of an interfacial layer can modify the Schottky barrier height ϕ_B and thereby modify the over-the-barrier current J_1; (2) the presence of an I layer introduces the transport prefactor Ξ_{ob} into the over-the-barrier current J_1, which may reduce J_1 since $\Xi_{ob} \leq 1$; (3) the presence of the I layer can reduce components J_2, J_3, and J_4 since it could reduce the defects that carry these currents; and (4) the presence of an I layer can cause the voltage developed in the semiconductor V_s to differ significantly from the total voltage V, especially if localized states in the I layer fill on the passing of a current.[1] This latter effect can reduce the strength of J_1 at a given bias since J_1 is driven by V_s (i.e., by $V_s = V/n_1$). The interfacial layer can also suppress the minority carrier hole current (for the n-type semiconductor of Fig. 6), but in an ideal MIS structure this does not occur since Eq. (11) is assumed to be valid.

In many devices the majority carrier, over-the-barrier current J_1

† This contribution usually is not important. However, for a specific configuration discussed in Section 4.2 it is very important.

dominates Eq. (22). If this over-the-barrier current were suppressed by some combination of increasing ϕ_B, hindering transport across the I layer ($\Xi_{ob} \to 0$), and voltage splitting (the n_1 factor), then the multistep tunneling component J_2, the recombination components J_3 and J_4, and the minority carrier diffusion-recombination component would remain. At least conceptually, J_2, J_3, and J_4 can be suppressed by improving the quality of the interface. The presence of a purposeful I layer may improve the quality of the interface, since it could stop cross diffusion, block deleterious chemical reactions, reduce lattice distortion at the interface, etc. The minority carrier diffusion-recombination current [the first component of Eq. (22)] can never be removed. It is present due to the quasi-Fermi level splitting seen in Fig. 6 and it is the same current which would flow in an ideal p^+-n homojunction made with the semiconductor of that figure.

3.2. SIS Structures

SIS Schottky-barrier-type junctions can be classified as either isotype or anisotype. If the degenerate, window semiconductor being used to form the barrier has the same conductivity type as the barrier-supporting semiconductor, the SIS structure is isotype. If the degenerate, window semiconductor being used to form the barrier has the opposite conductivity type as the barrier-supporting semiconductor, the SIS structure is anisotype.[1] Figure 7a depicts an isotype SIS structure and Fig. 7b depicts an anisotype.

To discuss transport in SIS structures we again turn to Eq. (22). In the case of isotype SIS junctions the first term of Eq. (22), the minority carrier diffusion current (carried by holes for the n-type barrier-supporting semiconductor of Fig. 7), is present as it is for all Schottky-barrier-type devices. The second term of Eq. (22), the majority carrier over-the-barrier, is also present in

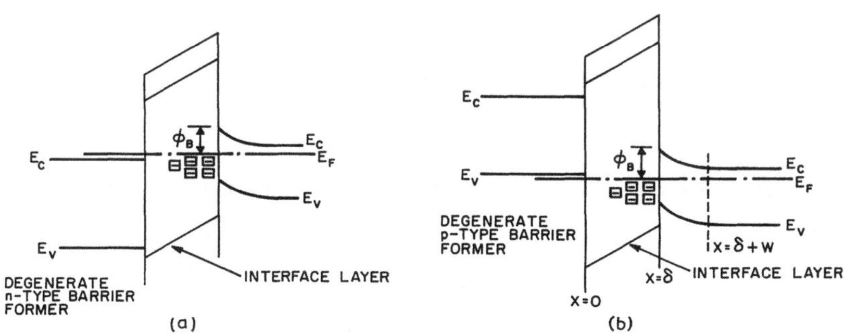

Figure 7. SIS junctions: (a) An isotype device; (b) an anisotype.

isotype SIS junctions. In addition, Fig. 7 shows that the multistep tunneling current and the two recombination current components of Eq. (22) can be present. As a consequence, all the various potential effects of an I layer discussed for MIS structures apply in the isotype SIS junction case.

Since easy minority carrier transport across the I layer is a necessity for many optoelectronic applications of SIS Schottky-barrier-type junctions, the condition of Eq. (11) is assumed to hold in our analysis of isotype SIS structures. Note that this has a unique implication for *isotype* SIS junctions: there must be states in the I layer which allow holes (for Fig. 7) to communicate very effectively through recombination and generation with the degenerate conduction band of the barrier-former.[1] If Eq. (11) is valid, then the photogenerated current density for the junction of Fig. 7 is given by Eq. (23).

For anisotype SIS junctions the first term of Eq. (22), the minority carrier diffusion current, is, of course, present. However, as can be seen from Fig. 7, the second term, the majority carrier over-the-barrier current, will be missing for anisotype SIS junctions due to the lack of final states in the barrier-former. The third term of Eq. (22) can still flow in anisotype junctions, if the electrons crossing the semiconductor barrier by multistep tunneling recombine at the semiconductor surface. The recombination components of Eq. (22), the fourth and fifth terms, are also present in anisotype SIS junctions. Although an I layer is no longer needed to suppress the over-the-barrier current in an anisotype structure, it can still have all the other effects discussed in Section 3.1.

Again easy minority carrier transport across the I layer is a necessity in many applications of effective SIS devices. Hence, we assume Eq. (11) is valid. If it is, then any photogenerated current present is given by Eq. (23).

4. SCHOTTKY-BARRIER-TYPE OPTOELECTRONIC STRUCTURES

This section surveys the research and development work that has been done on Schottky-barrier-type optoelectronic structures. As we will see, Schottky-barrier-type devices have shown promise as light emitting diodes (LEDs), photodiodes, and photovoltaic devices.

4.1. Schottky-Barrier-Type Light-Emitting Structures

Electroluminescence (EL) has been observed in a number of Schottky barrier structures for both forward and reverse bias. For example light emission has been observed from MIS structures employing III–V semiconductors such as forward-biased Au–oxide–(n)GaP diodes[14] and forward- and reverse-biased Au–oxide–(n)GaAs diodes[15,16] and from a number of

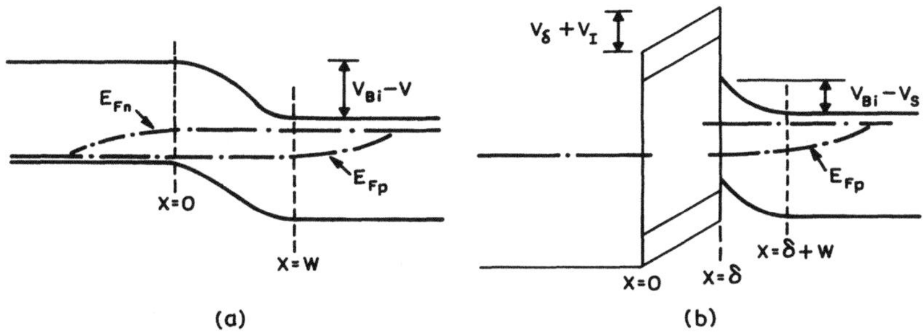

Figure 8. (a) A p–n homojunction and (b) the corresponding MIS device in forward bias. In the MIS device there is a voltage V_S developed in the semiconductor and a voltage V_I developed across the I layer such that $V = V_I + V_S$.

forward-biased MIS structures employing the II–VI semiconductors ZnSe(17–22) and ZnS.(18,23–27) Recently EL has also been reported for reverse-biased MS ZnSe and ZnS devices.(28) External quantum efficiencies† as high as 10^{-3} have been observed for ZnS MIS diodes under forward bias at room temperatures.(24,26) To put these current results in proper perspective, it must be noted that this is as much as two orders of magnitude below the external quantum efficiencies obtainable from p–n homojunction LEDs.

It is generally agreed that the source of light emission in all the forward-biased MIS structures is some form of injection electroluminescence. It is also observed that the I layer must be present for noticeable light emission in forward bias.(15,18,26) To understand the role of minority carrier injection and the I layer in light emission from these forward-biased MIS structures we turn to Fig. 8. Part (a) of this figure shows a p–n homojunction forward biased; part (b) shows the corresponding MIS junction fabricated using the same semiconductor, under forward bias.

In the forward-biased p–n homojunction electrons are injected into the p layer and holes are injected into the n layer. We focus on the n layer of the p–n junction, since this corresponds to the semiconductor of the MIS structure. The current density of injected holes at $x = W$ is given by Eq. (10). Since the holes have no trouble crossing the barrier region in a p–n homojunction, the hole quasi-Fermi level in Eq. (10) has the value V. In a direct gap semicon-

† External quantum efficiency EQE for light emission is defined as

$$\text{EQE} = (\text{photons per second out})/(I/e) \tag{24}$$

where I is the current flowing through the device and e is the charge on an electron. External quantum efficiency has the converse definition for light absorbing devices.

ductor, or in an indirect gap semiconductor with an optically active recombination path, a high percentage of these injected holes recombine by the emission of a photon of frequency ν resulting in injection electroluminescence. In a direct gap material $h\nu = E_G$; in an indirect gap material $h\nu < E_G$, since an intermediary step is required and only part of the energy may be released as light.[1] A high percentage of the total forward bias current can result in light emission in a properly designed p–n homojunction because the current can be designed to flow due to the just-described injection and recombination of holes in the n layer and due to the corresponding injection and recombination of electrons in the p layer.

In the corresponding forward-biased MIS junction seen in Fig. 8b, holes are also being injected into the n-type semiconductor at $x = \delta + W$ and this current density of injected holes is given by Eq. (10). Obviously, these injected holes can produce injection electroluminescence in the semiconductor. However, as we saw in Section 3, current flows in an MIS structure due to a number of other mechanisms: J_1, the over-the-semiconductor-barrier majority carrier current; J_2, the multistep tunneling-through-the-semiconductor-barrier current; and J_3 and J_4, the barrier region recombination currents.

The current density J_1 will not result in light emission† because electrons arriving in the metal due to this channel of current flow will thermalize mainly by electron–electron and electron–photon interactions.[26] For the same reason, the current density J_2 will not result in light emission, if these electrons also cross the I layer and enter the metal. The current density J_2 could produce electroluminescence, if the tunneling electrons, arriving at $x = \delta$, recombine with holes in an optically active transition. The recombination currents J_3 and J_4 could also result in electroluminescence, but only if the electrons recombine through optically active transitions.

In a forward-biased p–n junction all of the current, in principle, is available to produce electroluminescence. In MS and MIS Schottky-barrier-type diodes the fraction γ of current available to produce electroluminescence can vary from‡

$$\gamma = \frac{J_{\mathrm{dif}} e^{V/kT}}{J_{\mathrm{dif}} e^{V/kT} + \Xi_{\mathrm{ob}} A^* T^2 e^{-\phi_B/kT} e^{V/n_1 kT} + J_{02} e^{BT} e^{AV} + J_{03} e^{V/n_3 kT} + J_{04} e^{V/n_4 kT}} \tag{25}$$

† Proof of this is the fact that light is not observed for those forward-biased MS junctions where J_1 is clearly the dominant current.

‡ Note we have assumed the optimum situation for the hole quasi-Fermi level at $x = \delta + W$ in Eqs. (25) and (26); i.e., we have assumed hole transport is easily accomplished across the I layer giving $T_F^p(\delta + W) = V$ [Eq. (11)].

if J_2, J_3, and J_4 are nonradiative to†

$$\gamma = \frac{J_{\text{dif}}e^{V/kT} + J_{02}e^{BT}e^{AV} + J_{03}e^{V/n_3kT} + J_{04}e^{V/n_4kT}}{J_{\text{dif}}e^{V/kT} + \Xi_{\text{ob}}A^*T^2e^{-\phi_B/kT}e^{V/n_1kT} + J_{02}e^{BT}e^{AV} + J_{03}e^{V/n_3kT} + J_{04}e^{V/n_4kT}} \tag{26}$$

if J_2, J_3, and J_4 are all radiative.

As discussed in Section 3, the over-the-barrier current J_1 dominates in many forward-biased MS diodes and in many forward-biased MIS diodes; hence, Eqs. (25) and (26) show that light emission can be insignificant in these structures. However, these equations also show that an I layer can be chosen to suppress this over-the-barrier current to enhance light emission. The I layer can also be used to suppress J_2, J_3, and J_4, if they occur due to nonradiative processes. The I layer cannot inhibit hole transport (for the n-type semiconductor of Fig. 8) in an effective, forward-biased light-emitting junction.

In a study done on Au–(n)GaAs(E_G = 1.423V) MIS devices, it was found that there was no electroluminescence under forward bias for insulator thickness $\delta = 0$ and that the strongest forward bias electroluminescence for these devices was obtained for $20 \lesssim \delta \lesssim 40$ Å.[15] This was explained in terms of Eq. (26) and Eq. (11): the presence of the I layer decreases the majority carrier over-the-barrier current which dominates the dark current for $\delta = 0$ for GaAs (probably due to decreasing Ξ_{ob} and increasing ϕ_B in this case) and thereby enhances the injection of minority carrier holes into the n-type GaAs for any *current* level. These holes then cause injection electroluminescence. If the oxide is made too thick, Eq. (11) is no longer valid; and the relative rate of reduction in the hole component of the total current begins to exceed the relative rate of the advantageous reduction in the electron component. This causes an optimum insulator thickness, which is experimentally found to be $20 \lesssim \delta \lesssim 40$Å. The electroluminescence spectrum of these forward-biased structures was observed to have maxima at 0.98 and 1.24 eV; some samples also had an additional peak at 1.49 eV.[15]

A number of different light-emitting MIS structures have been fabricated using ZnS (E_G = 3.58–3.70 eV) including blue-emitting forward-biased diodes with EQE $\approx 10^{-3}$ at room temperature.[24,26] An interesting feature of many of these ZnS diodes is that the I layer has been relatively thick; in some cases it is reported to be thousands of angstroms in thickness.† Some of these blue light sources (those referred to as M–π–S) diodes have had I layers that have been produced by compensation of the (n)ZnS using implanted sulfur.[26] In others the I layer has been produced using the deposition of materials reported

† Transport across such "thick" I layers cannot be by direct tunneling but must be multistep tunneling, hopping, etc.[1]

to be as diverse as polystyrene, gelatine, and SiO.[27] In all cases, the presence of the I layer is found to be necessary for significant light emission from ZnS Schottky-barrier-type diodes.[18,26]

The ZnS MIS diodes that use I layers which are not based on ZnS yield the highest reported quantum efficiencies. An interesting feature of the highest EQE devices is that the insertion of the I layer between the metal and ZnS result in an *increase* in the forward current for a given bias in the range of significant electroluminescence. That is, the current of these MIS configurations is reported to exceed that of the MS configuration for a given bias even for "thick" insulator layers of thousands of angstroms.[27]

As we have seen, the reduction in MIS devices of the forward current present at a given voltage and a concomitant increase in electroluminescence for a given current is easily understood in terms of Eqs. (25) and (26). The presence of the I layer suppresses nonradiative channels of current flow, thereby reducing the total current and making radiative channels of current flow a larger component of the total current. However, this observation that use of nonnative I layer MIS devices with ZnS results in an increase in both forward current and electroluminescence seems contrary to explanations based on Eqs. (25) and (26).

In the detailed study of nonnative I layer MIS ZnS diodes reported in Ref. 27 it was deduced that there is an inadvertent I layer present in supposedly intimate metal–ZnS diodes. Examination of the forward J–V characteristics revealed that this inadvertent I layer caused voltage splitting (see Section 3). When purposeful I layers were introduced, more of the applied bias was found to be developed in the semiconductor which can increase currents J_1, J_2, J_3, and J_4 of Eq. (22) and, consequently, increase the total current at a given voltage. Apparently, the introduction of the purposeful I layer blocks the formation of the inadvertent I layer and its associated localized states which must store charge on the passing of a current and thereby develop a voltage V_I.[27]

This explanation tells us why the current increased at a given voltage with the purposeful I layer being present but it does not tell us why the electroluminescence increases. It is proposed in Refs. 18 and 27 that the origin of the electroluminescence is Auger recombination[1] of holes at $x = \delta$ (see Fig. 8b) and the subsequent radiative recombination of hot holes created by the Auger process. Figure 9 shows such a sequence suggested by Ref. 18: holes from the metal arrive at $x = \delta$, Auger recombine with electrons arriving at $x = \delta$, and produce hot holes in the valence band. These hot holes move into the bulk of the semiconductor and radiatively recombine as seen. It is assumed in Ref. 18 and 27 that electroluminescence increases in MIS ZnS diodes because the purposeful I layer enhances the useful Auger recombination levels seen in Fig. 9.

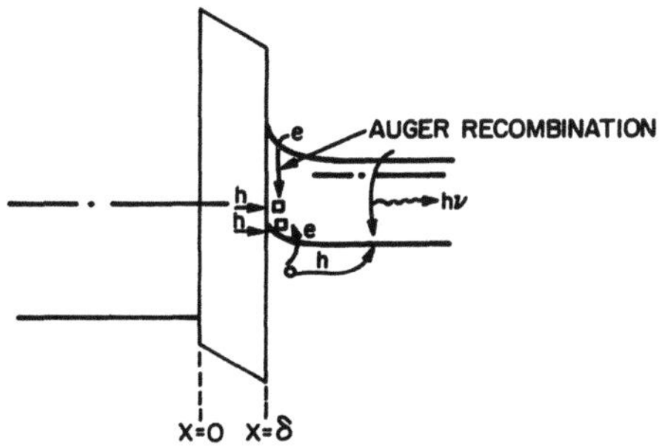

Figure 9. Radiative recombination of hot holes created by Auger recombination at $x = \delta$ according to model suggested in Ref. 18. For ZnS ($E_G \approx 3.6\,\text{eV}$), this recombination must be through a recombination center to yield visible light.

The introduction of this model employing Auger recombination at $x = \delta$ seems unnecessarily involved and cumbersome. For a simpler explanation one could turn to Eq. (26): although the presence of the purposeful I layer increases the total current, it may do so, not by increasing J_1 but by increasing some combination of radiative J_2, J_3, or J_4 current paths. Hence, the total current at a given voltage may increase but the electroluminescence will also increase.

In the forward-biased LED structures fabricated on ZnSe ($E_G = 2.67\,\text{eV}$) MIS configurations have again proven effective. In most of the devices examined, the light emission was mostly yellow or yellow-orange in color; however, there are also a number of reports of blue light emission for ZnSe MIS LEDs.[17–28] The former probably occurs in material where deep center emission dominates; the latter probably occurs in material where exciton decay dominates.[22] Again, the presence of the I layer in these devices enhances minority carrier injection, through Eqs. (25) or (26), thereby enhancing minority carrier injection EL. These injected minority carriers excite deep centers or create excitons or both causing the light emission.

Electroluminescence has also been observed in reverse-biased Schottky-barrier-type diodes. For MS diodes it has been demonstrated that electroluminescence can occur due to tunneling and subsequent recombination. For example, it has been seen that, for a *p*-type semiconductor, this can occur due to holes tunneling from the metal into the valence band where they then recombine with electrons which have tunneled from the valence band to the conduction band.[29]

Electroluminescence usually occurs, however, in reverse-biased Schottky-barrier-type devices because of one of the processes seen in Fig. 10 and 11. As may be seen from these figures, majority carriers may trigger the process in these reverse-biased EL mechanisms; hence, Eqs. (25) and (26) will not form the basis of this discussion.

Electroluminescence occurs in Fig. 10a due to carriers impact-exciting localized states. These localized states then relax by photon emission producing light. The process of EL in a reverse-biased diode could also involve

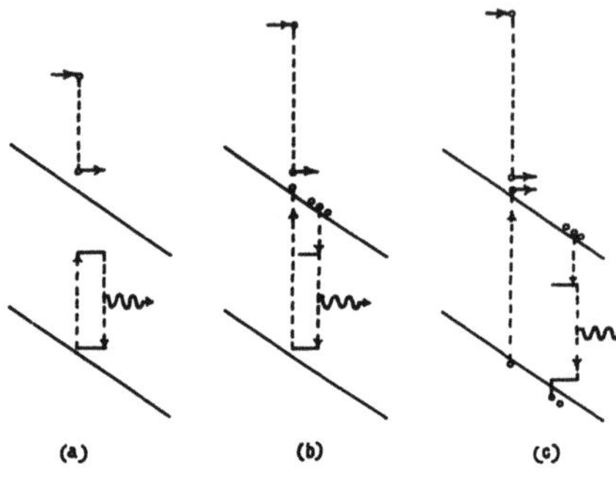

Figure 10. Three possible mechanisms for EL from reversed biased junctions. (a) Impact excitation of a localized level, (b) impact ionization of a localized level, and (c) avalanche multiplication and recombination through a localized level. (After Ref. 38, with permission.)

the mechanism of Fig. 10b. In this case the localized state is ionized by impact; subsequent capture of a carrier and its relaxation releases a photon. Figure 10c presents a third alternative: here avalanche multiplication produces carriers which recombine radiatively through a localized state. Figure 11 is another version of this where the recombination is band-to-band. The impact excitation process is believed to be responsible for the yellow-orange light observed from reversed biased ZnSe and ZnS MS devices.[28]

Electroluminescence in reverse-biased GaAs MIS diodes has been reported by several groups.[15,30] In the case of Ref. 30 the emission of white light visible to the naked eye was studied in Au, Al, and In MIS GaAs diodes which had anodically grown native oxide I layer with $\delta \approx 50\,\text{Å}$. The mechanism for this light emission in reverse bias is believed to be the avalanche electroluminescence of Fig. 11. The holes produced by this avalanche move toward the interface where they recombine with electrons in the semiconductor or in the I layer resulting in the emission of light.[30]

Even though the performance of Schottky-barrier-type light-emitting diodes studied thus far has not been competitive with that of *p–n* homojunction LEDs, Schottky-barrier-type devices are very interesting since they offer the possibility of developing LEDs from materials that are difficult to fabricate in *p–n* junction configurations. The work reported to date has been on MS and MIS light-emitting structures although the SIS configuration has a

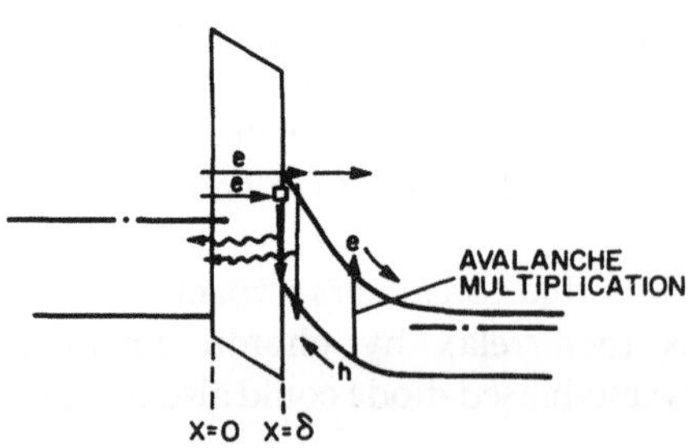

Figure 11. EL due to avalanche multiplication and band-to-band recombination, resulting in light emission.

great deal of potential in LED applications. The only work reported on nominally SIS type light emitting diodes used spray-deposited indium tin oxide (ITO) on p-type InP.[31] Light emission originating in the InP was observed, but examination of these devices revealed that they probably were buried n/p homojunctions. The SI layers served as a doping source of Sn to create the n layer in the (p)InP and then served as the contact to this n layer.[31] It is clear that the SIS structure will receive more attention as a light-emitting device in future research. Not only is the barrier-forming semiconductor transparent to light in the SIS configuration, but the anisotype junction has the advantage that the nonradiative current J_1 is absent (see Section 3).

4.2. Schottky-Barrier-Type Photodiodes

There are two classes of Schottky-barrier-type photodiodes. One class relies on absorbing the photons in the semiconductor of an MS or MIS structure and the other relies on absorbing the photons in the metallization of an MS structure. The former device class is designed to detect wavelengths shorter than

$$\lambda \leq 1.24/E_G \tag{27}$$

(where E_G is in electron volts); the latter device class is designed to detect wavelengths shorter than

$$\lambda \leq 1.24/\phi_B \tag{28}$$

Since $\phi_B > E_G$, the second type of device can detect radiation farther into the infrared. Both types of device are operated in reverse bias.

To understand the first device class we turn to Figs. 3 and 5. When light of wavelengths obeying Eq. (27) impinges in these structures, it creates electron–hole pairs in the semiconductor. If the junctions are reverse-biased, carriers created in the barrier region are swept out, giving a photocurrent. Carriers created in the base region can add to this since minority carriers can diffuse to the barrier and then be swept out. Equation (23) mathematically describes the situation. For a given set of bulk semiconductor properties the best detection of photons can be achieved *if the dark reverse current is small and controlled by minority carriers.*

In terms of our discussion in Section 3.1, this means that in cases where $J_1 > J_p(\delta + W)$ we must suppress J_1 vis-a-vis $J_p(\delta + W)$. In terms of the γ of Eq. (25),† we must have devices with $\gamma = 1$. Put succinctly: we must have

† Although we introduced γ in discussing light emission, it is a convenient parameter of general use. It is termed the minority carrier injection ratio when defined in terms of Eq. (25).

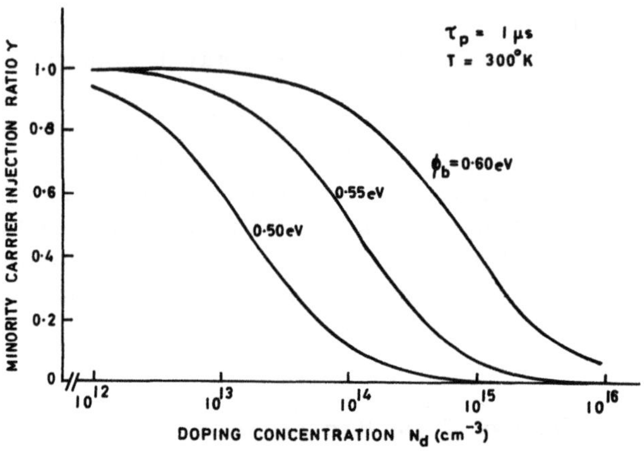

Figure 12. The γ of Eq. (25) (with $\Xi_{ob} = 1$ and $J_{02} = J_{03} = J_{04} = 0$) versus doping concentration N_D for MS (n-type) germanium diodes. The lifetime $\tau_p = 1$ μsec and $T = 300$ K. Three values of ϕ_B are used. Also, $\beta_5 \gg 1$ is assumed. (After Ref. 32, with permission.)

devices where the dark reverse current is controlled by minority carriers so that light-induced changes in minority carrier populations can significantly affect the reverse current. Figure 12 shows that our objective of $\gamma = 1$ can be achieved for Ge MS diodes without the need of an MIS structure to suppress J_1[32]; however, an MIS structure is needed to achieve this objective for Si. This difference between Si and Ge is essentially due to the low E_G for Ge; i.e., $J_{dif} \propto n_i^2 \sim e^{-E_G/kT}$ as may be seen from Section 3.1.

Schottky barrier MS photodiodes which rely on absorption in the semiconductor have been recently studied in detail for Ge[32,33] and GaSb.[34] The Ge [$E_G = 0.66$ eV] devices have been shown to be minority carrier devices (i.e., $\gamma = 1$) which serve as effective photodiodes.[32,33] The GaSb [$E_G = 0.72$] devices have the advantage that GaSb is a direct gap material whereas Ge is indirect gap[1]; hence, GaSb devices have stronger absorption. Metal–semiconductor GaSb photodiodes have been reported with quantum efficiency QE $\gtrsim 35\%$.[34]

Since more photocarriers are created in the barrier region for GaSb diodes, there is less reliance on diffusion of minority carriers into the barrier. This increased reliance on drift in the barrier directly as opposed to diffusion into the barrier should give better high-speed response. Because of the superior high-speed response of devices which collect the photocurrent by drift directly, metal/lightly doped n/n structures are always advantageous because the barrier will extend father into a lightly doped region.

To understand the second class of Schottky-barrier-type photodiode we turn to Fig. 13. When light of wavelength obeying Eq. (28) impinges on these

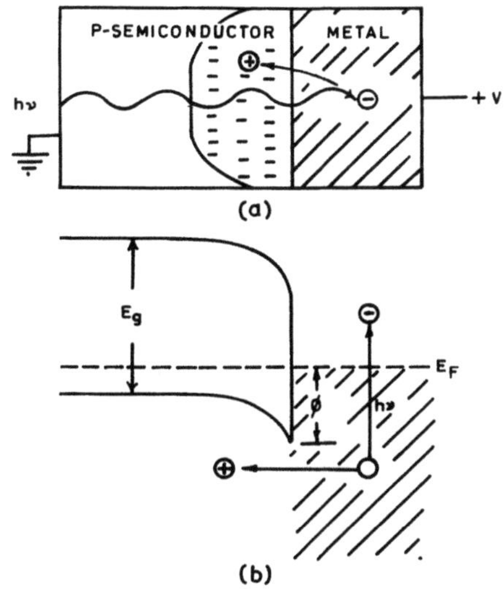

Figure 13. Class-two Schottky barrier photodiode relies on hot carrier creation in the metal and the sweeping out of that carrier by the barrier (field) region in the semiconductor. This process (referred to as internal photoemission) is seen for a *p*-type semiconductor. (After Ref. 35, with permission.)

structures, it creates hot carriers in the metal.† In the situation depicted in Fig. 13, hot electrons and holes are being created in the metal and some of the holes are ballistically emitted over the barrier. The holes emitted into the electric field region of the semiconductor in this figure are then swept out by the field setting up a photocurrent.[35] The photocurrent per photon flux gives the QE in this case. The expression for QE is.[35]

$$QE = C_1'(h\nu - \phi_B)^2/h\nu \tag{29}$$

Although the device in Fig. 13b is shown in thermodynamic equilibrium, these photodiodes are normally operated in reverse bias (Fig. 13a) since reverse bias lowers the effective barrier height below ϕ_B enhancing collection. This phenomenon of barrier lowering under reverse bias, characteristic of Schottky barriers, is referred to as Schottky barrier or image force lowering.[37]

The advantage of class two Schottky-barrier-type photodiodes is that they have a long-wavelength cutoff as dictated by Eq. (28). To fully exploit this cutoff advantage, one wants to fabricate devices with low barrier heights to allow majority carriers to cross easily from the metal. Consequently metal–semiconductor pairs are selected to yield low values of ϕ_B. Barrier shaping

† Wavelengths in the range

$$\lambda \leq 1.24/E_G$$

can also create electron–hole pairs in the semiconductor, if the metallization allows this light to pass.

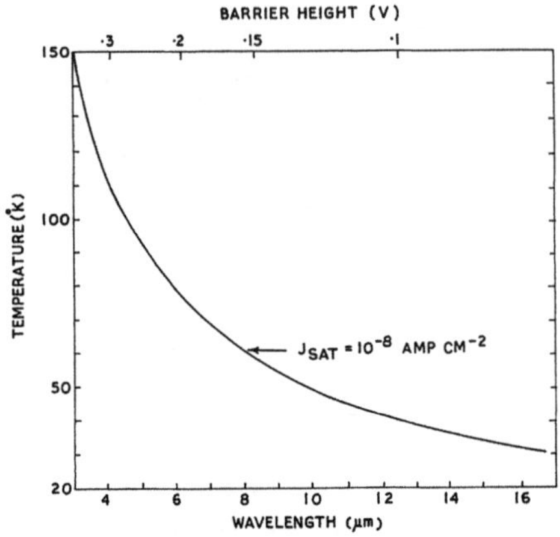

Figure 14. Cooling requirements for class-two Schottky barrier photodiodes. It is assumed that the reverse bias current density must be kept below 10^{-8} A/cm^2. (After Ref. 35, with permission.)

(using implantation) to enhance tunneling of majority carriers can also be used to effectively lower the barrier.[35]

Photodiodes of this class are majority carrier devices; i.e., thermionic emission (or thermally assisted field emission if barrier shaping has been used) dominates the dark current–voltage characteristics. Since the barrier heights are low, these majority carrier currents can be quite large. To reduce the dark reverse-bias current J_{SAT} to an acceptable level, cooling is needed in these devices. Assuming that J_{SAT} must be kept below 10^{-8} A/cm^2, Fig. 14 shows the cooling requirements versus barrier height and cutoff wavelength.[35]

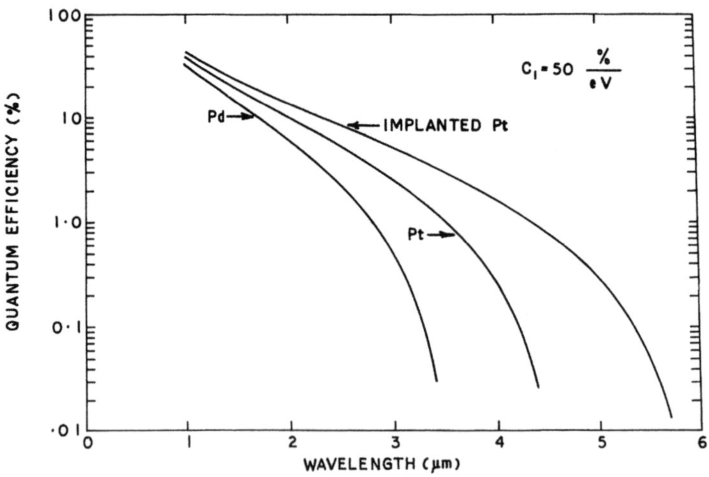

Figure 15. Quantum efficiency of various Schottky diodes based on experimentally observed ϕ_B and an assumed value of C_1'. Equation (29) was used; i.e., $\mathrm{QE} = C_1(h\nu - \phi_B)^2/h\nu$. (After Ref. 35, with permission.)

This class of Schottky barrier photodiode has been fabricated on p-Si using Pt and Pd metallizations. These metals were chosen since they give low barrier heights on p-Si and silicide stabilized interfaces.[35] As may be inferred from Fig. 4 and Eq. (28), Pt devices have a longer-wavelength cutoff than Pd devices. Using implantation to shape the barrier and effectively lower it has beeen demonstrated to be successful for these devices in extending the cutoff and in increasing QE for the 3.4- to 4.2-μm band.[35] These experimental results are summarized in Fig. 15.

4.3. Schottky-Barrier-Type Photovoltaic Devices

When a Schottky-barrier-type structure is designed as a photovoltaic device, it is operated in the fourth quadrant of a J–V plot. Consequently it is convenient to express the current produced, when a voltage V is developed and a light flux Φ_0 is impinging, as[1]

$$J = J_{\text{light}}(V, \Phi_0) - J_{\text{bucking}}(V, \Phi_0) \tag{30}$$

In ideal structures J_{light} is not a function of voltage and J_{bucking} is not a function of the light flux. Under these ideal conditions superposition is valid and J_{bucking} is given by Eq. (22) and J_{light} is given by Eq. (23).[1] Photovoltaic devices can be used specifically to produce power (solar cells) or as detectors.

4.3.1. MS and MIS Photovoltaic Devices

From Eq. (30) it may be seen that to optimize device performance, one wants to increase J_{light} and decrease J_{bucking} as much as possible. It follows from Eq. (23) that to increase J_{light} the barrier width W (if reliance is solely on drift in the barrier region) or the barrier width plus diffusion length $W + L_{n,p}$ (if reliance is principally on diffusion into the barrier region) must be made about the same as the depth of light absorption. Actually it is somewhat more complicated than that since the back contact can siphon off carriers. Thus the base width and back contact properties must also be considered and also optimized.[1] The back contact's influence on J_{light} is seen in the term involving the surface recombination speed in Eq. (23).

The fundamental question of whether to design for collection by diffusion or collection by drift is one that must be answered by considering the material used as the semiconductor and by considering the device application. If the semiconductor has a large diffusion length and if speed of response is not a consideration (it is not in solar cells but is in detectors), then collection by diffusion is attractive. It the semiconductor has a poor diffusion length or if speed of response is a consideration, then collection by drift is the approach to be used.[1]

It follows from Eq. (22) that to decrease J_{bucking} in Eq. (30) as far as possible for a given semiconductor, one wants to suppress the J_1, J_2, J_3, and J_4 of Section 1.2. Alternatively one wants the γ of Eq. (25) to be such that $\gamma = 1$. As we saw in Fig. 12, γ values of unity can be achieved in the MS configuration for large values of ϕ_B on relatively low band gap materials such as Ge. However, for larger band gap material such as Si, the MIS configuration is usually needed to suppress J_1 (and, perhaps J_2 and J_3 as discussed in Section 1.2) with respect to the minority carrier diffusion-recombination current to be able to produce devices with improved γ's and even, hopefully, devices with $\gamma = 1$. The MIS configuration is needed for these larger band gap materials because $J_{\text{dif}} \alpha e^{-E_G/kT}$ [see Eq. (13)]; i.e., the minority carrier diffusion-recombination current, which is always present in the dark $J-V$ and which is the only current which cannot be suppressed in the dark $J-V$, is very low in wide gap materials. It follows that the MIS configuration is useful for photovoltaic devices that are to be used as solar cells because the band gap of the semiconductor must be $E_G \sim 1-2$ eV in this application.[1]

In solar cell applications, to enhance cell efficiencies, the MIS configuration has further evolved into the MIS inversion layer (MISIL) cell seen in Fig. 16. The actual MIS junction areas are confined to being under the metal grid fingers which allows light to easily enter the absorbing semiconductor. However, the photocurrent-producing barrier region exists at the semiconductor surface across the complete device; under the oxide region, between grid fingers, the barrier is induced by charge built into the oxide.[38,39] In addition to the usual possible advantages of the MIS configuration, the MISIL cell also suppresses the majority carrier over-the-barrier component of J_{dark} by its geometry since this current can only flow beneath the metallizations. The band bending must be so extensive as to cause inversion in the semiconductor (hence the name inversion layer cell) because of the necessity of lateral photocurrent conduction to the grid fingers.

The MS and MIS configurations are very effective in evaluating the solar cell potential of materials since the low temperatures required in their fabrication have minimal effect on material properties. As an example, MS

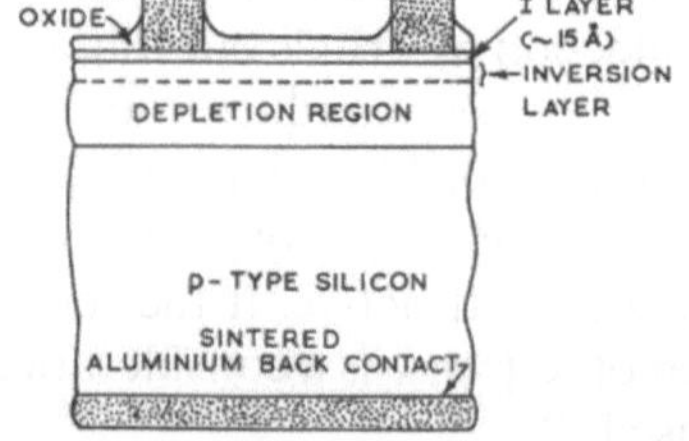

Figure 16. MIS inversion layer solar cell.

configurations have been used extensively to evaluate the potential of organic materials such as polyacetylene for solar cell applications.[40,41] Recently efficiencies as high as $\sim 1\%$ ($7\,\mathrm{mW/cm^2}$) have been reported for Al/$(CH)_x$/Au devices.[40]

The MIS configuration has been very successful on Si and GaAs,[1] yielding, in the case of Si, the highest open circuit voltages obtained to date with that semiconductor.[42] However, difficulties with stability of oxide charge have been observed at least for Al/SiO_x/Si solar cell structures.[43]

The MIS solar cell configuration has been used for a number of semiconductors other than Si and GaAs and a variety of I layer materials have been explored from native oxides to Langmuir films. In the case of Langmuir films, they have been used in Au/C4-anthracene Langmuir–Blodgett film/(*n*) CdTe MIS devices.[44]

4.3.2. SIS Photovoltaic Devices

From Eq. (30) it is again apparent that one must increase J_{light} and reduce $J_{bucking}$ to enhance the performance of SIS photovoltaic devices. As discussed in Section 3, isotype SIS devices are completely analogous to MIS devices except that J_{light} must flow due to recombination at the interface. Anisotype SIS junctions do not have this requirement that J_{light} must flow due to interface recombination as discussed in Section 3 and they have the advantage that thermionic emission of majority carriers ideally is blocked by the gap of the window material as also discussed in Section 3.

Using the *I* layer to reduce $J_{bucking}$ while maintaining J_{light} has been explored in a number of SIS configurations. In some cases the actual device structure produced has proven not to be the sought-after SIS configuration but a buried *p–n* homojunction contacted by the SI layer.[45] Isotype structures which probably are SIS devices fabricated using SnO_2 on (*n*)Si and indium–tin–oxide on (*n*)Si have yielded efficiencies well in excess of 10%.[46,47]

5. SUMMARY

When a Schottky-barrier-type device is to be used for light emission, it is designed to operate in the first, or perhaps third, quadrant of a *J–V* plot. In this mode, it converts electrical power into light. When a Schottky-barrier-type device is used as a photodiode, it is designed to operate in the third quadrant. When a Schottky-barrier-type structure is to be used as a photovoltaic device, it is operated in the fourth quadrant; i.e., the device converts light into electrical power. As a photovoltaic device, a Schottky-barrier-type structure can be designed to be a solar cell for optimized power generation or it can be designed as a photodetector.

Schottky-barrier-type devices are extremely convenient because they allow junction formation using only one semiconductor of one conductivity type and because they can be fabricated using low-temperature processing which minimizes degradation of semiconductor properties. Those MIS and SIS configurations which use charge in the oxide to modify the barrier height ϕ_B can be vulnerable, however, to instabilities in that oxide charge.

The coupling of light into (solar cells and detectors) or out of (LEDs) Schottky-barrier-type devices can be a problem for MS and MIS structures. The MISIL configuration minimizes this problem by reducing the metal coverage; the SIS configuration eliminates it by using the SI layers (with a transparent "window" S layer) as the barrier former.

REFERENCES

1. S.J. Fonash, *Solar Cell Device Physics*, Academic Press, New York (1981).
2. W. Schottky, *Naturwissenschaften* **26**, 843 (1938); N. Mott, *Proc. Cambridge Phil. Soc.* **34**, 568 (1938).
3. R. Gomer, *Field Emission and Field Ionization*, Harvard University Press, Cambridge (1961).
4. L. Pauling, *The Nature of the Chemical Bond*, Cornell University Press, Ithaca, New York (1960).
5. S. Kurtin, T.C. McGill, and C.A. Mead, *Phys. Rev. Lett.* **22**, 1433 (1969).
6. J.M. Andrews and J.C. Phillips, *Phys. Rev. Lett.* **35**, 26 (1975).
7. W. Spicer *et al.*, *Phys. Rev. Lett.* **44**, 420 (1980).
8. W.D. Johnston, H.J. Leamy, B.A. Parkinson, A. Heller, and B. Miller, *J. Electrochem. Soc.* **127**, 90 (1980); S.S. Chu, Y.T. Lea, A. Heller, and B. Miller, Abs. No. 580, 156th Electrochemistry Soc. Meeting, Los Angeles, (1980).
9. L. Brillson, *Phys. Rev. Lett.* **40**, 260 (1978); L. Brillson *et al.*, *Appl. Phys. Lett.* **36**, 326 (1980).
10. V. Heine, *Phys. Rev.* **138A**, 169 (1965).
11. V.I. Strikha, *Radio Eng. Electron. Phys.* (Eng. Transl.) **4**, 552 (1964).
12. E.H. Rhoderick, *Metal–Semiconductor Contacts*, Oxford University Press, London and New York (1978).
13. A.R. Riben and D.L. Feucht, *Int. J. Electron.* **20**, 583 (1969); *Solid-State Electron.* **9**, 1055 (1966).
14. H.C. Card and B.L. Smith, *J. Appl. Phys.* **42**, 5863 (1971).
15. A.A. Gutkin and V.E. Sedov, *Sov. Phys. Semicond.* **15**, 348 (1981).
16. B. Bayraktaroglu and H. Hartnogel, *Electron. Lett.* **14**, 470 (1978).
17. A.W. Livingstone, K. Turney, and J. Allen, *Solid-State Electron.* **16**, 351 (1973).
18. N. Lukyanchikova, T. Pavelko, G. Pehar, N. Tkachenko, and M. Sheinkman, *Phys. Status Solidi* (*a*) **64**, 697 (1981).
19. H. Watanabe, T. Chikamura, and M. Mada, *Jpn. J. Appl. Phys.* **13**, 357 (1974).
20. M. Ryall and J. Allen, *J. Phys. Chem. Solids* **34**, 2137 (1973).
21. M. Yamaguchi, A. Yamamots, and M. Kondo, *J. Appl. Phys.* **48**, 196 (1977); *Jpn. J. Appl. Phys.* **16**, 77 (1977).
22. X. Fan and J. Woods, *IEEE Trans. Electron Devices* **ED-28**, 428 (1981).
23. H. Katayama, S. Oda, and H. Kukimoto, *Appl. Phys. Lett.* **2** , 697 (1975).
24. N. Lukyanichova, G. Pekar, N. Tkachenko, Hoang M. Shin, and M. Sheinkman, *Phys. Status Solidi* (*a*) **41**, 299 (1977).

25. C. Lauther and J. Woods, *Phys. Status Solidi* (*a*) **50**, 491 (1978).
26. C. Lauther, S. Fujita, and T. Takagi, *Jpn. J. Appl. Phys.* **19**, 939 (1980).
27. N. Lukyanchikava, T. Pavelko, and G. Pekar, *Phys. Status Solidi* (*a*) **66**, 749 (1981).
28. N.T. Gordon, *IEEE Trans. Electron Devices* **ED-28**, 434 (1981).
29. P. Eastman, R. Haering, and P. Barnes, *Solid State Electron.* **7**, 879 (1964).
30. B. Bayraktaroglu and H. Hartnagel, *Electron. Lett.* **14**, 470 (1978).
31. L. Gouskov, H. Luquet, C. Gril, A. Oemry, and M. Savelli, *Rev. Phys. Appl.* **17**, 125 (1982).
32. D. Buchanan and H. Card, *IEEE Trans. Electron Devices* **ED-29**, 154 (1982).
33. E.Y. Chan and H. Card, *IEEE Trans. Electron Devices* **ED-27**, 78 (1980).
34. Y. Nagao, T. Hariu, and Y. Shibata, *IEEE Trans. Electron Devices* **ED-28**, 407 (1981).
35. P. Pellegrini, M. Weeks, and C. Ludington, *SPIE Proc.* **311**, (1981).
36. R. Fowler, *Phys. Rev.* **38**, 45 (1931).
37. S.M. Sze, *Physics of Semiconductor Devices*, 2nd edition, Wiley-Interscience, New York (1981).
38. R.B. Godrey and M.A. Green, *Appl. Phys. Lett.* **33**, 637 (1978).
39. R.E. Thomas, C.E. Norman, and R.B. North, Conf. Rec. 14th IEEE Photovoltaic Spec. Conf., p. 1350, IEEE, New York (1980).
40. J. Tsukamoto and H. Ohiyashi, *Synth. Metals* **4**, 177 (1982).
41. B. Weinberger, M. Akhtar, and S. Gau, *Synth. Metals* **4**, 187 (1982).
42. M.A. Green *et al.*, Conf. Rec. 15th IEEE Photovoltaic Spec. Conf., p. 1405, IEEE, New York (1981).
43. J.K. Kleta and D.L. Pulfrey, *IEEE Electron Device Lett.* **EDL-1**, 107 (1980).
44. G.G. Roberts, M.C. Petty, and I.M. Dharmadosa, to be published.
45. S. Ashok, S. Fonash, R. Singh, and P. Wiley, *IEEE Electron Devices Lett.* **EDL-2**, 184 (1981).
46. S. Ashok, P. Sharma, and S. Fonash, *IEEE Trans Electron Devices* **ED-27**, 725 (1980).
47. T. Feng, C. Feshman, and A. Ghosh, *Appl. Phys. Lett.* **34**, 198 (1978).

5

Schottky Barrier Photodiodes

S.C. Gupta and H. Preier

1. INTRODUCTION

In general, the detectors used for detecting electromagnetic radiation fall into one of two categories: (1) classical or thermal detectors and (2) quantum or photon detectors.

In thermal detectors, the incident radiation raises the temperature of the material. The raise in temperature causes changes in the temperature-dependent properties. Measuring one of these temperature-dependent parameters, the radiation can be detected.[1,2]

In photon detectors, the incident photons are absorbed within the material, by interaction with electrons, either bound to lattice atoms, or to impurity atoms or with free electrons. The interaction results in several photoeffects[3] such as photoconductive, photovoltaic, photoelectromagnetic, dember, or photon drag. Based on these photoeffects, there exist different types of photon detectors. All of these photon detectors are characterized by high detectivity and high speed of response when compared to thermal detectors. The resulting photosignal is dependent on wavelength of the incident radiation. It increases with increase of wavelength to a limit, which is characteristic of the detector material, beyond which it drops to zero. The two most widely employed photoeffects are the photoconductive and photovoltaic for detecting electromagnetic radiation.

Photoconductivity involves a process in which the electromagnetic radiation changes the electrical conductivity of the material. Intrinsic photoconductivity occurs when the electrical conductivity of the material is

---

S.C. Gupta and H. Preier • Fraunhofer-Institut für Physikalische Messtechnik, Heidenhofstrasse 8, D-7800 Freiburg, Federal Republic of Germany. S.C. Gupta's permanent address is Solid State Physics Laboratory, Lucknow Road, Delhi 110007, India.

increased as a result of free electron–hole pairs produced by incident radiation. Extrinsic photoconductivity results when incident radiation creates either free electrons and bound holes or free holes and bound electrons by interacting with bound electrons at impurity atoms. A photovoltaic detectors or photodiode depends on separation of photoexcited electron–hole pairs. It requires an internal potential barrier with built-in electric field. Several structures are possible to observe the photovoltaic effect. These include *p–n* junctions, *p–i–n* structures heterojunctions, and Schottky (metal–semiconductor) barriers. All practical photovoltaic detectors are intrinsic, though an extrinsic photovoltaic effect has also been reported.(3,4) The photodiode family has a faster response(5–7) and greater power conversion than photoconductors. We shall now briefly discuss the various photodiode structures.

A *p–n* junction is the most widely used photodiode structure for detection of radiation. It is normally prepared by diffusion. It consists of a thin *n*/*p* layer on *p*/*n*-type substrate of the same material. The thickness of the top layer is adjusted in such a way that most of the radiation is absorbed in the depletion region on both sides of the junction, such that the photoexcited electron–hole pairs can be separated by the internal electric field.

In *p–i–n* structures an intrinsic region is incorporated between the *p* and *n* sides of the photodiode. The surface region is made thinner than the optical absorption length, so that the absorption takes place in the intrinsic region, where the electron–hole pairs are produced. Because of the large electric field present there, photoexcited carriers can be very efficiently separated.

A heterojunction photodiode is formed by two dissimilar materials of different band gaps by epitaxial growth. The incident photons are absorbed completely in the low band gap material near the junction. The top layer of the wider band gap material can be quite thick. The main problem with this type of photodiode is the usual presence of a large density of defect states in the depletion region due to lattice mismatch causing a large dark current due to spontaneous electron–hole generation. Lattice matched structures in heterojunctions such as $GaAs$–$Ga_{1-x}Ga_{1-x}Al_xAs$, $PbTe_{1-x}Se_x$-$PbSn_xTe$, $PbSe_{1-x}S_x$-$Pb_{1-x}Sn_xSe$, etc., are of greatest interest.

A photodiode can also be realized at a Schottky barrier, formed at a metal–semiconductor junction. Like *p–n* junctions metal–semiconductor interfaces provides a potential barrier which separates photoexcited electron–hole pairs. Photoexcitation can occur within the semiconductor or at the metal–semiconductor interface. Schottky barrier photodiodes are of special interest for those materials in which a *p–n* junction cannot be formed.(3).

In recent years, more attention has been paid to fabricating Schottky barrier photodiodes(8–17) for detecting electromagnetic radiation. This is because of their advantages over *p–n* junction diodes; these advantages are fabrication simplicity, reliability, absence of high-temperature diffusion processes which can degrade carrier lifetime, and high speed of response.[17] The theoretical calculations[18] show that the Schottky barrier photodiodes of PbSnTe and PbSnSe can reach normalized detectivities of the order of 10^{11} cm $Hz^{1/2}/W$ and HgCdTe 10^{10} cm $Hz^{1/2}/W$ at the cutoff wavelength 12.4 μm. Schottky barrier photodiodes find application as ultraviolet and visible radiation detectors especially for laser receivers.[19,20] They are also used as infrared detectors and have applications in cameras.[21] Schottky barrier photodiodes are, therefore, of quite some interest.

In Section 2, the general parameters of Schottky barrier diodes are compared with those of the *p–n* junction diodes. Section 3 deals with criteria used for selecting the metal and semiconductor materials for fabricating the Schottky barrier photodiodes, Section 4 with fabrication technology, and Section 5 is on techniques used for the evaluation of the photodiode parameters. Section 6 deals with the application of these photodiodes, and in Section 7 conclusions are drawn.

2. GENERAL PARAMETERS OF PHOTODIODES

Photodiodes are specified by a number of parameters. These are the signal-to-noise ratio, the noise equivalent power, the detectivities, the area resistance product, and the response time.

2.1. Signal-to-Noise Ratio (S/N)

The signal-to-noise ratio is defined as the ratio of rms signal voltage (or current) generated as a result of a sinusoidally modulated radiant power falling upon a photodiode to the rms noise voltage (or current) of the photodiode, i.e., (V_s/V_n) or (I_s/I_n).

2.2. Noise Equivalent Power (NEP)

The noise equivalent power or NEP is the rms value of sinusoidally modulated radiant power P falling upon a photodiode which generates a signal equal to the noise level $V_s = V_n$ or $I_s = I_n$:

$$\text{NEP} = P(V_s/V_n)^{-1} = P(I_s/I_n)^{-1} \quad \text{(in watts)} \tag{1}$$

or in terms of the responsivity $\tilde{R} = V_s/P$ or I_s/P

$$\text{NEP} = V_n/\tilde{R} = I_n/\tilde{R} \tag{2}$$

2.3. Detectivity (D)

The figure of merit detectivity D is defined for psychological reasons. It was felt that good detectors should have larger figures of merit than poor detectors. This is not true for NEP—here poor detectors have large NEP. D was therefore defined as

$$D = \frac{1}{\text{NEP}} \qquad (\text{W}^{-1}) \tag{3}$$

2.4. Normalized Detectivity (D*)

The signal increases linearly with area of the photodiode. The noise varies as the square root of the area. The signal-to-noise ratio increases as the square root of the area. In order to compare two detectors, D is normalized with respect to unit area and unit bandwidth. A figure of merit called detectivity D^* is defined. It is given by the relation

$$D^* = D\sqrt{A}\sqrt{B} \qquad (\text{cm}\,\text{Hz}^{1/2}/\text{W}) \tag{4}$$

$$= \frac{\sqrt{A}\sqrt{B}}{\text{NEP}} = \frac{\sqrt{A}\sqrt{B}}{P}\cdot\frac{V_s}{V_n} = \frac{\tilde{R}}{V_n}\cdot\sqrt{A}\cdot\sqrt{B}$$

where A is the area of photodiode in square centimeters and B is the electrical bandwidth in hertz.

The source of radiation can be either monochromatic or blackbody. In the former case D^* is known as the spectral D^*_λ and is expressed as $D^*(\lambda, f, 1)$ where λ is the wavelength in micrometers, f is the modulation frequency in hertz, and 1 represents the 1-Hz bandwidth. It is important to state the modulation frequency because of the its influence on the amount of noise. In the latter case it is called blackbody D^*_T and is symbolized by $D^*(T, f, 1)$, where T is the blackbody temperature.

For small signal voltages V_s is given by

$$V_s = I_s R = q\eta A N_\lambda R \tag{5}$$

where I_s is the photoinduced current, R is the junction resistance at the operating point, q is the electronic charge, η is the quantum efficiency, and N_λ is the photon flux density of radiation of wavelength λ falling on the photodiode. The incident power P is given by

$$P = \frac{AhcN_\lambda}{\lambda} \tag{6}$$

where h is Planck's constant and c is the velocity of light.

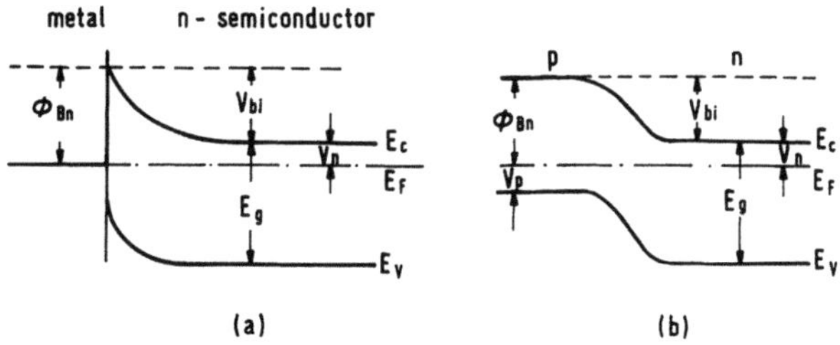

Figure 1. Energy band diagrams of a Schottky barrier consisting of (a) a metal and an *n*-type semiconductor and (b) a *p–n*-homojunction photodiode.

The noise in a photodiode is primarily determined by shot noise. The shot noise current I_{sn} is given by[(3)]

$$I_{sn}^2 = 2q\{I_p + I_0[\exp(qV/kT) + 1]\}B \tag{7}$$

for detectors (for an ideal Schottky barrier and a *p–n* junction, whose energy band diagrams are shown in Fig. 1a and Fig. 1b, respectively) in which the diode current–voltage *I–V* characteristic is expressed by

$$I = I_0[\exp(qV/kT) - 1] \tag{8}$$

and where I_p is the current due to background radiation, I_0 is the reverse-biased saturation current of the diode, k is the Boltzmann constant, and T is the temperature of the diode in degrees kelvin. When the diode is so shielded that the background photon flux becomes negligible, i.e., $I_p = 0$, then expression (7) becomes

$$\begin{aligned} I_{sn}^2 &= 2q\{I_0[\exp(qV/kT) + 1]\}B \\ I_{sn}^2 &= 2\left\{\frac{kT}{R_0}[\exp(qV/kT) + 1]\right\}B \end{aligned} \tag{9}$$

where $R_0 = (dI/dV)_{V=0}^{-1}$, is the dark resistance of the diode at zero bias voltage. The photodiodes are either operated at small negative bias or at zero bias. When $V = 0$, the shot noise is equal to the Johnson, or thermal, noise and becomes

$$\begin{aligned} I_{sn}^2 &= \frac{4kT}{R_0}B \\ V_n &= (4kTR_0B)^{1/2} \end{aligned} \tag{10}$$

and the expression (5) for $R = R_0$

$$V_s = q\eta A N_\lambda R_0 \tag{11}$$

By substituting expressions (6), (10), and (11) in (4), we get the spectral detectivity D_λ^* as

$$D_\lambda^* = \frac{q\eta\lambda}{2hc}\left(\frac{R_0 A}{kT}\right)^{1/2} \tag{12}$$

for the detection of radiation of the wavelength λ by a photodiode. The product R_0A (zero bias resistance times area of the photodiode) besides the quantum efficiency η is the most important parameter. It is determined by differentiating the current–voltage characteristic.

2.5. Detectivity Normalized Also with Respect to the Field of View (D^{**})

D^* is a function of the detector field of view. Jones had defined another figure of merit called D^{**}, which removes the need to specify the field of view. D^* and D^{**} are related as

$$D^{**} = (\Omega/\pi)^{1/2} D^* \tag{13}$$

where Ω is the detector field of view. For a circular symmetric field of view

$$\Omega = \pi \sin^2(\theta/2)$$

then

$$D^{**} = D^* \sin(\theta/2) \tag{14}$$

where θ is the total angle of the cone. For a hemispherical field of view $D^{**} = D^*$. The units of D^{**} are cm $\text{Hz}^{1/2}\text{sr}^{1/2}/\text{W}$.

2.6. Resistance Area Product

In case of photovoltaic detectors based on p–n-homojunction diodes and Schottky barrier photodiodes the I–V characteristic in the presence of radiation has the general form

$$I = I_0[\exp(qV/kT) - 1] - I_s \tag{15}$$

For a p–n-homojunction photodiode, the reverse saturation current I_0 is given by

$$I_0 = A(kT)^{1/2} n_i^2 q^{1/2} \left[\frac{1}{p}\left(\frac{\mu_n}{\tau_n}\right)^{1/2} + \frac{1}{n}\left(\frac{\mu_p}{\tau_p}\right)^{1/2} \right]^{-1} \tag{16}$$

where n_i is the intrinsic carrier concentration, n and p are the electron and hole majority carrier concentrations, μ_n and μ_p are the electron and hole mobilities, and τ_n and τ_p the electron and hole lifetimes in the p- and n-type regions, respectively. n_i is given by

$$n_i = 2\frac{(2\pi kT)^{3/2}}{h^3} \exp(-E_g/2kT) m_{dc}^{3/4} m_{dv}^{3/4} \tag{17}$$

where E_g is the band gap of the material and m_{dc} and m_{dv} are the density of states effective masses for conduction band and valence band, respectively. The lifetime is determined by the recombination of carriers.

There are three recombination processes normally considered. These are (i) Shockley–Read recombination, (ii) radiative recombination, and (iii) Auger recombination. In case of a Shockley–Read process,[22] the recombination of carrier is via recombination centers. These centers arise due to crystal imperfections, defects, and/or impurities in the crystals and have energy levels in the band gap. The recombination rate is proportional to the number of defect centers. These can, in principle, be controlled by appropriate crystal growth techniques. Theoretically it is difficult to calculate the lifetime of excited carriers due to Shockly–Read centers of general type. However, in case of a recombination center with two charge states and one defined energy level in the gap, the lifetimes are[23]

$$\tau_n = \tau_{n0}(1 + n_1/n) \tag{18}$$

and

$$\tau_p = \tau_{p0}(1 + p_1/p) \tag{19}$$

where

$$\tau_{n0} = \frac{1}{\sigma_n v_n N_T} \tag{20}$$

$$\tau_{p0} = \frac{1}{\sigma_p v_p N_T} \tag{21}$$

n_1 and p_1 are the electron and hole concentrations in the conduction and valence bands, respectively, when the Fermi level is at the trap level. σ_n and σ_p

are the capture cross sections for electrons and holes, v_n and v_p are the electron and hole thermal velocities, respectively, and N_T is the concentration of the traps. When a trap level lies at the center of the band gap, n_1 or p_1 is small. The carrier lifetime is independent of the free carrier concentrations. It varies slowly with temperature and depends mainly on the capture cross section and the trap concentration.

The radiative and Auger processes are intrinsic recombination mechanisms. An electron recombines directly with a hole. In the radiative process, the recombination energy is liberated as photon and in the Auger process, it is transfered to a third carrier (either an electron or a hole). Here, we shall give the lifetime of a minority carrier in n-type material only. The lifetime of a hole in n-type material for radiative or Auger process is given by[24]

$$\tau_p = n_i^2/[n + (n_i^2/n) + n_e]\bar{R} \tag{22}$$

where n_e is the excess pair density and $\bar{R}$ is the recombination rate either for a radiative process or for an Auger process. For radiative recombination R_R can be estimated from the expression given by Roosbroeck and Shockley[25]:

$$R_R = \frac{8\pi k^3 T^3}{h^3 c^2} \int_0^\infty \frac{n^{-2} \alpha U \, dU}{(e^U - 1)} \tag{23}$$

where c is the velocity of light, $\bar{n}$ is the refractive index, α is the absorption coefficient, and $U = E/kT$ with E the carrier energy. The lifetime in intrinsic meterial for $n_e \ll n_i$ is $\tau_{Ri} = n_i/2R_R$. It is the maximum value of the radiative lifetime. It varies with energy gap and temperature as $\exp(E_g/2kT)$. For extrinsic materials the lifetime varies inversely with the carrier concentration and it is independent of temperature.[24] Auger recombination is complementary to impact ionization. It has been investigated quite extensively by Beattie and Landesberg[26,27] for spherical bands located at the zone center. In this case, the recombination rate R_A is given by

$$R_A = \frac{8(2\pi)^{5/2} q^4 m_c |F_1 F_2|^2 n (kT/E_g)^{3/2}}{h^3 \epsilon_s^2 (1+u)^{1/2}(1+2u)} \exp - \left(\frac{1+2u}{1+u}\right)\frac{E_g}{kT} \tag{24}$$

where F_1, F_2 are overlap integrals, $u = m_c/m_v$ with m_c and m_v the effective masses in the conduction and valence band, respectively. For $n_e \ll n_i \ll n$, in extrinsic material, the Auger lifetime varies inversely as the square of the electron concentration. It depends on temperature and energy gap according to $E_g^{3/2} T^{3/2} \exp[uE_g/kT(1+u)]$. In intrinsic material, it varies with the energy gap and temperature as $E_g^{3/2} T^{-3/2} \exp[(1+2u)E_g/kT(1+u)]$.

The total lifetime τ is the parallel combination of the Schottky–Read (SR) lifetime τ_{SR} the Auger lifetime τ_A, and the radiative lifetime τ_R and is given by

$$\frac{1}{\tau}=\frac{1}{\tau_{SR}}+\frac{1}{\tau_A}+\frac{1}{\tau_R} \tag{25}$$

With the rapid development of the purification and the crystal growth techniques it might be possible to exclude τ_{SR}. The total lifetime is then determined by τ_A and τ_R. The relative importance of the radiative and Auger processes varies from material to material depending on the carrier concentration and the temperature. For example in $Pb_{1-x}Sn_xSe$, radiative recombination dominates below $2\times10^{15}\,cm^{-3}$ and in $Pb_{1-x}Sn_xTe$ below

TABLE 1. Values of Energy Band Gaps, Electron Affinities, Cut-off Wavelengths, Static Dielectric Constants, and Effective Richardson Constants of Various Semiconductors at the Indicated Temperatures

Semiconductor	Energy band gap (eV)	Temperature (K)	Electron affinity (eV)	Cut-off wavelength (μm)	Static dielectric constant	Effective Richardson constant (A/cm/K^2)
ZnS	3.66	300	3.9	0.34	8.3	40.8
ZnSe	2.67	300	4.09	0.46	9.2	20.4
CdS	2.42	300	4.8	0.51	8.9	24
GaP	2.26	300	4.0	0.55	11.1	42
ZnTe	2.25	300	3.5	0.55	10.4	10.8
AlAs	2.13	300	3.5	0.58	10.9	38.0
CdSe	1.74	300	3.93	0.71	10.2	15.6
AlSb	1.62	300	3.6	0.76	12.04	46.8
CdTe	1.5	300	4.5	0.83	10.6	13.2
GaAs	1.43	300	4.07	0.87	13.18	7.8
InP	1.35	300	4.4	0.92	12.35	9.2
Si	1.11	300	4.05	1.12	11.7	50
GaSb	0.7	300	4.06	1.77	15.69	5.9
Ge	0.67	300	4.13	1.85	16.3	26.4
PbS	0.42	300	4.6	2.95	172	12.84
	0.307	77	—	4.0	184	10.4
InAs	0.356	300	4.9	3.5	14.55	3.24
	0.404	77	—	3.1	—	2.76
InSb	0.228	77	4.59	5.4	17.78	1.8
PbTe	0.217	77	4.6	5.7	428	4.32
PbSe	0.176	77	—	7.0	227	5.1
$Pb_{0.8}Sn_{0.2}Te$	0.1	77	—	1.42	428	2.04
$Pb_{0.94}Sn_{0.06}Se$	0.1	77	—	12.4	227	3.24
$Hg_{0.8}Cd_{0.2}Te$	0.1	77	—	12.4	18.1	66.0

$1 \times 10^{16}\,\mathrm{cm}^{-3}$. For doping larger than these, Auger recombination dominates[28] in both of these materials.

For a metal–n-type semiconductor Schottky barrier with the energy level diagram shown in Fig. 1a, the reverse saturation current is[29]

$$I = AA^* T^2 \exp(-\phi_{Bn}/kt) \tag{26}$$

and

$$R_0 A = (kT^{-1}/qA^*) \exp(\phi_{Bn}/kT) \tag{27}$$

where $\phi_{Bn} = \phi_m - \chi_s$ is the barrier height for metal–n-type semiconductor junctions, ϕ_m is the work function of the metal, χ_s is the semiconductor electron affinity, and $A^*(= 4\pi qk^2 m^*/h^3)$ is the effective Richardson constant, where m^* is the effective electron mass. For free electrons $m^* = m_0$ (m_0 being the free electron mass), A^* becomes $A^* = 120\,\mathrm{A/cm/K^2}$. In semiconductors, A^* will differ from this value by the effective mass ratio m^*/m_0. Table 1 gives the calculated A^* for nearlv all useful semiconductors.

2.7. Response Time

The response time of a photovoltaic detector is normally determined by its RC time constant. The capacitance C per unit area of the junction is[29]

$$\frac{C}{A} = \frac{\epsilon_s}{W} \tag{28}$$

where ϵ_s is static dielectric constant and W is the space charge layer width. For an abrupt p–n-junction, W is[29]

$$W = \left[\frac{2\epsilon_s}{q}\left(\frac{n+p}{np}\right)V_{\mathrm{bi}}\right]^{1/2} \tag{29}$$

where $V_{\mathrm{bi}} = \phi_{Bn} - V_n$ is the built-in voltage (see Fig. 1). If $n \ll p$, the junction is one sided and

$$W \simeq \left(\frac{2\epsilon_s}{q}\frac{V_{\mathrm{bi}}}{n}\right)^{1/2} \tag{30}$$

This expression holds good for metal–n-type semiconductor Schottky barrier junction photodiodes also, and thus

$$RC \simeq RA(\epsilon_s qn/2V_{\mathrm{bi}})^{1/2} \tag{31}$$

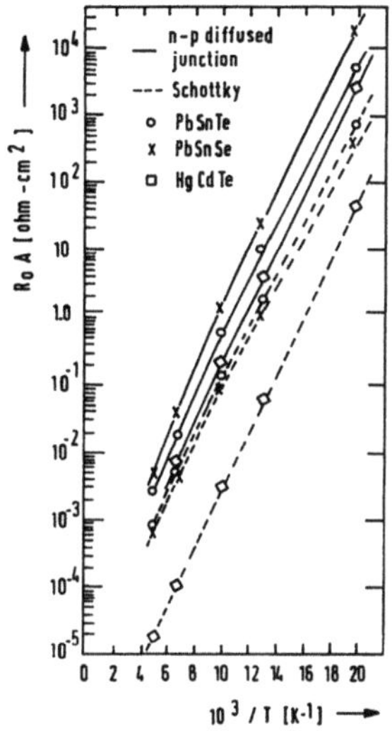

Figure 2. Calculated zero bias R_0A product vs. temperature for *n–p*-junction and Schottky diodes of PbSnTe, PbSnSe, and HgCdTe.

The comparison between the Schottky barrier and *n–p* diffused junction photodiodes for detection in the mid IR-spectral range has been made by Gupta *et al.*[18] Figure 2 shows the calculated zero bias R_0A product as a function of temperature in the temperature range 50–200 K for Schottky barrier diodes and *n–p* diffused Auger band-to-band recombination limited diodes of $Pb_{0.8}Sn_{0.2}Te$, $Pb_{0.94}Sn_{0.06}Se$, and $Hg_{0.8}Cd_{0.2}Te$. Table 2 gives the calculated values of R_0A, D^*, and C/A at 77 K for Schottky barrier and *n–p*-junction diodes of these materials. These results indicate that even though *n–p* diffused photodiodes have higher detectivities, the Schottky diodes of $Pb_{0.8}Sn_{0.2}Te$, $Pb_{0.94}Sn_{0.06}Se$, and $Hg_{0.8}Cd_{0.2}Te$ too have detectivities of the same order of magnitude. Figure 3 gives plots of $\ln(R_0AA^*)$ against ϕ_{Bn} or ϕ_{Bp} at 77 and 300 K. Knowing A^*, the maximum theoretical value of R_0A at these temperatures can be established from these plots for any material, because the maximum value of the barrier height is equal to the band gap. Once R_0A is

TABLE 2. Calculated Values of R_0A, D^*, and C/A at 77 K for Schottky Barrier and n–p-Junction Photodiodes

	D^* at $\lambda = 12.4\,\mu m$				
	Schottky barrier diode		*n–p* junction photodiodes		
Material	$R_0(\Omega\,cm^2)$	$(cm\,Hz^{1/2}/W)$	$R_0A(\Omega\,cm^2)$	$(cm\,Hz^{1/2}/W)$	C/A $(\mu F/cm^2)$
$Pb_{0.8}Sn_{0.2}Te$	1.9	2.1×10^{11}	9.8	4.8×10^{11}	0.56
$Pb_{0.94}Sn_{0.06}Se$	1.2	1.7×10^{11}	28.5	8.2×10^{11}	0.4
$Hg_{0.8}Cd_{0.2}Te$	0.06	3.7×10^{10}	4.1	3.1×10^{11}	0.11

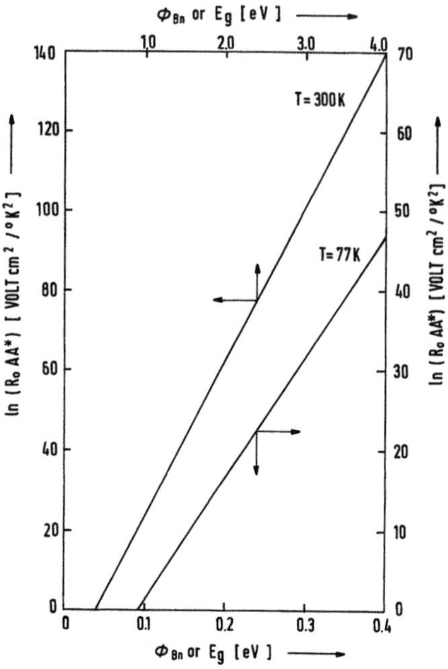

Figure 3. Plots of $\ln(R_0AA^*)$ vs. barrier height for Schottky barriers consisting of a metal and an n-type or a p-type semiconductor at 77 K and 300 K.

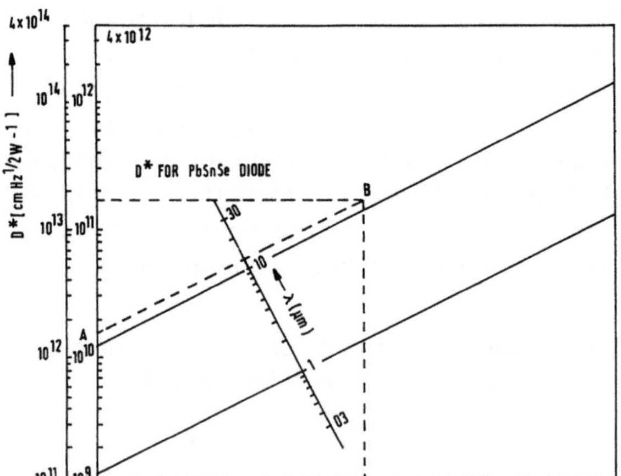

Figure 4. Nomograph for determining D^* at 77K for a wavelength λ and a given value of R_0A for $\eta = 1$.

known, D^* can be determined by using the monograph developed by Gupta *et al.*[30] Figure 4 shows the monograph for determining D^* at 77 K for a cutoff wavelength λ and a given value of R_0A for $\eta = 1$.

3. SELECTION OF MATERIALS

The selection of materials for fabricating Schottky barrier photodiodes involves metal systems and semiconducting materials.

3.1. Metal Systems

The choice of a metal for fabricating the Schottky barrier photodiode with a particular semiconducting material is determined by its work function ϕ_m. For an *n*-type semiconductor ϕ_m should be greater, while for a *p*-type semiconductor, it should be less than the electron affinity χ_s of the semiconductor. The barrier heights in these cases are given by

$$\phi_{Bn} = \phi_m - \chi_s \tag{32}$$

and

$$\phi_{Bp} = (\chi_s + E_g - \phi_m) \tag{33}$$

respectively. The barrier height of the metal–semiconductor junction should be as large as possible to maximize D^* [see expressions (12) and (27) in Section 2]. The maximum value of the barrier height for an ideal Schottky contact is about equal to the band gap of the semiconductor. If $\phi_B > E_g$, a layer of the semiconductor next to the surface is iverted in type, we have then a *p–n* junction within the material. However, in practice it is difficult to have an ideal Schottky contact and to have simple relationships like the expressions (32) and (33). This is due to interface states originating either from surface states[31,32] or from metal-induced gap states[33,34] and/or due to interface chemical reactions of metal and semiconductor atoms.[35–38] Cowley and Sze[32] have derived an expression for ϕ_{Bn} taking surface states into account. Then one obtains

$$\phi_{Bn} = \gamma(\phi_m - \chi_s) + (1 - \gamma)(E_g - \phi_0) - \Delta\phi \tag{34}$$

where ϕ_0 is the position of the neutral level of the interface states measured from the top of the valency band, $\Delta\phi$ is the barrier lowering due to image forces, and $\gamma = \epsilon_i/(\epsilon + q^2\delta D_s)$ where ϵ_i is the permittivity, δ the thickness of the interfacial layer, and D_s the density of interface states. Neglecting $\Delta\phi$, expression (34) reduces to expression (32) when $D_s \equiv 0$. In the literature, there

TABLE 3. Electronic Work Function of Some Important Metals

Metal	Work function (eV)	Metal	Work function (eV)
Pt	5.65	Zn	4.33
Ni	5.15	Al	4.28
Pd	5.12	Ag	4.26
Au	5.1	Pb	4.25
Cu	4.65	Ta	4.25
W	4.55	Cd	4.22
Cr	4.5	Ga	4.2
Hg	4.49	In	4.12
Sn	4.42	Zr	4.05
Ti	4.33	Cs	2.14

are numerous experimental data[39,40] on ϕ_m. There is a considerable variation among them. Michaelson[41] has analyzed these data and has attempted to correlate them theoretically with the atomic electron negativity. Table 3 gives the most preferred experimental values of work function compiled by him[41,42] for important metals.

The other important factors on which metal selection depends include its low diffusivity in the semiconductor and its ease of deposition (convenient temperature of deposition and good adherence), no (or a controllable) interface reaction, good electrical and thermal behavior, and adaptability to thermocompression bonding. The reactivity of a metal with the semiconductor is one of the most important factors affecting Schottky barriers. It has been shown by various workers[43–46] that pronounced chemical reactions occur at room temperature at a metal–semiconductor interface even under ultrahigh-vacuum conditions. It is, therefore, essential to know the reactive and nonreactive metals with regard to a particular semiconductor prior to selecting a metal system for it.

The usual way to deposit a metal on a semiconductor is by evaporation. The use of one particular metal may be restricted due to its high vaporization temperature. However, with the advent of ion implantation, this restriction may be lifted.

3.2. Semiconductor Materials

Any material is a potential photodetector if the number of free charge carriers can be altered by direct absorption of photons. The semiconducting materials fall into this category. The basic requirement is that absorbed photons excite a carrier from a nonconducting (valence band) or impurity state to a conducting state (conduction band). For intrinsic photodetectors the relation

$$h\nu \gg E_g \tag{35}$$

or

$$hc/\lambda \gg E_g \tag{36}$$

where ν is the frequency of the radiation and λ is the radiation wavelength or

$$\lambda \leq hc/E_g \tag{37}$$

must be satisfied.

The material is sensitive to all wavelengths less than λ_0. λ_0 is the long-wavelength limit or cutoff wavelength which is given by

$$\lambda_0(\mu\text{m}) = \frac{1.241}{E_g(\text{eV})} \tag{38}$$

Table 1 gives the energy band gaps and cutoff wavelengths for semiconducting materials at the indicated temperature.

If two or more materials may be used in the same wavelength region, then the selection of the material is made on the basis of barrier height obtained on it. The material with the larger barrier height will be preferred. This will cause a higher R_0A product and hence larger detectivity. The other important parameter in selecting the material is its dielectric constant, because the response time of the photodiode is normally determined by the RC time constant and is, therefore, proportional to ϵ_s. For example, the alloy systems PbSnTe and HgCdTe can be used in the same wavelength region. One reason why HgCdTe is preferred over PbSnTe is its shorter response time due to the smaller static dielectric contant (see Table 1). The other parameters which control the selection of one material over the other are compositional uniformity, chemical stability, inertness to the ambient, and ease of surface preparation.

4. FABRICATION TECHNOLOGY

The Schottky barrier junction photodetectors have been fabricated with various semiconductors in the cutoff range $\lambda_0 = 0.35\,\mu$m to $\lambda_0 = 14\,\mu$m using the metals as shown in Table 4. The metal system used with a particular

TABLE 4. Schottky Barrier Photodiodes

Schottky photodiode	Carrier density (cm^{-3})	Area (cm^2)	Quantum efficiency (%)	Responsivity (V/W or A/W)
W–*n*Ge	10^{15}	3×10^{-10}	—	2000 μV/W
(point contact)	10^{18}	3×10^{-10}	—	1000 μV/W
Ag–*n*Ge	0.1–10[a]	8.11×10^{-3}	43	—
Au–*n*Ge	0.1–1.0	8.11×10^{-3}	52	—
Cu–*n*Ge	0.1–1.0	8.11×10^{-3}	45	—
Ni–*n*Ge	0.1–1.0	8.11×10^{-3}	12–13	—
Pb–*n*Ge	0.1–1.0	8.11×10^{-3}	5–10	—
W–*n*Si	10^{19}	3×10^{-10}	—	1000 μV/W
(point contact)	10^{16}	3×10^{-10}	—	1–10 μV/W
Au–*n*Se	—	—	70	—
Au–*i*–Si	—	2×10^{-3}	70	—
Ag–*n*GaAs	3×10^{17}	7×8.10^{-3}	50	—
Cr–*n*GaAs	3×10^{17}	7×8.10^{-3}	25	—
Au–*n*GaAs	—	—	40	—
Pt–*n*GaAs	—	—	50	0.3 A/W
Ag–*n*ZnS	2×10^{15}	—	70	—
Au–*n*ZnS	2×10^{15}	—	50	—
Cr–*n*ZnS	2×10^{15}	—	50	—
Au–*n*GaSb	4×10^{16}	—	37	—
	1×10^{17}			
Pt–$nIn_xG_{1-x}As$	—	—	50	0.47 A/W
$x = 0.20$				
$x = 0.17$	—	—	50	0.38 A/W
Pb–*p*PbS	10^{17}	3.2×10^{-3}	55	1.6 A/W
	5×10^{17}	3.2×10^{-3}	62	—
Pb–*p*PbSe	10^{17}	3.2×10^{-3}	61	3.3 A/W
	9.6×10^{16}	10^{-3}	70	$(1–7)10^4$ V/W
	1.2×10^{17}			
In–$pPbS_{1-x}Se_x$	10^{17}	3.2×10^{-13}	62	2.1 A/W
$x = 0.15$				
Pb–$pPbS_{1-x}Se_x$ $x = 0.30$	10^{17}	3.2×10^{-13}	59	2.1 A/W
$x = 0.50$	10^{17}	3.2×10^{-13}	50	1.9 A/W
$x = 0.70$	5×10^{17}	3.2×10^{-13}	60	—
Pb–*p*PbTe	$(0.8–2.7)10^{17}$	$(2.7–3)10^{-3}$	50–60	$(2.2–5.7)10^5$ V/W
In–*p*PbTe	4×10^{17}	—	—	—
Cu–*p*PbTe	4×10^{17}	—	—	—
Pb–$pPb_{1-x}Sn_xSe$	5×10^{17}	3.2×10^{-3}	35	—
$x = 0.05$				
$x \simeq 0.062$	2.6×10^{17}	6.2×10^{-4}	53	—
$x \simeq 0.07$	2.6×10^{17}	6.2×10^{-4}	25	—
$x = 0.065$	2×10^{17}	3.2×10^{-3}	44	3.76 A/W
Pb–$pPb_{1-x}Sn_xTe$	10^{17}	—	—	10^4 V/W
$x = 0.21$	2.68×10^{17}	1.29×10^{-4}	49	—
Al–$pHg_{1-x}Cd_xTe$ ($x = 0.22$)	2×10^{16}	1.14×10^{-3}	—	—
Cr–$pHg_{0.68}Cd_{0.32}Te$	10^{15}	1.14×10^{-3}	—	—

[a]Resistivity in Ω cm.

Response time or capacitance (psec or nF)	Cut-off wavelength (μm)	Zero bias resistance area product ($\Omega\,cm^2$)	Normalized detectivity $D^>$ ($cm\,Hz^{1/2}W^{-1}$)	Field of view (deg)	Temperature (K)	Reference
0.10 psec	10	3×10^{-4}	4×10^{-11} W(NEP)	—	300	59
		3×10^{-4}	—	—		
—	1–2	14	—	—	300	14
—	1–2	30.5	—	—	300	14
—	1–2	3	—	—	300	14
—	1–2	1	—	—	300	14
—	1–2	4	—	—	300	14
—	10	9.4×10^{-5}	—	—	300	59
—	10	$> 3 \times 10^{-3}$	—	—	300	59
< 500 psec	0.63	—	—	—	300	19
—	0.38–0.8	—	10^{12}	—	300	37
—	< 0.36	2×10^6	—	—	300	60
—	< 0.6	—	—	—	300	60
0.03 nF	0.63	—	S/N = 1.2×10^7	—	300	20
—	< 0.9	—	—	—	300	61
—	< 0.35	—	—	—	300	62
—	< 0.35	—	—	—	300	62
—	< 0.35	—	—	—	300	62
—	< 1.6	2.8×10^2	—	—	300	17
—	1.11	—	—	—	300	61
—	1.0	—	—	—	300	61
3 nF	3.7	2.1×10^4	10×10^{11}	20	77	51
—	< 4	—	—	—	77	15
3 nF	6.9	30	2.7×10^{11}	20	77	51
—	< 7	12–88	$(1.7–1.8) \times 10^{10}$	56	77	8
3 nF	4.2	1.5×10^4	9×10^{11}	20	77	51
3 nF	4.4	4.4×10^3	3.3×10^{11}	20	77	51
3 nF	4.8	3.7×10^2	3×10^{11}	20	77	51
—	6.0	—	—	—	77	51
0.5–3 nF	< 5.7	$2 \times 10^2 - 2 \times 10^3$	6×10^{11}	1	77	63
5.4 nF	< 5.9	1.7×10^3	—	—	77	13
5.4 nF	< 5.9	73	—	—	77	13
—	10	—	—	—	77	15
—	10.1	1.6	5.2×10^{10}	26	77	9
—	11.5	0.6	2.4×10^{10}	45	77	9
—	10.6	1.3	7×10^{10}	20	77	50
—	12		1×10^{10}	—	77	10
3 nF	11.0	1.0	3×10^{10}	—	77	64
—	9.1	14.9	—	—	77	16
—	4.3	50.6	—	—	77	16

semiconductor material is selected on the basis of its low diffusivity and neutral electrical activity such that $\phi_m > \chi_s$ in n-type and $\phi_m < \chi_s$ in p-type semiconductors. The fabrication technology, in general, is alike for all the materials. It involves (1) preparation of surface of the material prior the deposition of the metal, (2) deposition of metal for the Schottky barrier junction formation on the active side, and (3) fabrication of an ohmic contact to the semiconductor.

The semiconductor surface is frequently prepared first by mechanically polishing and then chemically etching. However, this invariably produces a thin oxide layer. The exact nature and the thickness depend on the method of preparation. Also, if the chemical etching and the solvent cleaning are not carried out carefully, it gives rise to undesirable interfaces. These interfacial layers may also be caused by water or other solvents used for cleaning. Even if the surface were prepared by cleaving a crystal in order to have a fresh surface, an oxide layer would form by exposure to air by the time the metal is deposited over it. It has been reported(47) that a monolayer of silicon dioxide can be formed in air on clean Si surface in about 1 msec. Natural oxide in the 20–30 Å thickness range is normally present(48,49) on the surface of any GaAs sample kept under standard laboratory condition. Such adsorbed layers can usually be removed by heating the substrate or vacuum annealing(50,51) the substrate and then cooling it to room temperature prior to deposition. Reactive sputter etching has also been reported(52) in the literature for surface cleaning.

It has also been observed(53,54) that interfacial layers at metal semiconductor junctions are sometimes useful. Buchner *et al.*(55) have shown by using Auger depth profiling and $I-V$ measurements that Schottky barriers of Pb on $Pb_{0.8}Sn_{0.2}Te$ form only when the semiconductor surface is exposed to oxygen prior to the deposition of the Pb.

In most of the cases the Schottky contacts are made by evaporation of a specific metal. The lower melting point metals like Au, Pb, In, Al, etc. are usually evaporated by resistive heating from a boat, while the refractory metals like molybdenum, titanium, etc by electron beam heating in a vacuum chamber having a pressure of less than 10^{-6} torr. Other techniques reported in the literature are sputtering and electroplating.(56) The semiconductor is generally kept at an elevated temperature during the deposition process. This

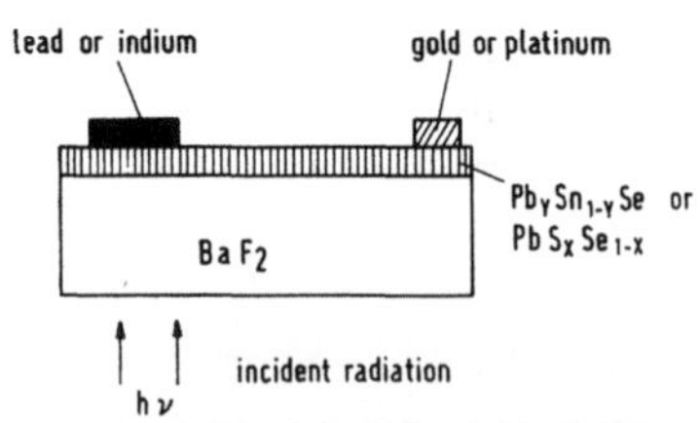

Figure 5. Configuration of Schottky barrier photodiodes of $Pb_ySn_{1-y}Se$ or PbS_xSe_{1-x}.

TABLE 5. *Fabrication Technology Details of Various Schottky Barrier Photodiodes Shown in Table 4*

Semiconductor	Surface preparation	Schottky metal	Substrate temperature (°C)	Vacuum pressure (mm)	Thickness (Å)	Metal ohmic contact	References
n-Ge	CP-4A solution for 8 min	Au, Cu, Ag, Pb and Ni	25	8×10^{-6}	50–1000	Sn	14
n-Si	Etched in HF	Au	—	2.3×10^{-7}	10–400	Au–Sb	19
n-GaAs	Methanol–bromine solution	Ag, Cr	150	2×10^{-6}	400–3000	Ag	60
n-ZnS	1% solution of bromine in mathanol	Ag, Au, Cr	100	2×10^{-6}	100–1500	In–Hg amalgam	62
n-GaSb	$3H_2SO_4 + 1H_2O_2$ solution	Au	—	10^{-8}	100–150	Au–Sn	17
n-$In_xGa_{1-x}As$ $x = 0.17$ $x = 0.20$	—	Pt	—	—	100	—	61
p-PbS	Vacuum annealed at 170°C for 30 min	Pb, In	25	—	—	Au	15, 51
p-PbSe	Vacuum annealed at 170°C for 30 min	Pb, In	25	—	—	Au	8, 51
p-$PbS_{1-x}Se_x$ $x = 0.15$ $x = 0.30$ $x = 0.50$ $x = 0.70$	Vacuum annealed at 170°C for 30 min	Pb, In	25	—	—	Au	15, 51
p-PbTe	Etched in HBr/Br_2 solution	In, Pb, Zn, Cu	Cooled	—	—	Au	13
	Etched in HBr/Br_2 solution	Pb	—	—	3000	Pt	63
p-$Pb_{1-x}Sn_xTe$ $x = 0.21$	Etched in HBr/Br_2 solution	Pb, In	25	—	100–500	Pt	10, 64
p-$Pb_{1-x}Sn_xSe$ $0.05 < x < 0.70$	Vacuum annealed at 170° C for 30 min	Pb	25	—	—	Au, Pt	9, 15, 50
p-$Hg_{1-x}Cd_xTe$ $0.2 < x < 0.38$	—	Al, Cr	—	10^{-6}	200–400	—	16

temperature is critical as the barrier properties in many cases depend on it. The area of the Schottky contact is often defined by a photomask, stainless steel mask, or by opening a window in an insulating layer deposited on the active side of the semiconductor prior to metal deposition.

The ohmic contacts are prepared by evaporation, sputtering, electroplating, or soldering. The metal is selected on the principle that its work function be less than the electron affinity of an *n*-type semiconductor or greater than the electron affinity of a *p*-type semiconductor. However, in the vast majority of cases ohmic contacts[40] work on the principle of having a thin layer of very heavily doped semiconductor immediately adjacent to the metal.

As an example the Schottky barrier photodiodes on IV–VI compound alloys are fabricated[50,51] as follows. Epitaxial films of either $Pb_ySn_{1-y}Se$ or PbS_xSe_{1-x} are grown on BaF_2 substrates using a modified hot wall technique.[57,58] These films are *p*-type with hole concentrations in the 10^{17}-cm^{-3} range. These films are vacuum annealed to 170°C for 30 min and cooled to room temperature prior to depositing the metal, in order to desorb an oxide layer from the film. Schottky barrier junctions are prepared by depositing lead or indium through a stainless-steel mask that defines the area of the contact. An ohmic contact is made by evaporating gold pads or by sputtering platinum. The photodiode so fabricated has the configuration shown in Fig. 5. Table 5 gives fabrication technology details of various Schottky barrier photodiodes. Figure 6 gives other configurations in which Schottky photodiodes can be fabricated.

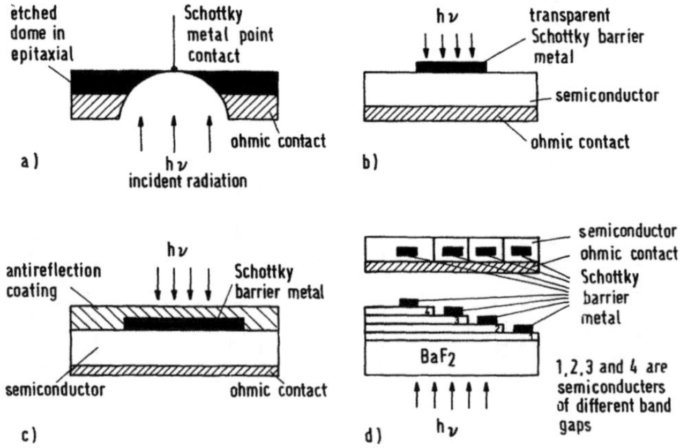

Figure 6. Various configurations of Schottky barrier photodiodes (a) point contact, (b) conventional photodiode, (c) conventional photodiode with antireflection coating, and (d) multispectral (four-color) detector.

5. TECHNIQUES FOR EVALUATING DEVICE PARAMETERS

The most important parameters[65] of a Schottky barrier photodiode are the barrier height and the reverse saturation current or zero-bias resistance. These parameters can be evaluated from current–voltage characteristics, capacitance–voltage characteristics, and/or photoelectric measurements.

5.1. Current–Voltage Characteristics

The I–V characteristic provides information about the nature of the contact. For ohmic contacts, the I–V characteristic is linear and in case of an ideal Schottky contact, the diode equation (8) should be obeyed:

$$I = I_0 \exp\left[\left(\frac{qV}{kT}\right) - 1\right] \tag{39}$$

$$I = AA^* T^2 \exp(-\phi_{Bn}/kT) \exp\left[\left(\frac{qV}{kT}\right) - 1\right] \tag{40}$$

From such I–V characteristics, parameters like I_0, R_0, ϕ_{Bn}, and A^* can be determined. If $\ln I$ is plotted against V in the forward direction a straight line should be obtained except for the region where $V < 3\,kT/q$. The extrapolation of this straight line to $V = 0$ gives the intercept $\ln I_0$. Knowing I_0, R_0 can be calculated from $R_0 = (kT/qI_0)$. If the calculated value of A^* is used (Table 1) ϕ_{Bn} can be obtained by using the relationship

$$\phi_{Bn} = \frac{kT}{q} \ln\left(\frac{AA^* T^2}{I_0}\right) \tag{41}$$

If the I–V characteristics are measured at various temperatures $\ln(I_0/AT^2)$ against $(1/kT)$ can be plotted yielding a straight line with an intercept on the vertical axis equal to $\ln A^*$ and a slope equal to $-\phi_{Bn}$.

5.2. Capacitance–Voltage Characteristics

The junction capacitance as a function of applied voltage is given by[29]

$$C_v = A\left[\frac{q\epsilon_s n}{2(V_{\text{bi}} - V)}\right]^{1/2} \tag{42}$$

If n is constant, a plot of $(1/C_v^2)$ vs. V will give a straight line. The intercept on voltage axis gives V_{bi} and the slope provides n. Knowing V_{bi}, the barrier height can be determined by (see Fig. 1a)

$$\phi_{Bn} = V_{bi} + V_n \tag{43}$$

where V_n is the distance between the Fermi level and the conduction band edge.

5.3. Photoelectric Measurements

This is the most direct method of determining the barrier height. When a monochromatic light $h\nu > \phi_{Bn}$ is incident upon a metal–semiconductor barrier, it generates photocurrent. The photoresponse $\tilde{R}$ is given by[29]

$$\tilde{R} \propto (n\nu - \phi_{Bn})^2 \tag{44}$$

A plot of $\sqrt{\tilde{R}}$ vs. photon energy will give a straight line. The intercept on the energy axis gives directly the barrier height. In Table 6, the barrier height values measured by these various techniques are compared. These values are also compared with the maximum theoretically possible value for an ideal junction as calculated from metal work function and electron affinity. As an example, in case of a metal–nGe junction, the maximum possible barrier height is 0.67 eV. However, for Au–nGe and Ni–nGe junctions, the calculation barrier height comes out to be more than 0.67 eV. Here a layer of the semiconductor next to the surface is inverted in type and then there is a p–n junction within the material. The junction will not be a true Schottky junction.

Although these methods are considered sufficient to characterize Schottky diodes, additional diagnosis is desirable in order to get some insight into performance, reliability, product yield, and aging. Techniques[52,55,66–68] developed for thin film technology like EBIC (electron beam induced current), ESCA (electron spectroscopy chemical analysis), and AES (Auger electron spectroscopy) may also be used for examining the Schottky barrier photodiodes.[55,68] Here we give a short description of the EBIC technique and the information it provides.

5.4. Electron Beam Induced Current Technique

This is a nondestructive technique. The junction device is scanned with the well-focused monoenergetic electron beam of a scanning electron microscope. The electrons incident on the device induce a current in an external circuit. This induced current is amplified and displayed on a cathode ray tube. There are two display modes. One is called intensity modulation: the induced current amplitude is represented as the level of brightness on the cathode ray

TABLE 6. Calculated and Measured Schottky Barrier Heights

Semiconductor	Metal (Ref.)	Barrier height (φ_B): Theoretical — Maximum possible (Eg) (eV)	Theoretical — Calculated (eV)	Experimental by various techniques — Current–voltage (eV)	Experimental — I_0/T vs. $1/T$ (eV)	Experimental — Capacitance–voltage (eV)	Experimental — Photoelectric measurement (eV)
*n*Ge	Au(14)	0.67(300 K)	0.97	0.58	—	0.58	0.60
	Cu(14)	—	0.52	0.51	—	0.50	0.54
	Ag(14)	—	0.13	0.54	—	0.56	0.584
	Ni(14)	—	1.02	0.49	—	0.49	0.482
	Pb(14)	—	0.12	—	—	0.56	0.51
*n*Si	Au(60)	1.11(300 K)	1.05	—	—	—	—
*n*GaAs	Ag(60)	1.43(300 K)	0.19	0.82	—	1.05	—
	Cr(60)	—	0.43	—	—	—	—
	Au(60)	—	1.03	—	—	—	—
	Pt(61)	—	1.58	—	—	—	—
*n*ZnS	Ag(62)	3.66(300 K)	0.36	—	—	1.9	—
	Au(62)	—	1.2	—	—	—	—
	Cr(62)	—	0.6	—	—	—	—
*n*GaSb	Au(17)	0.7(300 K)	0.94	—	0.6	—	—
*p*PbS	Pb(15, 51)	0.307(77 K)	0.657	—	—	—	—
	In(15, 51)	—	0.787	—	—	—	—
*p*PbSe	Pb(8, 51)	0.176(77 K)	—	—	—	—	—
p$PbSe_{0.5}S_{0.5}$	Pb(15, 51)	0.24(77 K)	—	—	—	—	—
*p*PbTe	Pb(12, 63)	0.217(77 K)	0.567	—	—	0.272	—
	In(13)	—	0.697	—	—	0.263	—
	Zn(13)	—	0.487	—	—	0.272	—
	Cu(13)	—	0.167	—	—	0.12	—
p$Pb_{0.94}Sn_{0.06}Se$	Pb(9, 15, 50)	0.1(77 K)	—	—	—	—	—
p$Pb_{0.8}Sn_{0.2}Te$	Pb(10, 64)	0.1(77 K)	—	—	—	—	—
p$Hg_{0.78}Cd_{0.22}Te$	Al(16)	0.136(77 K)	—	0.101	0.104	0.071	—
p$Hg_{0.68}Cd_{0.32}Te$	Cr(16)	0.288(77 K)	—	0.272	0.275	0.225	—

tube screen. The second is known as deflection modulation. Here the intensity of the spot remains constant. The induced signal is plotted in the y direction.

This technique can be used to examine surface inversion layers,[69] the junction depth,[70] device damage due to fabrication,[71] nonuniformity in surface preparation,[66] antireflection coatings,[66] and also diffusion lengths.[66]

6. APPLICATIONS

Table 4 gives the summary of Schottky barrier photodiodes fabricated of various semiconductors in the wavelength range $0.35\,\mu m$ to $14\,\mu m$. Photodiodes fabricated for various optical frequency ranges find applications as ultraviolet and visible radiation detectors, especially for laser receivers[19,20] because of their high-frequency response. Recent developments in phosphosilicate fibers for optical communications have accelerated the research in quantum detectors in the infrared wavelength region $1.0–1.6\,\mu m$. The basic requirements for the detectors are fast response time, high conversion officiency and narrow band detection. Germanium, gallium antimonide, and the alloys of aluminum antimonide and gallium antimonide and of indium arsenide and gallium arsenide are appropriate materials for detectors in this region. Schottky barrier photodiodes of germanium and gallium antimonide have been fabricated and studied for this purpose and are expected to find application in optical communication systems.[14,17]

Schottky barrier photodiodes made of lead salt alloy systems can utilize their adjustable band gap feature, to adjust their maximum responsivity to any infrared radiation in the range 3 to $14\,\mu m$.[10,50,51] Multispectral sensors[15] have been fabricated by epitaxial growth of thin layers of different lead salts on BaF_2 substrates.[15] It has been demonstrated that Schottky barrier photodiodes based on CdHgTe can be fabricated.[16] In spite of their smaller detectivity if compared with p–n-junction devices they may find future use because of their fast response and simple fabrication procedure.

7. CONCLUSIONS

The various aspects of Schottky barrier photodiodes have been reviewed. These include selection of metal and semiconductor meterials, fabrication technology, and characterization or evaluation of photodiode parameters. The properties of Schottky barrier and p–n junction photodiodes have been compared. Theoretical detectivities of $Pb_{0.8}Sn_{0.2}Te$, $Pb_{0.94}Sn_{0.6}Se$ and $Hg_{0.8}Cd_{0.2}Te$ Schottky barrier photodiodes have been calculated and were found to be of the same order of magnitude as p–n junction diodes.

The fabrication technology of Schottky barrier photodiodes is simpler and offers advantages over *p–n* junction diodes. However, this technology is relatively new. It is necessary to come to a clear metallurgical understanding of the metal–semiconductor interfaces which is required for perfecting the metallization procedure. Photolithographic, bonding, and packaging procedures are still in the development state. In view of the advanced *p–n* junction technology the future will tell if Schottky barrier photodiodes will find wide application.

ACKNOWLEDGEMENTS. The authors would like to acknowledge the support of this work by the Alexander von Humboldt Foundation (AvHF) and by the Fraunhofer Society. S.C. Gupta is thankful to the AvHF for financial assistance during his stay at the Fraunhofer Institute.

REFERENCES

1. E.H. Putley, in *Semiconductors and Semimetals* (R.K. Willardson and A.C. Beer, eds.), Vol. 5, pp. 259–285, Academic Press, New York (1970).
2. E.H. Putley, in *Topics in Applied Physics, Optical and Infrared Detectors* (R.J. Keyes, ed.), Vol. 19, pp. 71–100, Springer-Verlag, Berlin, (1980).
3. P.W. Kruse, in *Topics in Applied Physics, Optical and Infrared Detectors* (R.J. Keyes, ed.), Vol. 19, pp. 5–69, Springer-Verlag, Berlin (1980).
4. S.M. Ryvkin, *Photoelectric Effects in Semiconductors*, Consultants Bureau, New York (1964).
5. R.P. Riesz, High-speed semiconductor photodiodes, *Rev. Sci. Instrum.* **33**, 994–998 (1962).
6. C.A. Burrus and W.M. Sharpless, Planar *p–n*-junction germanium photodiodes for use at microwave modulation frequencies, *Solid-State Electron.* **13**, 1283–1287 (1970).
7. D.H. Seib and L.K. Aukerman, in *Advances in Electronics and Electron Physics* (L. Marton, ed.), Vol. 34, pp. 95–221, Academic Press, New York (1973).
8. D.K. Hohnke and H. Holloway, Epitaxial PbSe Schottky-barrier diodes for infrared detection, *Appl. Phys. Lett.* **24**, 633–635 (1974).
9. D.K. Honke, H. Holloway, K.F. Yeung, and M. Hurley, Thin-film (Pb, Sn)Se photodiodes for 8–12 μm operation, *Appl. Phys. Lett.* **29**, 98–100 (1976).
10. S.G. Parker, Expitaxial deposition of $Pb_xSn_{1-x}Te$ on $Pb_xSn_{1-x}Te$ substrates in a closed system, *J. Electrochm. Soc.* **123**, 920–924 (1976).
11. T.K. Chu, A.C. Bouley, and G.M. Black, Preparation of epitaxial thin film lead salt infrared detectors, Proc. SPIE—Int. Soc. Opt. Eng. 285 (Infrared Detect. Mater.) (1981), p. 33.
12. M. Drinkwine, J. Rozenbergs, S. Jost, and A. Amith, The lead/lead sulfide selenide $PbS_{0.5}Se_{0.5}$ interface and performance of lead/lead sulfide selenide ($PbS_{0.5}Se_{0.5}$) photodiodes, Proc. SPIE—Int. Soc. Opt. Eng. 285 (Infrared Detect. Mater.) (1981), p. 36.
13. J. Baars, D. Basset, and M. Schulz, Metal–semiconductor barrier studies of PbTe, *Phys. Status Solidi(a)* **49**, 483–488 (1978).
14. E.Y. Chan and H.C. Card, Infrared optoelectronic properties of metal–germanium Schottky barriers, *IEEE Trans. Electron. Devices* **ED-27**, 78–83 (1980).
15. R.B. Schoolar, J.D. Jensen, G.M. Black, S. Foti, and A.C. Bouley, Multispectral PbS_xSe_{1-x} and $Pb_ySn_{1-y}Se$ photovoltaic infrared detectors, *Infrared Phys.* **20**, 271–275 (1980).
16. D.L. Polla and A.K. Sood, Schottky barrier photodiodes in $pHg_{1-x}Cd_xTe$, *J. Appl. Phys.* **51**, 4908–4912 (1980).

17. Y. Nagao, T. Hariu, and Y. Shibata, GaSb Schottky diodes for infrared detectors, *IEEE Trans. Electron Devices* **ED-28**, 407–411 (1981).
18. S.C. Gupta, B.L. Sharma, and V.V. Agashe, Comparison of Schottky barrier and diffused junction infrared detectors, *Infrared Phys.* **19**, 545–548 (1979).
19. M.V. Schneider, Schottky barrier photodiodes with antireflection coating, *Bell System Tech. J.* **45**, 1611–1638 (1966).
20. W.M. Sharpless, Evaluation of a specially designed GaAs Schottky barrier photodiode using 6328 Å radiation modulated at 4 GHz, *Appl. Opt.* **9**, 489–494 (1970).
21. F.D. Shepherd, Recent advances in Schottky IR-photodiodes and projected camera capabilities, International electron device meeting, Washington D.C., 7 December 1981.
22. W. Shockley and W.T. Read, Statistics of the recombination of holes and electrons, *Phys. Rev.* **87**, 835–842 (1952).
23. A.G. Milnes, *Semiconductor Devices and Integrated Electronics*, Van Nostrand Reinhold Company, New York (1980).
24. J.S. Blakemore, *Semiconductor Statistics*, Pergamon Press, Oxford (1962).
25. W. van Roosbroeck and W. Shockley, Photon radiative recombination of electrons and holes in germanium, *Phys. Rev.* **94**, 1558–1560 (1954).
26. A.R. Beattie and P.T. Landesberg, Auger effect in semiconductors, *Proc. R. Soc. London Ser. A* **249**, 16–29 (1959).
27. A.R. Beattie and P.T. Landesberg, One-dimensional overlap functions and their application to Auger recombination in semiconductors, *Proc. R. Soc. London Ser. A* **258**, 486–495 (1960).
28. H. Preier, Comparison of the junction resistance of (PbSn)Te and (PbSn)Se infrared detector diodes, *Infrared Phys.* **18**, 43–46 (1978).
29. S.M. Sze, *Physics of Semiconductor Devices*, John Wiley and Sons, New York (1969).
30. S.C. Gupta, B.L. Sharma, and V.V. Agashe, Nomographs for evaluating parameters of Schottky barrier IR-detectors, *Infrared Phys.* **19**, 673–675 (1979).
31. J. Bardeen, Surface states and rectification at a metal semiconductor contact, *Phys. Rev.* **71**, 717–727 (1947).
32. A.M. Cowley and S.M. Sze, Surface States and barrier height of metal–semiconductor systems, *J. Appl. Phys.* **96**, 3212–3220 (1965).
33. V. Heine, Theory of surface states, *Phys. Rev. A* **138**, 1689–1696 (1965).
34. S.G. Louie, J R. Chelikowsky, and M.L. Cohen, Ionicity and the theory of Schottky barriers, *Phys. Rev. B* **15**, 2154–2162 (1977).
35. L.J. Brillson, Transition in Schottky barrier formation with chemical reactivity, *Phys. Rev. Lett.* **40**, 260–263 (1978).
36. K. Zdansky and Z. Sroubek, in *Physics of Semiconductors* (B.L.H. Wilson, ed.), Conference Series No. 43, pp. 761—764, Institute of Physics, London (1979).
37. R.H. Williams, V. Montgomery, and R.R. Varma, Chemical effects in Schottky barrier formation, *J. Phys. C: Solid State Phys.* **11**, L735–L738 (1978).
38. M. Schlüter, Chemical trends in metal–semiconductor barrier heights, *Phys. Rev. B* **17**, 5044–5047 (1978).
39. V.S. Fomenko, *Handbook of Thermionic Properties*, Plenum Press, New York (1966).
40. E.H. Rhoderick, *Metal–Semiconductor Contacts*, Clarendon Press, Oxford (1978).
41. H.B. Michaelson, Relation between an atomic electro negativity scale and the work function, *IBM J. Res. Devp.* **22**, 72–80 (1978).
42. H.B. Michaelson, *Work Function of the Elements, Handbook of Chemistry and Physics* (R.C. Weast, ed.), 58th ed. CRC Press, Cleveland, Ohio, pp. E81–E82 (1977–1978).
43. R.Z. Bachrach and A. Bianconi, Interface states at the Ga–GaAs interface, *J. Vac. Sci. Technol.* **15**, 525–528 (1978).

44. L.J. Brillson, Chemical reaction and charge redistribution at metal–semiconductor interfaces, *J. Vac. Sci. Technol.* **15**, 1378–1383 (1978).
45. I. Lindau, P.W. Chye, C.M. Garner, P. Pianetta, C.Y. Su, and W.C. Spicer, New phenomena in Schottky barrier formation on III–V compounds, *J. Vac. Sci. Technol.* **15**, 1332–1339 (1978).
46. L.J. Brillson, Chemical reactions and local charge redistribution at metal–CdS and CdSe interfaces, *Phys. Rev. B* **18**, 2431–2446 (1978).
47. V.L. Rideout, Review of the theory, technology and application of metal–semiconductor rectifiers, *Thin Solid Films* **48**, 261–291 (1978).
48. F. Lukes, Oxidation of Si and GaAs in air at room temperature, *Surf. Sci.* **30**, 91—100 (1972).
49. A.C. Adams and B.R. Pruniax, Gallium arsenide surface film evaluation by ellipsometry and its effect on Schottky barriers, *J. Electrochem. Soc.* **120**, 408–414 (1973).
50. R.B. Schoolar and J.D. Jensen, Narrowband detection at long wavelengths with epitaxial $Pb_ySn_{1-y}Se$ films, *Appl. Phys. Lett.* **31**, 536–538 (1977).
51. R.B. Schoolar, J.D. Jensen, and G.M. Black, Composition-turned PbS_xSe_{1-x} Schottky-barrier infrared detectors, *Appl. Phys. Lett.* **31**, 620–622 (1977).
52. R. Longshore, M. Jasper, B. Summer, and P. LoVecehio, Evaluation of $Pb_{0.8}Sn_{0.2}Te$ detector fabrication using surface analysis, *Infrared Phys.* **15**, 311–315 (1975).
53. H.C. Card, E.S. Yang, and P. Panayotatos, Peaked Schottky-barrier solar cells by Al–Si metallurigical reactions, *Appl. Phys. Lett.* **30**, 643–645 (1977).
54. J. Basterfield, J.M. Shannon, and A. Gill, The nature of barrier height variations in alloyed Al–Si Schottky barrier diodes, *Solid State Electron.* **18**, 290–291 (1975).
55. S. Buchner, T.S. Sun, W.A. Beck, N.E. Byer, and J.M. Chen, Schottky barrier formation on (Pb, Sn)Te, *J. Vac. Sci. Technol.* **16**, 1171–1173 (1979).
56. B.L. Sharma and S.C. Gupta, Metal–semiconductor Schottky barrier junctions, Part 1-Fabrication, *Solid State Technol.* **23**, 97–101 (May 1980).
57. J.D. Jensen and R.B. Schoolar, Surface charge transport in PbS_xSe_{1-x} and $Pb_{1-y}Sn_ySe$ epitaxial films, *J. Voc. Sci. Technol.* **13**, 920–925 (1976).
58. M. Bleicher, H.D. Wurzinger, H. Maier, and H. Preier, *n*-type PbS and $PbS_{1-x}Se_x$ layers prepared by the hot-wall epitaxy, *J. Mater. Sci.* **12**, 317–322 (1977).
59. D. Tsang and S.E. Schwarz, Detection of 10 km radiation with point-contact Schottky diodes, *Appl. Phys. Lett.* **30**, 263–265 (1977).
60. R.D. Baertsch and J.R. Richardson, An Ag–GaAs Schottky-barrier ultraviolet detector, *J. Appl. Phys.* **40**, 229–236 (1969).
61. G.E. Stillman. C.M. Wolfe, A.G. Foyt, and W.T. Lindley, Schottky barrier $In_xGa_{1-x}As$ alloy avalanche photodiodes for 1.06μm, *Appl. Phys. Lett.* **24**, 8–10 (1974).
62. J.R. Richardson and R.D. Baertsch, Zinc sulfide Schottky barrier ultra-violet detectors, *Solid State Electron.* **12**, 393–397 (1969).
63. E.M. Logothetis, H. Holloway, A.J. Varga, and E. Wilkes, Infrared detection by Schottky barriers in epitaxial PbTe, *Appl. Phys. Lett.* **19**, 318–320 (1971).
64. R.A. Chapman, M.R. Johnson, and H.B. Morris, Metal–semiconductor diode infrared detector having semi-transparent electrode, U.S. Patent 3, 980, 915 (September 14, 1976).
65. B.L. Sharma and S.C. Gupta, Metal–semiconductor Schottky barrier junctions: Part II—Characterization and applications, *Solid State Technol.* **23**, 90–95 (June 1980).
66. M. Lanir, A.H.B. Vanderwyck, and C.C. Wang, EBIC characterization of HgCdTe crystals and photodiodes, *J. Electron. Mat.* **8**, 175–189 (1979).
67. R.W. Grant, J.G. Pasko, J.T. Longo, and A.M. Andrews, ESCA surface studies of $Pb_{1-x}Te$ devices, *J. Vac. Sci Technol.* **13**, 940–947 (1976).
68. A. Christon and K. Sleger, in *GaAs and Related Compunds, St. Louis* 1976 (L.F. Eastman, ed.), Conference Series No. 33b, pp. 191–200, Institute of Physics, London (1977).

69. J.J. Lander, H. Schreiber, Jr., T.M. Buch, and J.B. Mathews, Microscopy of internal crystal imperfections in Si *p–n* junction diodes by use of electron beam, *Appl. Phys. Lett.* **3**, 206–207 (1963).
70. W. Czaja, Response of Si and GaP *p–n* junctions to a 5- to 40-keV electron beam, *J. Appl. Phys.* **37**, 4236–4248 (1966).
71. T.E. Everhart, O.C. Wells, and R.K. Matta, A novel method of semiconductor device measurements, *Proc. IEEE* **52**, 1642–1647 (1964).

6

Microwave Schottky Barrier Diodes

Y. Anand

1. INTRODUCTION

Point contact diodes have been in use for many decades for mixer and detector application from uhf through millimeter-wave frequencies. The first published paper on the subject appeared in 1874 when Braun reported the asymmetrical nature of conduction between metal points and crystals. Point contacts are relatively unsophisticated devices consisting of a metal whisker making pressure contact with the semiconductor chip, normally tungsten for silicon and phosphorus bronze for germanium and gallium arsenide. The point contact diodes are generally encapsulated in axial lead glass, axial prong ceramic, cartridge-type ceramic (1N21 and IN23), or metal coaxial enclosures. In the early 1960s Schottky barrier diodes were introduced for similar applications. The Schottky diode, also a metal–semiconductor rectifying junction, is formed by depositing a variety of metals on *n*-type or *p*-type semiconductor materials by chemical deposition (electroplating), evaporation or sputtering; *n*-type silicon and *n*-type gallium arsenide (GaAs) are the most commonly used materials. Owing to the need for higher cutoff frequency, GaAs devices are preferred at millimeter-wave frequencies, since electrons have a higher mobility in GaAs than in silicon.

Noise figure lower than 5 db (double sideband) is now possible at 94 GHz using epitaxial GaAs Schottky barrier diodes. These improvements have resulted from the recent advances made in the areas of semiconductor material growth and purity, photolithographic techniques to achieve small junction areas for Schottky diodes, and planar processing of semiconductor surfaces.

Y. Anand • M/A-COM Gallium Arsenide Products, Inc, South Avenue, Burlington, Massachusetts 01803.

Major achievements were the fabrication of stable evaporated or sputtered contacts by Archer and Atalla[1] (1963), silicide formation by Kahng and Lepselter[2] (1965), multilayer high-temperature refractory metallization processes to improve the reliability and RF burnout performance,[3,4] the fabrication of honeycomb diodes,[5] and beam leaded diodes.[6]

Recent research has continued in the following directions: (a) Abrupt epitaxial GaAs material grown by metal–organic chemical vapor deposition (MOCVD)[7] and molecular beam epitaxy (MBE),[8] and low ohmic contact resistance metallization which are necessary to achieve low series resistance and consequently noise figure; (b) theoretical analysis of low-noise microwave and millimeter-wave mixers by Held and Kerr,[9] which has shown that the conversion loss and noise of a mixer can be predicted very accurately from measurable device and circuit parameters and; (c) the use of strip transmission line in waveguide mixers[10,11] has resulted in new types of mixers which have exhibited extremely low-noise performance in cryogenic mixers approaching the theoretical limit at millimeter-wave frequencies. Monolithic receiver work is currently being pursued in many research laboratories. This may result in low-cost optimized microwave and millimeter receivers for many scientific and communication applications.

2. DIODE DESIGN CONSIDERATIONS

2.1. Equivalent Circuit

The mixer (or Schottky diode) may be regarded as a nonlinear conductance g shunted by a capacitance C_j in series with a series resistance R_s as shown in the equivalent circuit (Fig. 1); the conductance is the nonlinear barrier conductance at the rectifying contact, and the capacitance is the barrier capacitance. At low frequencies the capacitance does not affect rectification, but at microwave frequencies its shunting action reduces the rf voltage across the barrier. Since it is impossible to tune out this capacitance at microwave

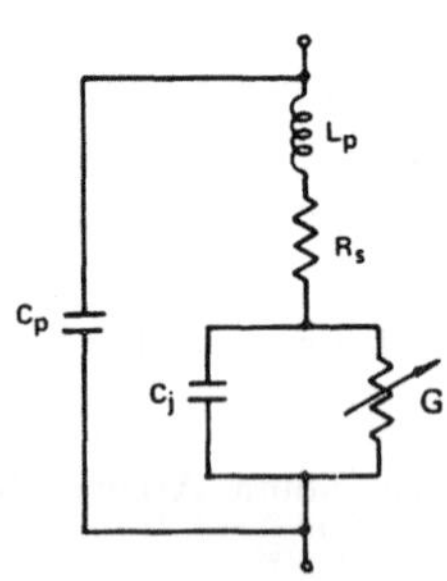

Figure 1. Equivalent circuit of varistor diode.

frequencies with an external inductance due to the presence of R_s, it reduces the rectification efficiency.

2.2. Frequency Conversion

Superheterodyne receivers employing semiconductor mixer diodes are widely used at microwave frequencies. The mixer, in its usual application, converts a microwave frequency signal to a signal centered at a lower intermediate frequency (if). Frequency conversion is a process which converts a signal at a low power level from one frequency to another by combining it with a larger local oscillator voltage in a nonlinear device, the mixer diode. This diode generates, in general, an infinite number of sum and difference frequencies. The difference frequency between signal (rf) and local oscillator (LO) frequencies is an intermediate frequency (if) at a low power level. The mixer diode also generates harmonics of the local oscillator and signal frequencies. Local oscillator harmonics at high power levels beat with the intermediate frequency to produce sum and difference frequencies called harmonic sidebands. Interaction between low-level frequencies (for example, signal harmonics and if frequency) can be ignored since the power levels of the resultant frequencies are extremely low.

The image frequency is generated as a result of direct beat between the rf signal and the local oscillator second harmonic and also from the beat between local oscillator and intermediate frequency. If the mixer input circuit is broadband, i.e., the same impedance is presented to both signal and image terminals, the image power will be absorbed mostly in the source impedance. If the input circuit bandwidth is smaller, the image frequency power can be completely reflected back to the diode: further conversion to intermediate frequency takes place which improves or degrades the conversion efficiency depending upon the image reflection phase. If the phase of the reflected image signal is adjusted to increase the available if power an image enhancement mixer results.

2.3. Basic Mixer Diode RF Parameters

A fundamental limitation on the sensitivity of a microwave receiver employing a mixer arises from the fact that in the frequency conversion process only a fraction of the available rf signal power is converted into power at the intermediate frequency. This overall conversion loss is dependent primarily on the diode junction properties and secondarily on the diode's package parasitics (i.e., mismatch of signal power, R_s, C_j) and on mismatching at the input and output ports of the mixer. An additional limitation in performance arises from the fact that the mixer itself generates noise (noise temperature ratio) when it is driven by an rf signal (local oscillator). Thus, the conversion loss and the noise temperature ratio are the essential parameters of the microwave mixer diode. The mixer diode is completely characterized by the following parameters:

a. conversion loss;
b. noise temperature ratio;
c. receiver noise figure;
d. rf impedance;
e. if impedance;
f. rf burnout;
g. local oscillator power required for mixing.

2.3.1. Conversion Loss Theory

The conversion loss of a mixer diode is dependent on several factors, including both the package and the device chip, and can be considered to be the summation of several losses. The first component of total diode conversion loss can be called the "matching loss" as it is dependent on the degree of impedance match obtained at both the rf signal and if ports. Less than optimal match at either of these ports will result in a reduction in the available rf signal at the diode and the inefficient transfer of the if signal. The "matching loss" can be expressed as

$$L_1(\mathrm{dB}) = 10\left[\log\frac{(S_1+1)^2}{4S_1} + \log\frac{(S_2+1)^2}{4S_2}\right] \tag{1}$$

where S_1 and S_2 are rf and if VSWRs, respectively.

The second loss represents a loss of signal power due to the diode's parasitic elements and, hence, will be called the diode "parasitic loss." The parasitic elements are the junction capacitance, C_j, and the series resistance, R_s. The "diode parasitic loss" is the ratio of the input rf signal power to the power delivered to the junction resistance R_j:

$$L_2(\mathrm{dB}) = 10\log\frac{(P_{\mathrm{in}})}{(P_{\mathrm{out}})} \tag{2}$$

Expressing this loss in terms of diode parameters,

$$L_2(\mathrm{dB}) = 10\log\left[1 + \frac{R_s}{R_j} + (\omega C_j)^2 R_s R_j\right] \tag{3}$$

where R_j is the time average value, as established by the local oscillator drive. The minimum value of L_2 occurs when R_j is equal to $(1/\omega C_j)$:

$$(L_2)_{\min}(\text{dB}) = 10\log(1 + 2\omega C_j R_s) \tag{4}$$

Since the value of R_j is strongly dependent on the local oscillator drive level, the value of L_2 is a function of LO drive. R_s is also a slight function of drive level. Further increase in the drive results in increased L_2 due to the dissipation in R_s, while decreasing drive also gives insertion loss increase due to the shunting effect of the junction capacitance.

The third type of loss is the actual conversion loss at the diode junction. This depends mainly on the voltage-vs.-current characteristics of the diode and the circuit conditions at the rf and if ports.

The nonlinear behavior of the diode is represented by a time-varying conductance g which is dependent on the dc characteristics of the diode and local oscillator voltage waveform across the diode.[12] Torrey and Whitmer[13] described the mixer as a linear passive three-port network having terminals at signal, image, and intermediate frequencies and described the network parameters in terms of Fourier coefficients of the time-dependent conductance. Conversion loss and impedance values can be calculated for the various image terminations by means of linear network theory. The minimum conversion loss at diode junction, L_3, for a broadband mixer (image properly terminated) can be expressed in terms of incremental conductances[14]:

$$(L_3)_{\min} = 2\frac{1 + \{1 - [2g_1^2/g_0(g_0 + g_2)]^2\}^{1/2}}{1 - \{1 - [2g_1^2/g_0(g_0 + g_2)]^2\}^{1/2}} \tag{5}$$

where g_0, g_1, and g_2 are the incremental conductances and are defined by:

$$\begin{aligned} g_0 &= I_s I_0(\alpha V_0)\alpha \\ g_1 &= I_s I_1(\alpha V_0)\alpha \\ g_2 &= I_s I_2(\alpha V_0)\alpha \end{aligned} \tag{6}$$

where I_0, I_1, and I_2 are modified Bessel functions. V_0 is the amplitude voltage of local oscillator drive.

Equation (6) is valid under the following assumptions: (a) diode series resistance is negligible, (b) no dc bias is applied to the diode, (c) contribution due to nonlinear capacitance of the diode and higher-order harmonics of local oscillators is neglected, and (d) that the diode follows the relation

$$i = I_s[\exp(\alpha V) - 1] \tag{7}$$

where I_s is the saturation current and $\alpha = \eta kT/q$, where k is Boltzmann's constant, T is the absolute temperature, and q is the elementary charge. η is an empirical factor equal to 1 for ideal diodes and, for experimental diodes, in the range between 1.02 and 1.2 for Schottky and 1.2 to 2.5 for point contact diodes.

$$g(t) = 1/R_j = \frac{\partial i}{\partial V} = I_s \exp(V_0 \cos \omega_0 t) \tag{8}$$

where $V = V_0 \cos \omega_0 t$, voltage of the local oscillator. In Eq. (5) $(L_3)_{\min}$ approaches a value of 2 as the expression

$$1 - 2g_1^2/g_0(g_0 + g_2)^{1/2}$$

approaches zero. Thus, for an ideal mixer, theoretical minimum conversion loss is 3 dB under broadband conditions. Thus, a maximum of half the power is delivered to the if port and the remaining power is dissipated at the image termination. Further, the above theory also predicts conversion loss of 0 dB for open or short circuited image terminations.[12] These mixers are called image enhancement mixers.

The overall conversion loss, L, is the sum of the three described losses:

Conversion loss = matching loss + parasitic loss + junction loss

or

$$L = L_1 + L_2 + L_3 \qquad \text{(all in dB)} \tag{9}$$

Barber,[15] Saleh,[16] Egami,[17] and Held and Kerr[18,19] extended this work, using a nonsinoidal voltage, including effects of nonlinear capacitance, harmonics of local oscillator, and nonideality of Schottky diodes. Held and Kerr[9] have shown that an accurate prediction of the mixer performance is possible if the electrical properties of the diode and the impedance across the diode terminals is known at the signal, the image, the if, and all harmonics of the local oscillator.

Conversion loss of a mixer can be calculated accurately using the computer program by Siegel and Kerr[20] which performs a nonlinear analysis to determine the diode conductance and capacitance waveforms produced by the local oscillator. A small signal linear analysis is then used to determine the conversion loss, if impedance, and input noise temperature of the mixer. It has been shown[20,9] that the parametric effects of the voltage-dependent capacitance of a Schottky barrier diode may either degrade or improve the mixer performance depending on the diode and circuit parameters.

The following conclusions can be drawn from the mixer theories: (a) R_sC_j product should be kept minimum as possible. This implies that the cutoff frequency of the diode ($f_c = 1/2\pi R_s C_j$) should be as high as possible (b) Minimum conversion loss for broadband mixer is 3 dB, and for minimum conversion loss for image enhancement mixer; the diode terminals should be short circuited at the image and harmonics of the local oscillator. (c) The average junction resistance should be equal to the reactance of the diode at zero bias ($R_j \approx 1/\omega C_j$).

2.3.2. Noise Temperature Ratio

In varistor mixers, there are three main sources of noise. These are thermal noise, which is present in any conductor at thermodynamic equilibrium; shot noise, generated under the influence of an electric field; and an additional component which is inversely proportional to frequency, usually referred to as $1/f$, or flicker noise. The other possible noise sources are: noise due to phonon scattering, correlated shot noise,[2] and, in gallium arsenide, intervalley scattering.[22] In room temperature mixers shot and thermal noise predominate at if frequencies greater than 10 MHz, with scattering noise contributing typically 5%–10% of the overall mixer noise.[9]

2.3.2.1. *Thermal and Shot Noise.* Thermal noise for the Schottky barrier is given by an expression

$$\overline{i^2} = 4kTG_BB \tag{10}$$

where k is Boltzmann's constant, G_B is the conductance of the diode, B is the bandwidth, and $\overline{i^2}$ is the mean square noise current.

The shot noise in a Schottky barrier is derived from the considerations identical to those presented by Uhlir[23] and Van der Ziel[24] for p–n junctions. Under forward bias condition, there is net flow of electrons from the semiconductor to the metal, giving rise to dc current I. Equal and opposite components of saturation current also flow in the barrier; these do not produce a net current in the external circuit, but do produce a shot noise. Full shot noise is ascribed to the three components; the resulting shot noise current is given by

$$\overline{i_N^2} = 2(1 + 2I_S)Bq \tag{11}$$

In terms of ac conductance G_B of the diode, the noise temperature ratio t_B of the barrier is defined by

$$t_B = \frac{\overline{i_N^2}}{4kTG_BB} \tag{12}$$

t_B is the ratio of the diode mean square noise current to the mean square thermal noise current of a passive conductance. The current–voltage relationship of the Schottky barrier diode is given by an expression

$$i = I_s \exp(qV/\eta kT) - 1 \tag{7'}$$

The small signal conductance G_B is found by differentiating (7) with respect to V, and the result for $\eta = 1$ is

$$G_B = \frac{q}{kT}(I + I_s) \tag{13}$$

Substituting (11) and (13) into (12) yields the barrier noise temperature ratio

$$t_B = \frac{1}{2}\left(1 + \frac{I_s}{I + I_s}\right) \tag{14}$$

The noise temperature ratio, t, of the composite device, consisting of Schottky barrier with noise temperature t_B and series resistance with the thermal noise, is given by the expression

$$t = \frac{R_j t_B + R_s}{R_j + R_s} \tag{15}$$

where R_j is the dynamic resistance of the barrier (reciprocal of G_B). Values t and t_B less than one have been measured experimentally for Schottky barrier diodes. In the limit when the Schottky barrier diode is due entirely to shot noise, from Eq. (14)

$$t_B = \frac{1}{2}\left(1 + \frac{I_s}{I_s + 1}\right) \tag{16}$$

The saturation current, I_s, for platinum–silicon (n-type) or palladium–GaAs (n-type) Schottky barrier diodes is 2×10^{-14} A and the rectified current I is usually 0.1–1 mA under dc or local oscillator bias condition. Thus, for ordinary dc forward biases,

$$t_B \approx \tfrac{1}{2} \tag{17}$$

Under optimum local oscillator excitation, correlation effects reduce the shot noise to much smaller values. At the same time, however, conversion of the source and image thermal noise, together with the series resistance thermal noise, results in a noise temperature ratio close to 1.0. Experimentally obtainted t values are slightly less than 1 for Schottky diodes and slightly greater than 1 for point contact diodes.

2.3.2.2. *Flicker or* $1/f$ *Noise.* Flicker noise has a current-dependent power spectrum which varies inversely with frequency and is present in all devices at low frequencies when a current flows. Its causes, though not fully understood, are mainly surface effects.(25) Schiff(26) attributed it to an instability of the metal–semiconductor contact due to thermal effects. Weiskoph(27) regards it as being caused by surface ions whose motion influences the contact potential. Jantch(28,29) proposed a model of slow surface states in an attempt to explain this phenomenon.

Investigations of noise in silicon, gallium arsenide, and germanium point-contact and Schottky barrier diodes made under both rf and dc excitation show that flicker noise is the dominating factor at low if frequencies have a spectral distribution(30,31)

$$\overline{i^2} = \frac{I^m}{f^n} B \tag{18}$$

where I is the dc current and m and n are constants, approximately 2 and 1, respectively.

2.3.3. Overall Receiver Noise Figure

The most important criterion of mixer performance is the overall noise figure of the receiver. The noise at the output of a receiver is the sum of the noise arising from the input termination (source) and noise contributed by the receiver itself (i.e., due to if amplifier and mixer diode). The noise factor is the ratio of the actual output noise power of the device to the noise power which would be available if the device were perfect and merely amplified by the thermal noise of the input termination without contributing any noise of its own. It is given by the following relation:

$$\text{NF} = \frac{S_i/N_i}{S_0/N_0} \tag{19}$$

where S_i is the available signal power at the input of receiver, N_i is the available noise power at the input of receiver, S_0 is the available signal power at the output of receiver, and N_0 is the available noise power at the output of receiver. The noise figure is the noise factor in decibels.

$$\text{NF(dB)} = 10 \log_{10}\left(\frac{S_i/N_i}{S_0/N_0}\right) \tag{20}$$

The overall noise figure of the receiver depends on conversion loss L and noise temperature ratio of mixer diode t and also on the noise figure of the if amplifier, F_{if}. It is given by Friis[32]:

$$NF = L(t + F_{if} - 1) \tag{21}$$

2.3.4. Mixer Noise Temperature

Mixer noise is also expressed in terms of mixer input noise temperature, T_M, which is defined as the temperature of the input termination on an equivalent noise free mixer which would produce the same output noise power as the actual mixer with a noise free input termination.[9] The noise temperature of a receiver, which is a mixer followed by an if amplifier, is given by

$$T_R = T_M + LT_{if} \tag{22}$$

where T_{if} is the noise temperature of the if amplifier. The sensitivity of a radiometer, such as a radio telescope, is given by the radiometric equation[9]

$$\delta T = \frac{K_s T_R}{(\tau B)^{1/2}} \tag{23}$$

where δt is the smallest detectable signal, τ is the observation time required to detect δt, K_s is the sensitivity constant which is of the order of 1, and B is the bandwidth of the receiver.

2.3.5. RF Impedance

The rf impedance of varistor rectifiers is a property of prime importance in the design of mixers. Impedance mismatch at radio frequency not only results in signal loss due to reflection but also affects that if impedance at the if terminals of the mixer, an effect that becomes more serious with rectifiers of low conversion loss as shown in Eq. (11). The rf impedance of a varistor diode can be measured by VSWR method.

The rf impedance is a complicated function of package geometry, size and shape of package parts, composition of the semiconductor, and junction parameters. In order to establish a good match between a semiconductor chip and rf line, a matching transformer is generally an essential part of a microwave package.

2.3.6. IF Impedance

A matter of prime importance in the design of coupling circuits between the mixer and if amplifier is the if impedance that is seen on looking into the if terminals of a varistor mixer. The pertinent if impedance (Z_{if}) is the impedance at the output terminals of the mixer when the rectifier is driven by a local oscillator. It is a function of the local oscillator power level and depends on the rf properties of the mixer and circuit connected to the rf terminals of the mixer. IF impedance of a mixer diode in terms of incremental conductances for a broadband case is given by the following:

$$\text{if impedance} = Z_{\text{if}} = \frac{1}{g_0}[1 - 2(g_1)^2/g_0(g_0 + g_2)]^{1/2} \tag{24}$$

where g_0, g_1, and g_2 are incremental conductances given by Eq. (6).

An accurate measurement of if impedance is essential for measuring noise temperature ratio t and conversion loss L of a mixer diode.

2.3.7. Receiver Sensitivity

The following equation for the sensitivity of a receiver shows the parameters which affect system sensitivity:

$$S = -114 + \text{NF}_0 + 10\log_{10} B + 10\log_{10}(S/N) \tag{25}$$

where, S is the receiver sensitivity in dBm, NF_0 is the receiver overall noise figure in dB, B is the receiver bandwidth in MHz, and S/N is the minimum acceptable receiver signal-to-noise ratio in dB.

2.3.8. Doppler Shift

Doppler radars utilize the fact that microwave energy reflected by a moving target is shifted in frequency. The amount of frequency shift is directly proportional to the target's velocity relative to the radar's transmitter. A similar effect of audible frequencies occurs when an automobile horn is moving with respect to an observer. The sound pitch is higher when the horn is moving towards the observer and decreases as it moves away from him. The Doppler shift frequency f_d is given by

$$f_d = 2v\left(\frac{f_0}{c}\right)\cos\phi \tag{26}$$

where, f_0 is the transmitter frequency, C is the velocity of light (3×10^8 msec), v is the velocity of the target (msec), and ϕ is the angle between the microwave beam and the target's path. Note: $\cos \phi = 1$, for moving directly forward or away from the radar beam, velocity v in a vectorial sense will determine the sign of the Doppler shift frequency.

2.3.9. Typical Doppler Radar System

A typical Doppler radar system consists of an rf (i.e., microwave) section, a signal processing section, and a bias supply.

In order to design a Doppler radar system, one must first know

1. The maximum range at which the target is to be detected. (This determines the overall sensitivity required of the transceiver.)
2. The maximum and minimum target speeds that the system is to measure. (This determines the characteristics of the amplifier.)

The Doppler systems for police radars and intrusion alarms usually operate with a "zero if" because the transmitter source (Gunn oscillator) is also used as the local oscillator for the mixer. With this technique, signal amplification occurs at the Doppler shift frequency. For example, at a transmitter frequency 10.525 GHz, a vehicle traveling 50 mph cause a Doppler shift of 1568 Hz. The amplifier bandpass for the police radar might be 50 to 5000 Hz. to include the range of target speeds expected.

The maximum range of a radar system can be determined by the following equation:

$$R_{\max} = \left(\frac{P_t G_a K}{\mathrm{NF}} \right)^{1/4} \tag{27}$$

where P_t is the transmitted power, G_a is the antenna gain, NF is the receiver noise figure, and K is a constant. Thus, the effective range of a radar system is inversely proportional to the fourth root of the overall receiver noise figure.

2.4. Basic Detector RF Parameters

Microwaves can be detected by direct rectification of the rf signal at a nonlinear semiconductor metal contact. The sensitivity is mediocre in comparison to the superheterodyne receiver.

Detector operation is based on the slope and curvature of the voltage-vs.-current characteristics of the diode in the neighborhood of the bias point. The output voltage of the detector is directly proportional to the input rf power (square of the input voltage). The concept of noise figure is not applicable to a square-law detector.

A comprehensive analysis of square-law detectors and their quantitative comparison are given elsewhere.[33] The sensitivity of a low-level detector depends upon the following:

1. rectification efficiency, output impedance, and noice properties of the diode;
2. input impedance, bandwidth, and noise properties of the amplifier;
3. rf matching structure.

The rectification efficiency of the diode is usually stated as either current sensitivity, $\beta = \Delta I / P_{\text{in}}$, where ΔI flows in the detector diode output circuit, or voltage sensitivity, $\gamma = \Delta V / P_{\text{in}}$, ΔV being the time average increase in voltage across the detector output, and P_{in}, the applied microwave power. In unbiased detector operation, the self-bias current is extremely small, resulting in low current sensitivity. Generally detector diodes are externally forward biased (0.25–30 μÅ) in order to improve their performance. In 1944, Berringer[34] introduced a "figure of merit" concept to characterize the video receivers:

$$M = \frac{\beta R_v}{(R_a + R_v)^{1/2}} \tag{28}$$

where R_v is the dynamic resistance of the diode and is called video resistance, and R_a is a constant resistance representing the noise contribution due to amplifiers. The term $R_a \approx 1200\Omega$ no longer seems to be valid for present low-noise transistor amplifiers. Figure of merit does not consider shot and flicker noise introduced by the bias current. Therefore, this method of characterization is of limited value in describing Schottky barrier diodes.

2.4.1. Video Resistance (R_v)

R_v is the real part of the diode's small signal impedance. This parameter is dependent on the dc bias current and diode's series resistance. Video detectors are presently being characterized by signal sensitivity types of measurement, i.e., amount of available signal power (in decibels referred to 1 mW) required to produce a specified signal-to-noise ratio. The various terms recently used are minimum detectable signal (MDS),[34] tangential signal sensitivity (TSS),[34] noise equivalent power (NEP),[33] and nominal detectable signal (NDS).[35]

$$R_V = R_s + R_j \tag{29}$$

where R_j is the active small signal junction resistance and R_s is the series resistance. For a Schottky diode, using diode current–voltage relationship:

$$i = I_s \exp(qV/\eta kT) - 1 \tag{7'}$$

$$R_j = \left(\frac{\partial i}{\partial v}\right)^{-1} = \frac{\eta kT}{q} \frac{1}{I + I_s}$$

$$I_s \approx 10^{-9}\,\text{A} \ll I$$

$$R_j = \frac{\eta kT}{qI} = \frac{0.26}{I} \tag{30}$$

for $\eta = 1$, $T = 300\,\text{K}$, and I is in mA. Generally Schottky detector diodes are biased approximately to 5 microamperes (μA) of current, and for these cases

$$R_v \approx R_j \approx 5000\,\Omega \tag{31}$$

It follows from (28), since $R_a \ll R_v$ for low-noise amplifiers, that

$$M = \frac{\beta R_v}{\sqrt{R_v}} = \beta\sqrt{R_v} = \frac{\gamma}{\sqrt{R_v}} \tag{32}$$

2.4.2. Voltage Sensitivity

The voltage sensitivity of a detector diode is a ratio of the open circuit video signal voltage to the rf input power:

$$\gamma = \frac{V_{\text{oc}}}{P_{\text{in}}} \tag{33}$$

where V_{oc} is the open circuit video voltage and P_{in} is the rf power incident on the detector. Voltage sensitivity is expressed in units of millivolts per milliwatt. To ensure that the detector is in the square low range, γ is usually measured at -20 to $-30\,\text{dBm}$.

2.4.3. Current Sensitivity β

Current sensitivity, (β), for a detector diode is the ratio of the short circuit video current to the rf input power:

$$\beta = \frac{I_{\text{sc}}}{P_{\text{in}}} \tag{34}$$

where I_{sc} is the short circuit video current. The units of β are milliamps per milliwatt, γ and β are related as follows:

$$\gamma = \beta R_v \tag{35}$$

In terms of diode parameters and physical constants

$$\beta = \frac{q}{2\eta kT}\left(\frac{1}{1 + R_s/R_j + \omega^2 C_j^2 R_s R_j}\right) \tag{36}$$

where c_j is the junction capacitance.

2.4.4. Minimum Detectable Signal (MDS)

The minimum detectable signal (MDS) is defined as the smallest signal which may be observed on an oscilloscope when its position along the trace is unknown. This corresponds to a signal-to-noise ratio of approximately unity and is a subjective measurement.

2.4.5. Tangential Signal Sensitivity (TSS)

The tangential signal sensitivity (TSS) is a direct measure of the signal-to-noise voltage in a detector receiver. The measurement is carried out with a pulse signal, the level of which is adjusted so that the highest noise peaks observed on an oscilloscope in the absence of signal are at the same level as the lowest noise peaks in the present of signal. The signal level thus determined gives the TSS value. TSS corresponds to a signal-to-noise ratio of approximately 2.5. Although the measurement is highly subjective and depends upon the operator, it is still most commonly used by the industry.

2.4.6. Nominal Detectable Signal (NDS)

The nominal detectable signal (NDS) is defined as the microwave power required to produce an output power equal to the noise power. This corresponds to a signal-to-noise ratio of unity.

$$\mathrm{NDS} = \frac{2}{\gamma}\left\{kTR_v\left[\left(t_w + \frac{T_0}{T}(F_v - 1)\right)B + B_x \ln\frac{f_h}{f_1}\right]\right\}^{1/2} \tag{37}$$

where t_w is the white noise temperature ratio, F_v is the noise figure of the video amplifier, and $B_x \ln f_n/f_1$ represents flicker noise of the diode.

If $F_v = 1 + (R_a/R_v)$, then

$$\text{NDS} = \frac{2}{M}(kTB)^{1/2} \tag{38}$$

The TSS is found empirically to be 4 dB above NDS, under ordinary conditions. The parameters TSS, MDS, and NDS all depend on the amplifier bandwidth, usually varying as the square root of the bandwidth. Thus, the value at which measurements are made must be quoted in specifying the detector; the usual value is a 1 MHz video bandwidth.

2.4.7. Noise Equivalent Power (NEP)

The NEP provides a measure of the threshold sensitivity characteristics of a detector and is independent of the associated video amplification circuitry. NEP is defined as the microwave input power required to produce an output signal-to-noise ratio of unity, for a bandwidth of 1 Hz. This measure is appropriate for modern video amplifiers, which are of extremely low noise, and for applications in which the detector output frequency is very low.

$$\text{NEP} = \frac{2nkT}{q}\left(\frac{4kTt_w}{R_j}\right)^{1/2}\left(1+\frac{R_s}{R_j}\right)^{1/2}\left[1+\left(\frac{f}{f_c}\right)^2\right]\left(1+\frac{f_n}{f_v}\right)^{1/2} \tag{39}$$

where f_n and f_v are "noise corner" and video frequency, respectively. If may be shown that

$$\text{TSS} = \text{NEP} + 4 + 5\log_{10} B \tag{40}$$

where the parameters are in decibels and B is the bandwidth, using the fact that TSS corresponds to a current signal-to-noise ratio of about 2.5. The NEP is typically around -80 dBm, so that the addition of the term $5\log_{10} B_w$ decreases the TSS, e.g., if $B = 1$ kHz, and NEP $= -84$ dBm, then TSS $=$ -65 dBm.

2.4.8. Video Bandwidth

Although the detector may have a wide bandwidth capability, the circuit in which the detector diode is used will determine the video bandwidth of the overall detector. The typical detector circuit has its low-frequency video response limited by the inductance of the RF choke and the series coupling capacitor to the video amplifier. The high-frequency video response is limited by the amplifier input impedance and the rf bypass capacitance. The upper frequency, 3 dB roll-off point, is given by

$$f(3\,\mathrm{dB}) = \frac{R_v + R_a}{2R_v R_a C_t} \tag{41}$$

where R_a is the amplifier input resistance and C_T is the sum of amplifier input capacitance and rf bypass capacitance.

2.4.9. Superheterodyne vs. Single Detection

Mixers and detectors both downconvert microwave signals so that they may be displayed or processed further. Low noise amplification (about 100 dB) is more readily achieved at vhf and below, than at microwave frequencies. Most mixer (superheterodyne) systems use if amplification at an intermediate frequency (30–200 MHz) and then use a second downconverter such as a video detector to obtain the modulating signal that was imposed on the microwave carrier. Such a superheterodyne detection system, a microwave receiver with 10-dB noise figure and 1 MHz if bandwidth, would have a maximum sensitivity of − 104 dBm.

A single detection system using only video amplification can achieve a tangential sensitivity (TSS) of perhaps −60dBm for a 1 MHz video bandwidth compared with the − 104 dBm for the heterodyne system. However, the single detection system has the advantage of simplicity, low cost, and potentially wide bandwidth.

2.5. Mixer Configurations

Five basic types of mixers have been developed for various applications. The five types are as follows:

a. Single-ended,
b. Single-balanced,
c. Double-balanced,
d. Image-rejection,
e. Image-enhanced (or recovery).

2.5.1. Single-Ended Mixer

The single-ended mixer is the simplest type of diode mixer and has the advantage of wide rf bandwidth. It consists mainly of rf and if ports, with both local oscillator and rf signals applied together at rf terminals as shown in Fig. 2a. The downconverted if signal is decoupled from the mixer diode by a

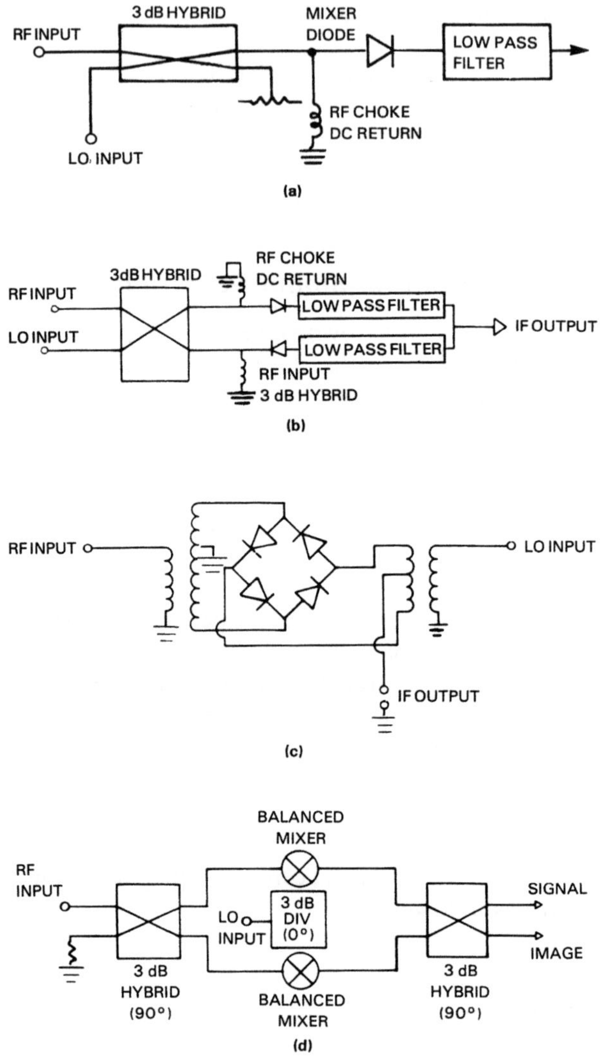

Figure 2. (a) Single-ended mixer; (b) single-balanced mixer; (c) double-balanced mixer; (d) image rejected mixer.

low-pass filter. This low-pass filter also reflects the rf and LO signals back into the mixer. The dc return, which offers a low-resistance path for the diode rectified current, also acts as a high-pass filter which prevents loss of if signal out the rf port. The drawbacks of the single-ended mixer are (a) it does not suppress local oscillator noise and (b) it has poor LO/rf isolation.

2.5.2. Single-Balanced Mixer

Two single-ended mixer circuits can be combined into a balanced mixer configuration commonly known as a single-balanced mixer, as shown in Fig. 2b. The rf and LO power is supplied to two mixing diodes by using a 3-dB hybrid. Either a 90° or 180° hybrid can be used for this mixer. The choice of which hybrid to use will generally depend on LO to rf isolation versus VSWR trade-off. The balanced mixer offers the advantage of canceling AM noise generated by the local oscillator and better LO rf isolation, but its bandwith us limited by the characteristics of the hybrid.

2.5.3. Double-Balanced Mixer

A double-balanced mixer essentially consists of two single-balanced mixers connected in parallel with each other, but 180° out of phase. A typical double-balanced mixer is shown in Fig. 2c. The double-balanced mixer is superior to the single-balanced mixer by its design symmetry, such that it suppresses the even harmonics of both LO and rf signals and still maintains high isolation between the if, rf, and LO ports. The performance of a double-balanced mixer depends strongly on the degree of diode-to-diode matching in the diode ring and on the symmetry of the transformer. The drawbacks are twofold: a high-level oscillator power requirement to drive four diodes and circuit complexity which prevents their use for high-frequency applications.

2.5.4. Image Rejection Mixer

A typical image-rejection mixer is shown in Fig. 2d. This mixer uses two single-ended balanced mixers combined through the use of rf and if hybrids. The signals are combined in such a way that the desired signal will appear at one port of the if hybrid and the image at the other port. The degree of image rejection depends on the amplitude and phase balance of the two mixers. This type of mixer provides good input VSWR and a 3-dB improvement in power handling over other designs. But the drawbacks of this mixer are that the LO power required is 3 dB higher than for a balanced mixer, and the conversion loss is higher due to additional loss of the rf and if hybrids.

2.5.5. Image Enhanced or Image Recovery Mixer

The same method that rejects the image of an incoming signal can be used for image recovery of an internally produced signal. For 180°-type mixers, the generated image appears at the rf input port. For narrow-band applications, a filter at the rf input port will reflect the image power back into the mixer[67]

TABLE 1. Mixer Comparison Guide

Mixer type	Conversion loss	VSWR LO, rf	LO/rf isolation	LO power required	Spurious rejection	Harmonic supression	Third-order intercept
Single-ended	Good	Good, poor	Fair	+3 dBm	Poor	Poor	—
Balanced							
(90°)	Good	Good, good	Poor	+5 dBm	Fair	Fair	+13 dBm
(180°)	Good	Fair, fair	Very good	+3 dBm	Fair odd: fair	Even: good	+13 dBm
Double-balanced	Very good	Poor, poor	Very good	+10 dBm	Good	Very good	+18 dBm
Image reject	Good	Good, good	Good	+7 dBm	Fair	Even: good odd: fair	+15 dBm
Image recovery	Excellent	Good, good	Very good	+7 dBm	Fair	Even: good odd: fair	+15 dBm

[a]After Reynolds and Rosenzweig. Reprinted with permission from *Microwaves*, May 1978. Copyright Hayden Publishing Co., Inc.

Fig. 2e). By adjusting the electrical length between the filter and mixer, image signal can be reflected back to the mixer diode for further conversion into if signal. Two mixers can also be combined such that the image frequency produced is reactively terminated with the proper phasing to minimize conversion loss. Both techniques have been used to build mixers with conversion loss improvement of 1 to 2 dB lower than single-ended and single-balanced mixers. The performance of various types of mixers is summarized in Table 1.

3. PROPERTIES OF SCHOTTKY BARRIER DIODES

A Schottky barrier diode is a rectifying metal–semiconductor junction formed by plating, evaporating, or sputtering a variety of metals on *n*-type or *p*-type semiconducting materials. The properties of a forward-biased Schottky barrier diode are determined by majority carrier phenomena as opposed to *p*–*n* junction diodes whose properties are primarily determined by minority carriers. As a result, the Schottky diodes can be switched rapidly from forward to reverse bias without minority carrier storage effects.

The typical current-vs.-voltage curve of a Schottky barrier diode resembles that of a p–n junction with the following exceptions. The reverse breakdown voltage of a Schottky barrier diode is lower and the reverse leakage current higher than a p–n junction diode using the same resistivity n-type material. The forward voltage at a specific foward current level is generally lower for a Schottky barrier diode than for a p–n junction. For example, at 2 mA a low barrier silicon Schottky diode will have forward voltage of 0.3 V while a p–n junction diode will have a voltage of 0.9 V.

Silicon, gallium arsenide, indium phosphide, and germanium have been used as semiconductor materials for the fabrication of microwave mixer and detector diodes. Generally, n-type silicon Schottky diodes are used up to K_a-band frequencies. For higher frequencies, n-type GaAs Schottky diodes are preferred. This is because higher mobility of electrons in GaAs results in lower series resistance and thus a lower noise figure at millimeter-wave frequencies. Indium phosphide is also an excellent material for millimeter Schottky diodes, but further work is needed to fabricate reliable, reproducible, and low-noise devices at microwave- and millimeter-wave frequencies.

3.1. Diode Theory

In 1938, W. Schottky[37] described the nature and theory of the idealized rectifying contact between a metal and a semiconductor. Design theory can be only roughly applied to the point contact diodes as the basic mechanism affecting the barrier properties is not well understood. Recent reviews of the properties and applications of Schottky diodes have been given by Attalla,[38] Sze,[39] Watson,[40] Viola and Mattauch,[41] Schneider,[42] and Wrixon and Kelley.[43]

For simplicity, we will consider the diode rectification theory developed by Schottky[37] which is generally valid for both silicon and gallium arsenide diodes at microwave frequencies. The current-vs-voltage characteristics of an ideal Schottky diode is given by an expression

$$I = I_s\left(\exp\frac{qV}{\eta kT} - 1\right) \tag{7''}$$

where I_s is the saturation current, q is the electron charge, T the absolute temperature, k Boltzmann's constant, η the ideality factor, and V the voltage across the diode junction. η is equal to 1 for an ideal diode.

$$I_s = AA^*T^2e^{-q\phi/kT} \tag{42}$$

where I_s is the saturation current, A is the area, A^* is the Richardson constant, and ϕ is the barrier height. The barrier height is an important parameter, because it determines the local oscillator power necessary to bias the diode into its nonlinear region.

3.2. DC Parameters

The total capacitance of a packaged Schottky barrier is given by

$$C_T = C_j + C_0 + C_p \tag{43}$$

where C_j is the metal–semiconductor junction capacitance, C_0 is the overlay capacitance across the oxide layer, and, C_p is the package capacitance. The overlay and package capacitance are either eliminated or minimized above K_a-band frequencies.

3.2.1. Junction Capacitance

In the case of large area planar Schottky barrier diodes, the junction capacitance is given simply by the one-sided abrupt junction analysis.[39] Device capacitance is given by the following relation:

$$C = A[q\epsilon_r\epsilon_0 N_d/2(V_d - V)]^{1/2} \quad \text{(farads)} \tag{44}$$

and is shown as a function of diode diameter for the use of millimeter wave GaAs mixer diodes in Fig. 3. Here A is the device area, q is the electronic charge, ϵ_0 is the dielectric constant or relative permittivity, ϵ_r is the permittivity of free space, N_d is the donar density in n-layer, V_d is the diffusion or metal–semiconductor contact potential, and V is the applied voltage. In order to achieve high conversion efficiency it is desirable to have the capacitive

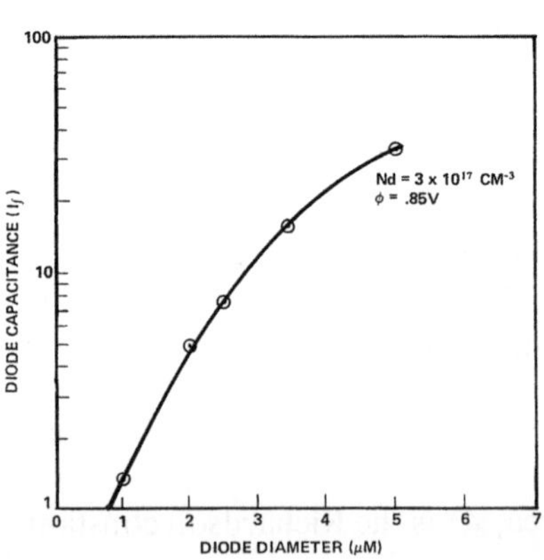

Figure 3. Diode capacitance (ff) vs. diode diameter (μm).

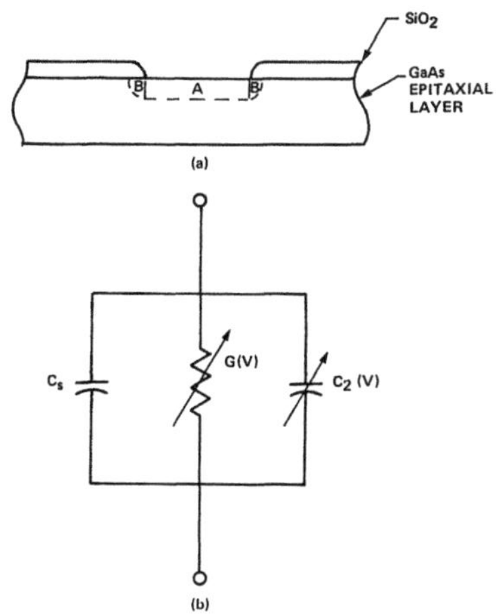

Figure 4. Equivalent circuit of GaAs Schottky diode.

reactance of the junction large with respect to the diode small signal conductance; furthermore, the rf impedance of the signal and LO circuitry is generally between 50 and 100Ω. Therefore, it is necessary, as the operating frequency increases, to proportionally decrease device capacitance. This is accomplished by reducing the device area.

This simple model, however, becomes inappropriate as the diode radius approaches the zero-bias space-charge region depth shown in Fig. 4a, since edge capacitance becomes so large as to no longer be negligible. In this case, the device capacitance is modeled as two capacitors in parallel. These are due to parts A and B of the space-charge region shown in Fig. 4a. While the junction capacitance due to part A can be expressed by the above equation and is voltage dependent, the capacitance due to part B is voltage independent and must be expressed as a constant. This is, on first consideration, surprising, but on further analysis one notes that capacitance is, in general, given by the simple parallel plate geometrical relation:

$$C = \frac{\epsilon_r \epsilon_0 A}{d} \tag{45}$$

where ϵ_r, ϵ_0, and A are given above and d is the plate separation. As the applied bias increases d increases, but so also does the space-charge region edge area, A, increase by the same factor. This effectively cancels out the effect of applied bias on the edge capacitance giving rise to the device junction capacitance model consisting of a voltage-dependent capacitor shunted by a voltage-independent capacitor as shown in Fig. 3b. It seems as though the obvious approach to high-frequency devices would be to decrease the device area. This action, however, is not without adverse effect on device operation because device series resistance increases with decreasing Schottky barrier diode anode area. If circular anode devices are considered separately one finds that the series resistance increases very rapidly with anode diameter, decreasing below approximately 2 μm.

3.2.2. Overlay Capacitance

The capacitance contributed by the metal contact overlaying the passivating dielectric layer in Schottky barrier diodes may be important. Assuming negligible space-charge penetration (a realistic or conservative assumption for SiO_2 on the semiconductor), the overlay capacitance is

$$C_0 = \frac{\epsilon_r \epsilon_0 A}{w_0} \tag{46}$$

wherein the parameters are the dielectric constant, area, and thickness of the oxide. This parasitic capacitance must be kept to a minimum, particularly at frequencies in the X-band and above. Overlay contacts are not generally used above 40 GHz frequencies because they degrade the overall performance of a mixer diode.

According to Fig. 5

$$A = \text{area} = \pi((R_i + \Delta)^2 - R_i^2) = \pi(R_i + \Delta + R_i)\Delta$$
$$= 2(R_i + \Delta)\Delta$$

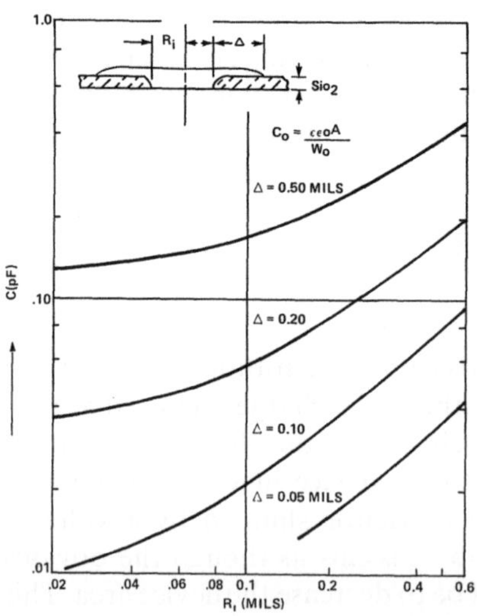

Figure 5. Overlay capacitance for SiO_2 width of 0.1 μm.

$$\epsilon_0 = 8.8 \times 10^{-14}\text{F/cm}$$

$$\epsilon_r = \text{for SiO}_2 = 4$$

$$C_0 = \frac{\epsilon_0 \epsilon_r}{W_0} = \frac{8.8 \times 10^{-14} \times 4 \times \pi(2R_i + \Delta)\Delta}{W_0}$$

The curves in Fig. 5 gives the different values of overlay capacitance for different values of R_i, Δ, and W_0. A thick layer of oxide such as a 2-μm layer really reduces the overlay capacitance of a Schottky diode.

3.2.3. Series Resistance

Several models have been proposed to explain device series resistance and individual contributions thereto. An early work which is quite useful is by Kennedy[44] and treats the thermal, or electrical, spreading resistance of a cylindrical semiconductor device. Dickens[45] presented a rather involved and complete spreading resistance model based on an oblate spheroidal coordinate system, while Clifton *et al.*[46] used the result of Dickens along with a few extensions to predict the series resistance of Schottky barrier diodes having 2- to 3-μm diameters. All methods assume a cylindrical anode placed coaxially on a cylindrical semiconducting substrate.

Recently, Carlson, Schneider, and McMaster[11] and Kelley and Wrixon[43] have calculated series resistances for millimeter and submillimeter devices using different anode geometries, shape, and size of the ohmic contacts.

The series resistance R_s is voltage and frequency dependent and its major

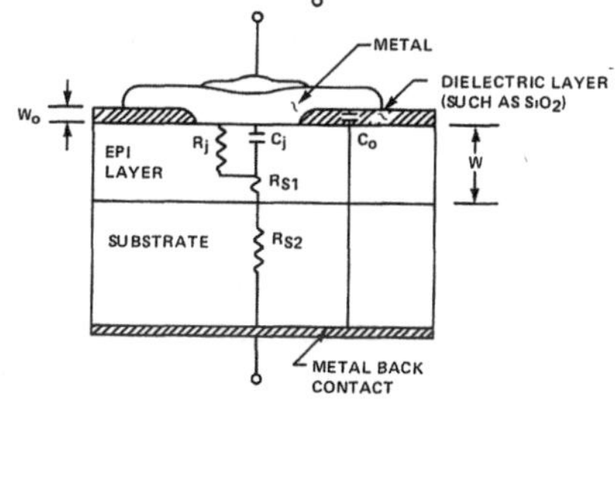

Figure 6. Schottky diode chip equivalent circuit.

contribution is caused by the resistance of the semiconducting substrate and the undepleted epitaxial layer. The series resistance in the epitaxial layer and the semiconducting substrate depends on the junction geometry, the frequency, and to a lesser extent, on the applied voltage. For simplicity, we discuss below the low-frequency model which neglects skin effect. The total series resistance as shown in Fig. 6 consists of R_{s1} (due to the epitaxial layer) plus R_{s2} (due to the substrate).

The epitaxial layer contribution to the resistance is given by

$$R_{s1} = \frac{\rho}{A} = \frac{2w}{(q\mu_e N_d)A} \tag{47}$$

where ρ is the resistivity of the epitaxial layer, w is the thickness of the epitaxial layer, A is the area of the Schottky junction, μ_e is the electron mobility in the epitaxial layer (this assumes the layer is n-type), and N_d is the donor density in the epitaxial or active layer. The resistance contributed by the substrate may be modeled by using the resistance of a contact dot the size of the junction on a semi-infinite semiconductor substrate:

$$R_{S2} = \frac{\rho}{2d} = 2\rho_s\left(\frac{A}{\pi}\right)^{1/2} \tag{48}$$

where ρ_s is the substrate resistivity and d is the active junction diameter $= 2(A/\pi)^{1/2}$.

3.2.4. Figure of Merit

The cutoff frequency, that is, figure of merit by definition, of a Schottky barrier diode is maximized by minimizing the $R_s C_j$ product. Furthermore, mixer conversion loss can be shown to be directly proportional to the product of diode series resistance, R_s, and junction capacitance, C_j. By converting these parameters to semiconductor properties of the active junction, the following figure of merit for a Schottky barrier diode can be obtained:

From equations (44) and (47):

$$L \alpha R_s C_j \alpha \frac{W \epsilon^{1/2}}{N_d^{1/2} \mu} \tag{49}$$

where ϵ is the electric permittivity of the semiconductor, W is the active layer thickness, N_d is the carrier concentration in the active region, and μ is the carrier mobility in the active region.

3.3. Semiconductor Materials

Schottky barrier diodes are generally fabricated from n-type Si or n-type GaAs epitaxial material, GaAs being preferred for millimeter-wave frequencies because of its greater mobility (five times greater than Si).

There has been considerable interest in using some of the other, even higher mobility, III–V compounds, but there have been practical difficulties. One of the most promising materials is InSb with mobility nine times higher than GaAs. Schottky barrier diodes have been fabricated on InSb by Korwin-Pawlowski and Heasell,[47] Kelly and Wrixon,[48] and McColl and Millea,[49] but have exhibited enormously high series resistance in addition to relatively low shunt resistance. One of the drawbacks of the InSb is its low band gap which requires that it be cooled to produce a rectifying barrier. Christou *et al.*[50] have also fabricated Schottky diodes on InP, but these devices need refinements before they can be used practically in mixer circuits. Presently, GaAs is considered to be the most suitable material for the fabrication of millimeter frequency Schottky barrier diodes because of its high mobility, ready availability, wide bandwidth, and room temperature operation.

3.4. Epitaxial GaAs

The epitaxial layers to be used must have (a) the correct crystallographic orientation, (100) and not (111), in order to avoid the uncertainties and patchy results characteristic of using a polarized crystal face, (b) a very low crystallographic imperfection density in order to ensure minimization of avalanche noise, (c) excellent surface morphology to allow for submicron device dimensions, and most important, (d) exactly the correct epitaxial layer doping level, abrupt interface and thickness to ensure the lowest possible device series resistance and the highest possible cutoff frequency.

Item (d) above is known to be of paramount importance in the success of the resulting diodes. The epitaxial layer thickness, however, must be no greater than 2000 Å and in many cases significantly less than this amount for optimum device design. In addition to the correct epitaxial layer properties mentioned, it is necessary to impose requirements on the GaAs substrates in order to minimize the primary parasitic resistance term, R_s. To minimize the contribution of the substrate R_s it is necessary that the resistivity of the substrate be as low as possible, preferably less than 0.001 Ω cm.

3.5. Barrier Height Lowering

The barrier height, ϕ_{Bn}, is an important parameter, as it determines the local oscillator power required to bias the diode in its nonlinear region. In many systems especially at millimeter-wave frequencies, local oscillator power is at premium and, therefore, low barrier Schottky diodes are desired.

Metal–semiconductor contacts can be ohmic or rectifying. Those made on heavily doped semiconductors are generally ohmic. The ohmic behavior of metal–semiconductor contacts is based on carriers tunneling through thin barriers.[51,52] The contacts on lightly doped material result in rectifying behavior with an energy barrier ϕ_{Bn} existing between the metal and the semiconductor. The barrier ϕ_{Bn} is the difference between the metal work function ϕ_m and the electron affinity X of the semiconductor: $\phi_{Bn} = (\phi_m - X)$. In principle, the sum of barrier heights for a metal on n-type and p-type semiconductor material equals the band-gap energy: $E_g = q(\phi_{Bn} + \phi_{Bp})$. Therefore, one is able to predict the barrier height for an idealized Schottky diode on a given semiconductor material. However, in practice, surface states[53] and image forces[54] determine the ultimate barrier: $\phi_{Bn} = (\phi_m - X - \Delta - \Delta\phi)$, where Δ and $\Delta\phi$ are the contributions due to surface states and electrostatic image forces, respectively. Generally in silicon Schottky diodes Δ and $\Delta\phi$ are second-order effects. But in the case of GaAs, the barrier height is mainly controlled by the surface states and is generally independent of barrier metal.

Schottky diodes have been fabricated with several metals and alloys[39,55–58] using p- and n-type silicon and n-type gallium arsenide, with barriers ranging from 0.15 to 0.90 V. See Table 2 and Fig. 7 for details.

TABLE 2. Experimental Values of Metal–Semiconductor Barrier Height in Volts

Metal	ϕ_B Si (n-type) (111) (V)	ϕ_B Si (p-type) (111) (V)	ϕ_B GaAs (n-type) (100) (V)
Au	0.81		0.90
Ni	0.3	0.8	
Mo	0.60		
Ti	0.35, 0.50	0.61	
W	0.69		
Ag	0.69		0.88
Pt	0.85		0.86
Pt–Ni Alloy	0.6[a]		
Pd	0.72	0.15–0.25[b]	
Au–Ge 300°C			0.27
Au–Ge 200°C			0.35

[a] Zero bias Schottky.[65]
[b] Recent low-barrier and high-burnout Schottky.[64]

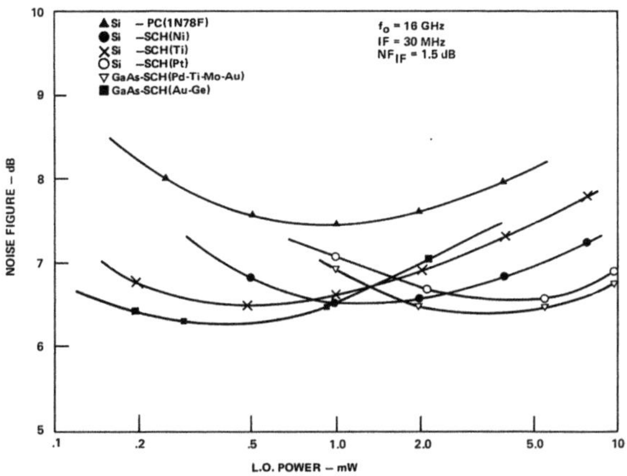

Figure 7. Noise figure vs. LO power of K_u-band mixer diodes.

The energy band diagram of a lightly doped semiconductor in contact with a metal under forward bias conditions is shown in Fig. 8a. The depletion region is wide and, therefore, electrons reach the metal by thermionic emission over the potential barrier ϕ_{Bn}. For a heavily doped semiconductor, the energy barrier is sufficiently thin for electrons to tunnel through the barrier (field emission), resulting in an ohmic contact (see Fig. 8b. Moroney and Anand[58] showed that the formation of Au–Ge alloy Schottky diodes on *n*-type GaAs results in high-quality microwave mixer and detector diodes with low barriers. The Au–Ge yields a barrier of 0.27 to 0.9 V depending on fabrication temperatures of 300 to 25°C, respectively. The application of Au–Ge to lightly doped *n*-type GaAs material at 500°C results in ohmic contact as reported by Braslave *et al.*[59] An explanation for this ohmic behavior to lightly doped material is the formation of gallium vacancies during

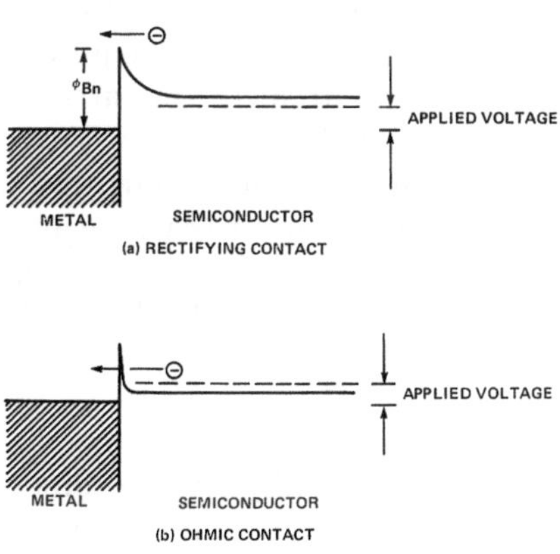

Figure 8. Energy diagrams of metal *n*-type semiconductor under forward bias voltage. (a) Semiconductor lightly doped; rectifying contact. (b) Semiconductor heavily doped (degenerate); ohmic contact. (After Anand and Moroney. Reprinted with permission from *Proc. IEEE.* **59** 1971.)

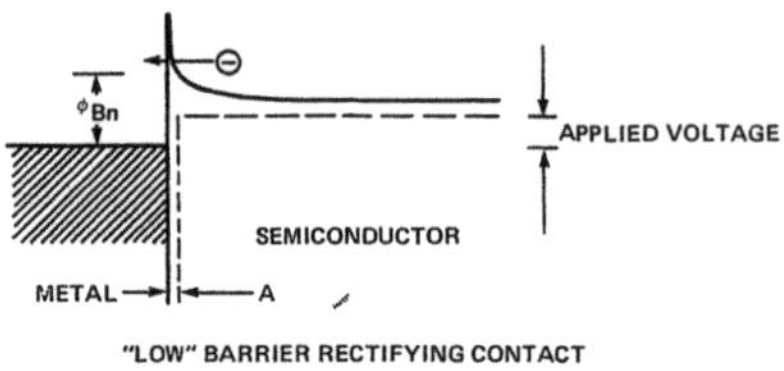

Figure 9. Energy diagram of metal *n*-type semiconductor under forward bias voltage. Semiconductor lightly doped with a heavily doped or disordered layer < 10 Å thick at surface. (After Anand and Moroney. Reprinted with permission from *Proc. IEEE* **59**, 1971.)

alloying with subsequent germanium atoms occupying these gallium sites and becoming donors. For the case where a low barrier results, it is supposed that a thin layer (10 Å) of degenerate material exists at the surface that modifies the barrier shape (see Fig. 9). A low forward bias is required to raise the majority carrier over the first part of the barrier by thermionic emission but the remaining barrier (part A) is sufficiently thin for tunneling to occur.

Anand, Christou, and Dietrich[60] used ion implantation techniques to obtain low-barrier-height GaAs Schottky barrier mixer diodes. By utilizing ion implantation, a shallow n^+ surface layer can be used to control the field at the surface and consequently reduce the effective barrier height of a Schottky barrier on *n*-GaAs.[61] Recently, Christou, Anderson, Davey, and Anand[62] deposited Ge films epitaxially to obtain low barrier height Schottky diodes at millimeter frequencies. Ge/GaAs–titanium Schottky barrier diodes exhibited a low noise figure of 7.0 dB (S.S.B.) (single side band) at 36 GHz frequency. The barrier height was lowered from $\phi = 0.75$ V (standard diode) to $\phi = 0.45$ V (Ge/GaAs–Ti–Mo–Au) diode and local oscillator requirement was reduced from 2 to 0.75 mW. Further lowering of the barrier height can be accomplished by optimizing the Ge film thickness and doping level. Investigations are underway to improve Ge film properties using MBE or MOCVD process.

The ion implant technique was also used to lower the barrier height of Pt–*n*(Si) and Ni–*n*(Si)[63] Schottky barrier diodes. Recently, (platinum–nickel)–*n*(Si)[64] Schottky barrier diodes have been developed, which exhibit high rf burnout performance of platinum schottky diodes but barrier height of 0.6 V and thus require 0.5 MW of local oscillator power for mixer operation. Extremely low barrier silicon schottky diodes known as "zero bias Schottky diodes" ($\phi \approx 0.15$ V) were introduced, especially for low-level detection applications at microwave- and millimeter-wave frequencies.[65,66] These devices have barrier height lower than point contact diodes and do not require external dc bias for their operations. The low barrier height is attained by forming a metal-silicide on silicon through heat treatment.

3.6. Fabrication

The Schottky barrier diodes are generally fabricated by a planar technique. A silicon dioxide (SiO_2) or silicon nitride layer (Si_3N_4) (10,000 Å)

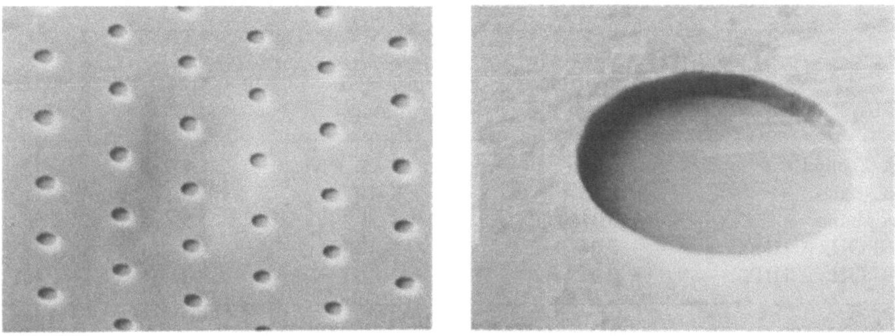

Figure 10. SEM pictures of etched windows in SiO_2 layer.

for 1–36 GHz, and 4000 Å for above 36 GHz) is deposited on the semiconductor wafer and windows are etched in the Si_3N_4 by photolithographic techniques. Schottky junctions are formed by evaporation sputtering or plating techniques. Metal on the oxide is removed by a second photo step. Junction diameters as small as 5 μm are made by this technique (see Fig. 10). This approach maintains the required clean interface between the metal edge and the semiconductor and exhibit low $1/f$ noise. The $1/f$ noise is affected by surface states, therefore controlled processing in the formation of Schottky chips is required to minimize this parameter. Initially, Schottky diodes were developed to replace the cartridge type point contact diodes (1N21, 1N23, and 1N78). Therefore, a multidot array structure (honeycomb) (see Fig. 10) was used and a whisker was used to form contact with one of the rectifying junctions. This technique is still widely used for millimeter-wave Schottky diodes. In order to reduce the parasitic capacitance, overlay capacitance approach is not used. Millimeter-wave Schottky diodes are fabricated either by electroplating (Ni–Au or Pt–Au) or by a "photo-lift-off" technique to remove the excess metallization over the oxide.

Recently, thermal compression bondable Schottky diodes are in great demand to K_a-band frequencies. A single chip consisting of several rectifying contacts of different diameters is used and a 0.75 mil to 1.0 mil gold wire thermal compression bonded to one of the junctions (actually overlay structure) to the package. The diodes fabricated by this technique are rugged and highly reliable for adverse environments such as space and military applications.

These devices are available in glass and ceramic packages up to X-band frequencies, coax and ceramic pill packages are used up to K_a-band frequencies. The capacitance and series resistance of GaAs devices is given in Table 3. GaAs beam lead Schottky diodes, bondable and multijunction

TABLE 3. RF Characteristics of GaAs Schottky Barrier Diodes

Frequency	Junction capacitance (pF)	Series resistance (ohms)	Noise figure (if = 30 MHz) (dB)	TSS (2 MHz) (dBm)
L-X band	0.12	8–20	5.5	−55
K_u-band	0.10	8–12	6.0	−55
K_a-band	0.04–0.06	4–8	7.5	−50
60 GHz (CHIPS)	0.04	5–10	8.0	−48
90 GHz (CHIPS)	0.02	5–10	≈9.0	−45

[a] Single side band.

Schottky chips are available up to 90 GHz for stripline and microstrip applications.

Below *X*-band frequencies, various configurations of beam lead diodes such as Star and Quads are available for balanced and doubly balanced mixers.

4. MICROWAVE PERFORMANCE

4.1. Mixer Diodes

Image enhancement or image reflected mixers have been studied experimentally at *X*-, and K_u- and K_a-band frequencies in single-ended and balanced mixer configurations.[67–72] Recently, Utsumi[70] achieved a minimum noise figure of 3.3 dB (S.S.B.) at 11.7 GHz using a single-ended mixer. Mixers are generally used in a broadband condition (image absorbed). Conversion loss and noise temperature, T_M of GaAs mixer diodes versus rf frequency are shown in Fig. 11 and 12.

Recently, low noise mixers and complete receivers showing excellent performance at millimeter frequencies have been built at many laboratories (Kerr,[10] Linke *et al.*,[73] Carlson *et al.*,[11] Keen *et al.*,[74] and Carlson and Schneider.[75] Typical single side band mixer noise temperature and conversion loss which can be consistently achieved at room temperature are 500 K and 5.3 dB, respectively, at 100 GHz. The noise temperatures that can be attained from cryogenically cooled mixers are substantially lower. Recently, Weinreb[76] achieved minimum noise temperature of 86 K and conversion loss of 5.2 dB for cryogenic mixer at 90.8 GHz. The low-noise performance of GaAs Schottky diodes at 100 GHz is due to recent refinements in mixer mounts[73,77] and availability of good quality epitaxial GaAs material. Above 100 GHz, there is rapid degradation of mixer performance and this is caused by the

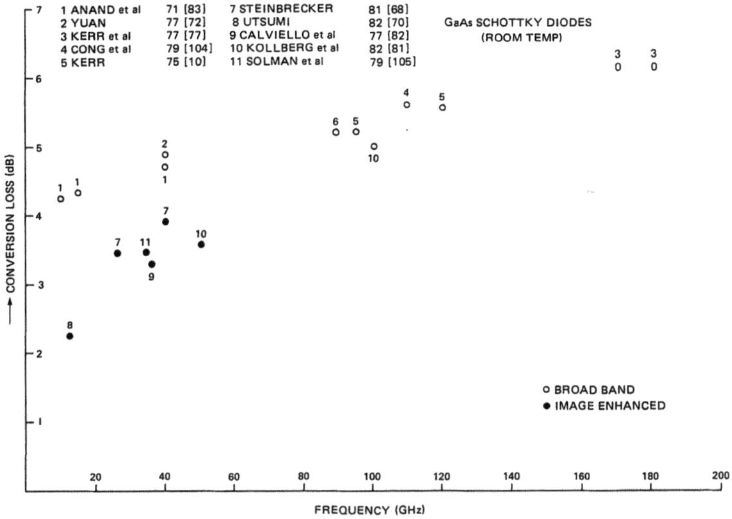

Figure 11. Measured SSB conversion loss of Schottky barrier mixers vs. rf frequency.

increased skin resistance of the diode, the increased series resistance of the diode for the reduced junction size, and the greater losses in smaller waveguide and coupling circuits.

Noise figure versus local oscillator power for K_u-band point contact, silicon and GaAs Schottky diodes are given in Fig. 7. Low barrier (Au-Ge)–GaAs and (Ti)–silicon exhibit a low noise figure even at extremely low power

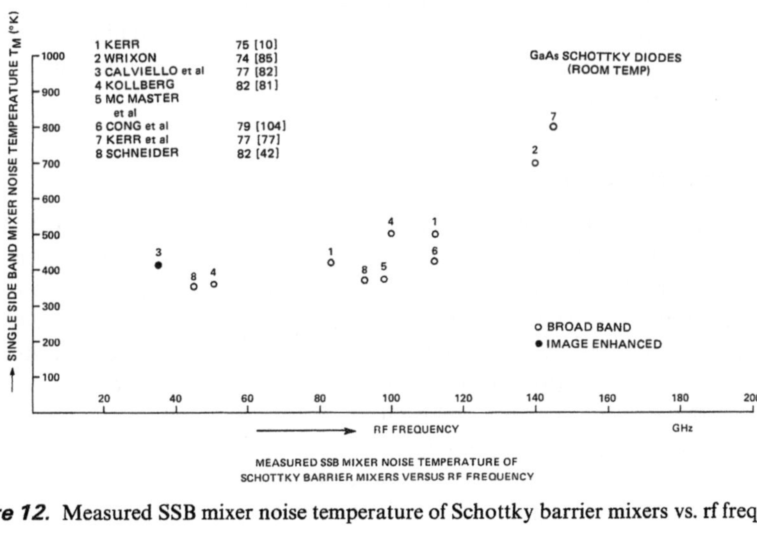

Figure 12. Measured SSB mixer noise temperature of Schottky barrier mixers vs. rf frequency.

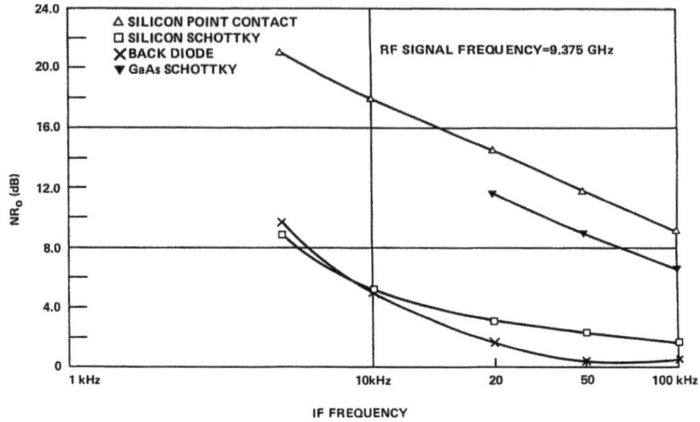

Figure 13. The variation of noise temperature ratio with frequency of X-band mixer diodes.

levels. Silicon Schottky diodes exhibit a minimum noise figure of 5.5 dB up to K_u-band frequencies. Above K_u-band frequencies, GaAs Schottky barrier diodes have better performance due to higher cutoff frequency.

Low-frequency ($1/f$) noise has been studied under dc (Weiskopf,[20] Miller,[30] and rf excitations (Sprinks *et al.*,[30] Eng,[78] and Anand.[79,80] Under dc excitation $1/f$ measurements give only a relative indication of mixer performance. Typical results for X-band point contact, back diode, silicon and GaAs Schottky barrier diodes under rf excitation are shown in Fig. 13. At if frequency of 10 kHz, silicon Schottky and back diodes exhibit a 15 dB lower noise figure than point contact diodes. But back diodes require special circuits to accommodate their low if impedance. GaAs Schottky diodes exhibit higher $1/f$ noise compared to silicon Schottky diodes. This is due to higher density of surface states in GaAs Schottky diodes and, therefore, at present they are not suitable for doppler applications.

4.2. Detector Diodes

The TSS is the most widely used criterion for video detector performance. It indicates the ability of the detector to detect a signal against a noise background and also includes the noise properties of the detector diode and video amplifier. The TSS of S-, X-, K_u-, and K_a-band diodes was calculated, as shown in Fig. 14 along with the experimental results. Detector diodes are often biased in the forward direction in order to reduce the video resistance and to improve the bandwidth and sensitivity.

Recently, extremely low barrier ($\phi = 0.15$ V) silicon Schottky diodes have been introduced especially for video detector applications (Anand[65] and Kerr and Anand[66]). These devices do not require external dc bias for video

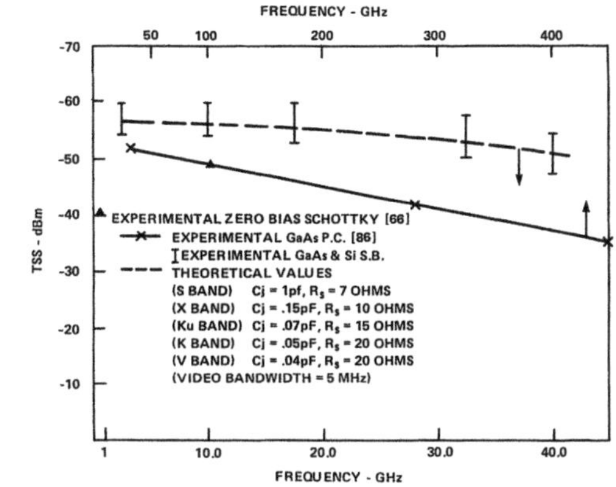

Figure 14. TSS vs. rf frequency for point-contact and Schottky barrier diodes. (After Anand and Moroney. Reprinted with permission from *Proc. IEEE* **59**, 1971).

detector operation and are known as the zero bias Schottky diodes. Compared to other point contact and Schottky diodes, zero bias Schottky diodes exhibit superior $1/f$, sensitivity, dynamic range, and TSS even at WR-10 band (75–110 GHz) frequencies.[66] Performance of the zero bias and conventional Schottky diode detectors at 95 GHz is shown in Fig. 14 with a PM 1038-V12 analyzer. Broadband performance of zero bias Schottky's at WR-10 band is shown in Fig. 15. Figure 16 shows the TSS versus dc bias for various types of diodes. With no dc bias, zero bias Schottky diodes have better TSS than other

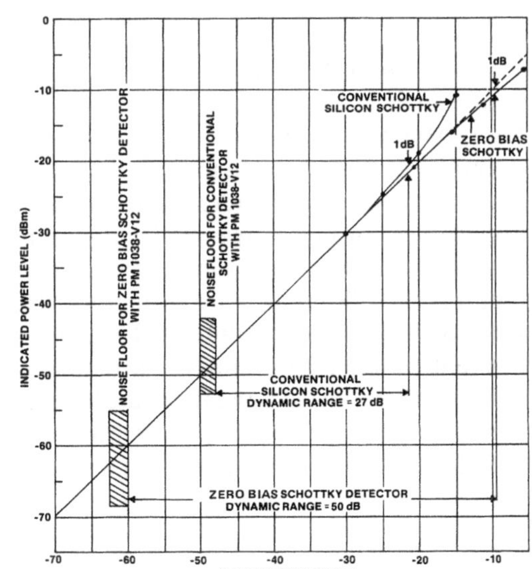

Figure 15. Performance of the zero bias and conventional Schottky-diode detectors at 95.5 GHz with a PM 1038-V12 analyzer. (After Kerr and Anand. Reprinted with permission from *Microwave J.*, December 1981).

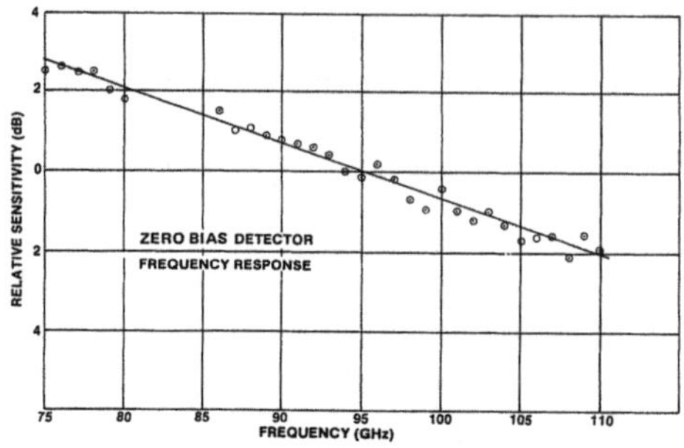

Figure 16. Frequency response of the zero bias Schottky detector. (After Kerr and Anand. Reprinted with permission from *Microwave J.*, December 1981).

Schottky diodes. With applied dc bias most other Schottky diodes exhibit improvement while zero bias Schottky diodes degrade in performance.

The rf performance of various types of X-band silicon Schottky barrier diodes is summarized in Table 4.

5. RF PULSE AND CW BURNOUT

5.1. Introduction

Burnout of microwave mixer and detector diodes is a continuing maintenance problem for almost all systems. Burnout is generally caused by large rf power reaching the diode either from other radar sets or from the radar set's own transmitter. These diodes usually exhibit deterioration in detection and conversion properties. Burnout is generally defined as either a 1–3-dB degradation in diode noise figure at 30 MHz if frequency or a catastrophic failure such as an open or short circuit. The Schottky barrier diodes designed for X-band and higher frequencies have a small active region (see Table 5—for example 45 μm^2 at X-band and 3 μm^2 at 94 GHz), and, therefore, very susceptible to rf and accidental electrostatic burnout. These diodes find many applications in high-power radar and communication systems.

In duplex radar systems, transmit–receive (TR) gas tubes, solid-state limiters, and TR limiters are commonly used to protect the microwave diode. Peak power passes through the TR tube for a finite length of time before the incident power is substantially attenuated. In radar systems where the TR switch is not followed by a solid-state limiter, rf energy (spike height times width) depends on the particular TR switch as well as the magnetron pulse rise time. In addition, harmonics, as well as random bursts of rf energy can occur in some systems. "Flat" leakage is the energy following the spike and continues for the length of the magnetron pulse. For periods of 1 μsec or longer ($\tau > 1$ μsec) this energy can be considered to be equivalent to cw for X-band or higher frequency

TABLE 4. RF Characteristics of Various Barrier Silicon Schottky Diodes (8–12 GHz)

Characteristics	Zero bias	Low barrier	Medium barrier	High barrier
		Mixer characteristics		
LO required for min NF	−5 dBm	−1 dBm	+1 dBm	+3 dBm
Z_{IF} At LO for min NF	250 Ω	350 Ω	400 Ω	450 Ω
		Detector characteristics		
	No dc bias	DC biased	DC biased	DC biased
TSS (2 MHz BW)	−52 dBm	−55 dBm	−55 dBm	−55 dBm
R_V	2–8 kΩ	2–8 kΩ	2–8 kΩ	2–8 kΩ)
Upper limit of square law detection	−20 dBm	−20 dBm	−20 dBm	−20 dBm
		General characteristics and special applications		
$1/f$ noise	Low	Low–moderate	Low–moderate	Low–moderate
RF burnout	Low–moderate	Moderate	Moderate	Excellent
Sensitivity to static discharge	Sensitive	Sensitive	Sensitive	Not sensitive
Special applications	Wide video bandwidth and excellent pulse fidelity excellent on microphonics and can replace tunnel and back diodes in certain applications	For general use and for intrusion alarm mixer and detector applications with low LO power	Good general purpose mixer diode with good rf burnout	Modulator, high-speed switch and low-power limiter. High-burnout application. lower inter mod. high-power detector and subharmonic pump mixer

diodes. Therefore, for complete burnout characterization, both the cw and the peak power or "spike" capability must be considered.

In low-power radar systems, mixer diodes encounter cw or rf pulses with pulsewidths greater than 100 nsec. These pulses occur in low-to-medium-power radar systems where protective limiter-type devices are not used. In this case, direct pickup by antennas either from friendly radars or enemy jamming

TABLE 5. DC Characteristics of GaAs Schottky Barrier Diodes

Frequency (band) (GHz)	Active area (μm^2)	Junction capacitance (pF)	Series resistance (ohms)
9 (X-band)	45	0.12	8–15
16 (K_u-band)	30	0.10	8–12
36 (K_a-band)	12	0.06	4–8
94 (V-band)	3	0.02	5–10

may also be responsible for degrading the performance of radar system mixer diodes.

The unwanted high-voltage video pulses or ac ripple in the if circuit, harmonics of high-power magnetron of a radar system,[8] and electrostatic discharges in handling these devices are also often responsible for the catastrophic failure or degradation of the mixer diodes. RF-induced burnout in Schottky barrier diodes has been the subject of many investigations.[83,88,99] The recent increase in synthetically produced clothing, rugs, and shoes has made accidental electrostatic failure a problem of great concern. On a dry, winter day, for example, a person could acquire a static potential of 3000 to 4000 V, a potential sufficient to cause catastrophic failure of Schottky barrier diodes.

During the last decade, significant progress has been made in understanding the burnout mechanism and improving the power-handling capabilities of Schottky barrier diodes.[88–99] It was experimentally observed that the early dc pulse burnout test methods using video pulses such as with Torrey line or a

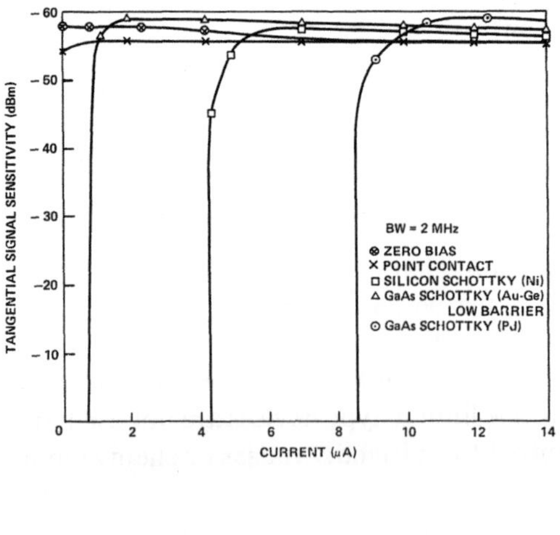

Figure 17. Tangential sensitivity vs. dc bias of K_u-band Schottky barrier diodes.

mercury relay switch and a charged coaxial line do not give valid measurement of the diode's burnout resistance in a radar receiver. This is because the diode encounters a different environment in a radar receiver test than the Torrey line test setup. A new rf pulse burnout system, using two *p-i-n* switches, has been introduced and discussed in detail.[102]

Lepselter *et al.*[91] have succeeded in improving the reliability and burnout performance of low-frequency silicon Schottky diodes (up to 4 GHz) by introducing a diffused guard-ring structure at the Schottky barrier edge. This technique, however, increases the total capacitance of the device. Furthermore, it is extremely difficult to fabricate devices with diffused guard-ring structures with a geometry suitable for *X*-band and higher frequencies. Another type of guard ring was formed by proton implantation around the periphery of the metal dot in GaAs Schottky barrier diodes for K_u-band frequencies.[83] The protons converted the *n*-type GaAs to semi-insulating ($10^6 \Omega$cm) material. Diodes fabricated by this technique did not exhibit parasitic losses and showed a 3-dB improvement in burnout. Test data[92] showed that tripod inverted mesas have improved cw burnout at *S*-band frequencies.

5.2. Factors Affecting RF Burnout

At least two significant factors affect the burnout resistance of point contact and Schottky barrier mixer diodes. The first is rf impedance matching. High burnout point contact diodes and Pt–silicon Schottky diodes have a self-protecting behavior at high-power levels by appearing as a mismatch and reflecting most of the incident power back towards the rf source. This behavior

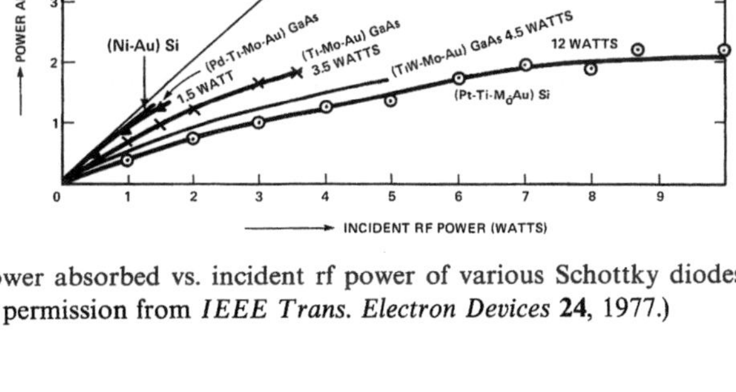

Figure 18. Power absorbed vs. incident rf power of various Schottky diodes. (After Anand. Reprinted with permission from *IEEE Trans. Electron Devices* **24**, 1977.)

could be due to variation of the barrier capacitance.[93] Generally, low-barrier Schottky diodes are better matched, and therefore, absorb more incident rf energy and have marginal rf burnout performance as shown in Fig. 18 and Table 6. Recently, (Pt–Ni)–silicon (n-type) diodes have been introduced which exhibit lower barrier height $\phi = 0.6$ V and high rf burnout of 12–15 W ($\tau = 1$ μsec, 1 kHz) ans 80–100 W ($\tau = 3$ nsec, 1 kHz).[64]

A second burnout factor involves resistance to the nanosecond pulses. Schottky diodes tend to avalanche in localized areas around the junction periphery due to high electric fields and fail catastrophically due to a localized alloyed site through the metal–semiconductor junction. The point contact also fails at junction periphery, although at varying rates. The first degradation occurs in $1/f$ noise[101] due to excessive leakage current and then is followed by an overall degradation in noise figure due to loss of nonlinear characteristics. Use of high eutectic temperature barrier metal systems generally results in higher burnout resistant Schottky diodes. This is also true to some extent for cw and microsecond rf pulses. However, in the long pulse condition, diodes fail in the center of the junction, indicating a thermal dissipation phenomenon.

Recently, the burnout resistance of silicon and gallium arsenide schottky barrier diodes have been improved considerably by using (a) high-temperature semiconductor Schottky barrier metal, (b) a metal diffusion barrier between

TABLE 6. RF Burnout of X-Band Mixer Diodes (1N23 Package) to 3ns and 1 μsec RF Pulses

Metallization scheme	Barrier height, ϕ_B (V)	3 $n$sec pulses (W)	1 μsec pulses (W)	Local oscillator power requirements for optimum mixer performance (mW)
Nickel (n-type silicon)	0.55	5–10	0.2–0.4	0.75
Nichrome–Au (n-type silicon)	0.55	10–15	0.2–0.4	0.75
Palladium (p-type silicon)	0.45	10–15	0.2–0.4	0.5
Ti–Mo–Au (N and p-type silicon	0.50	15–25	1.0–2.0	0.5
Moly–Au (n-type silicon)	0.65	40–60	2.0–3.0	1.0
Pt–Ti–Mo–Au (n-type silicon)	0.85	80–(> 100)	8–15	2.0
(Pt–Ni)–Ti–Mo–Au (n-type silicon)	0.6	10	8–15	0.75

the Schottky barrier and gold overlay, and (c) optimizing the thickness and resistivity of the epitaxial layer. RF burnout resistance of the diode can also be improved by (a) increasing the junction area, (b) increasing the thickness of epitaxial layer, and (c) resistivity of the epitaxial layer. But, these have adverse affects on the noise figure. Thus, noise figure and rf burnout resistance have conflicting requirements and, therefore, generally a compromise is made to optimize the overall performance of the diode.

5.3. Experimental Results

The experimental data include rf pulse generated by four separate techniques. In the first technique, a solid-state rf burnout system simulates the pulse through a limiter or TR tube.[102] By means of two *p-i-n* diode switches (normally off) and a delay line between them, rf pulses 3–100 nsec long are obtained and controlled by a precision attenuator.

In the second method, a traveling-wave tube (TWT) is used to amplify a low-level rf signal pulse. The 1-nsec data and part of the 10-nsec data were obtained by this technique.[103]

A third approach is to use the radar transmitter with a selected TR tube. The problem is the availability of a TR tube with proper pulse height and width. The 3-nsec data were obtained with a reject TR tube having maximum peak power leakage of 8 W.

A fourth scheme is a Varian Extended Interaction Oscillator (VA 245) and a special modulator was used to generate 5-nsec and 0.5-μsec pulses at 95 GHz.[100] Local oscillator power of 1–2 mW was introduced by a 10-dB directional coupler to the diode, and output voltage, reflected power, and noise figure degradation were monitored.

To simulate the electrostatic buildup of a charge on a human body, an electronic circuit was used. The circuit consists of a high-voltage power supply and a RC network with values of R and C chosen to be 1000 Ω and 100 pf.[88] A switch operated either to charge the capacitor to variable peak voltages or to directly apply the voltage to the diode under test.

RF pulse burnout of various types of X- and K_u-band silicon and GaAs mixer diodes is shown in Fig. 19. The bars represent approximately 10% and 90% failure points. An arrow at the top of the bar signifies that pulse power was limited and the 90% failure point is approximated. The failure point criterion is an increase of 1.0 dB in diode noise figure (30 MHz if). Duty cycle is typically 0.001 or less and the time of test approximately 0.5 min. K_u-band Schottky and point-contact diodes used in the tests had noise figures of 6.0–6.5 dB and 7.0–7.5 dB, respectively. We have optimized the Schottky barrier diode design for minimum noise figure. The silicon point-contact "high-burnout" diode is made from specially selected materials. The experimental results show that

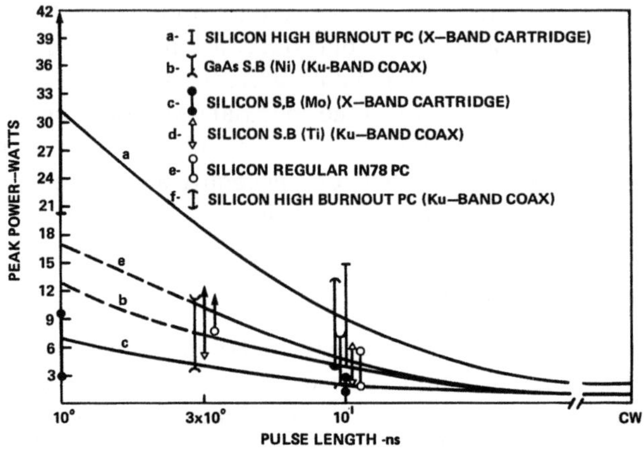

Figure 19. RF burnout vs. pulse length for point contact and n-type Schottky barrier diodes. (After Anand and Moroney. Reprinted with permission from *Proc. IEEE* **59**, 1971.)

with decreasing pulsewidth, peak power handling capability increases but not at a sufficient rate for constant energy. This is clearly shown in Fig. 20 for the X-band diodes at several different pulse lengths. Thus, rf burnout of a mixer diode depends on both pulsewidth and rf peak power, which, however, are not directly interchangeable. Power absorbed versus incident rf power ($\tau = 1$ μsec)

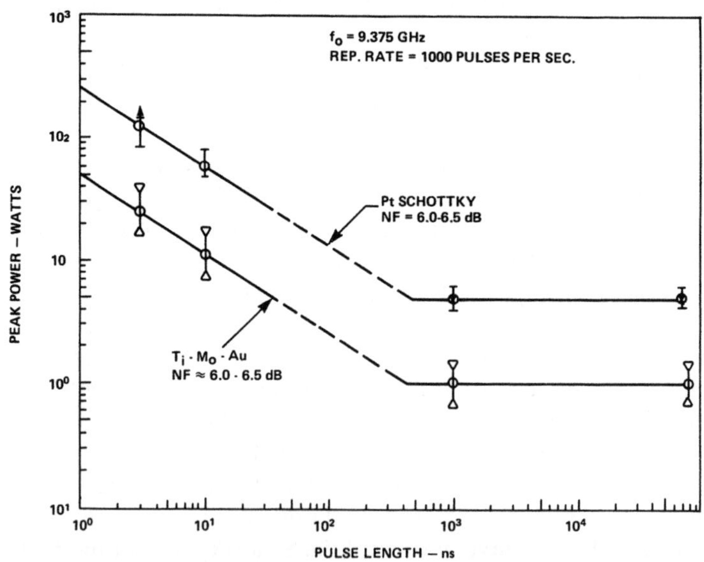

Figure 20. RF burnout vs. pulse length for X-band silicon Schottky barrier diodes. (After Anand. Reprinted with permission from *IEEE Trans. Electron Devices* **24**, 1977.)

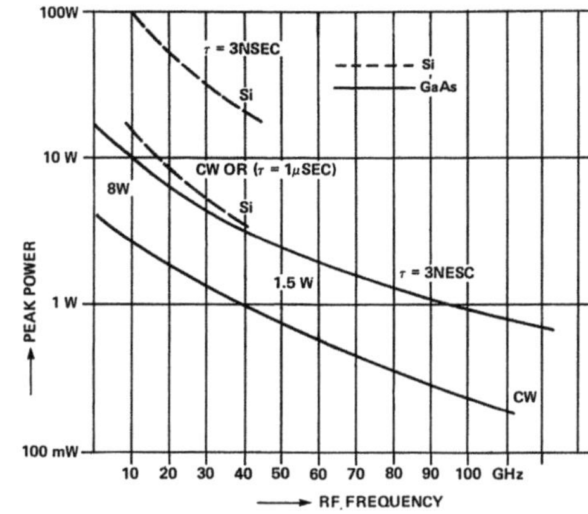

Figure 21. RF burnout resistance vs. rf frequency of Schottky barrier diodes.

of various silicon and GaAs Schottky barrier diodes at 9.375 GHz are shown in Fig. 18. Pt–silicon Schottky diodes exhibit higher reflection capability at high power levels, which may be due to variation of the barrier capacitance.(93) Further analysis shows that rf burnout has an accumulative effect on the diode performance. Ninety X-band titanium Schottky barrier diodes, selected from the same wafer with similar dc and rf characteristics, were subjected to 3-nsec rf pulses at 9.375 GHz. Forty of those diodes were tested by the described method, while the other 50 diodes were step-stressed for 5-min time intervals with increasing rf power until 1-dB degradation of noise figure (NF) occurred. The results show that step-stress test of 5-min interval is more severe than the standard step-stress test of 0.5-min duration.(3)

Recently, Anand and Christou(98–100) have improved the rf burnout resistance of K_a- (26–40 GHz) and W- (75–100 GHz) band GaAs Schottky diodes. The diodes were fabricated with different high-temperature refractory metals. GaAs Schottky diodes exhibited burnout resistance of 2 W ($\tau = 1\ \mu$sec, 1 kHz) at 36 GHz and 0.6 W ($\tau = 0.5\ \mu$sec, 1 kHz) at 95 GHz. RF burnout resistance of silicon and gallium arsenide Schottky diodes versus rf frequency is shown in Fig. 21 and indicates that the rf burnout resistance decreases rapidly with rf frequency. Silicon Schottky diodes exhibit higher burnout resistance than comparable gallium arsenide Schottky diodes.

5.4. Physical Analysis of RF Pulsed Silicon Schottky Barrier Failed Diodes

Experimental failure studies conducted on X-band point contact and Schottky diodes(96,97) using a scanning electron microscope and auger spectroscopy techniques have shown that cw or microsecond rf pulse burnout

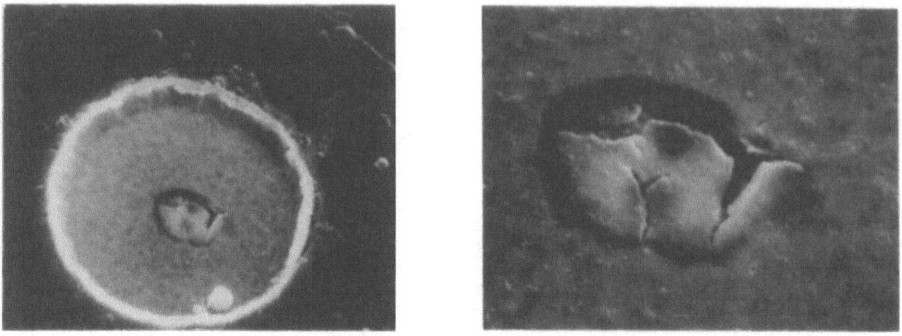

Figure 22. Typical burnout in Pt–Ti–Mo–Au diode resulting from microsecond pulses. Rf power = 8 W, $\tau = 1 \mu s$. (Left) 7600 ×; (right) 21,600 ×.

is due to thermal dissipation and causes diffusion of gold in silicon, particularly at the center of the Schottky junction as shown in Fig. 22. The overlay and barrier metals were removed by chemical etching to observe the failure points. The diodes failed at the center of the junction due to excessive heating. With longer pulses, the heat reaches equilibrium, and the center of the diode is hotter than the outer portion so that burnout usually occurs near the center. In contrast, diodes subjected to nanosecond pulses were characterized by edge burnout as shown in Fig. 23. Since the heat generated in the diode does not have time during the short nanoseconds pulse to spread or reach equilibrium, the diode fails at the periphery due to high electrical fields. Failure points at the junction periphery are clearly shown in Fig. 23 for titanium–molybdenum–gold (Ti–Mo–Au) and platinum–titanium–molybdenum–gold (Pt–Ti–Mo–Au) Schottky barrier diodes. Figure 23a shows a scanning electron microscope (SEM) picture of Ti–Mo–Au Schottky that failed at 20 W peak rf power using 10-nsec wide pulses [1000 pulses per second (pps)]. An

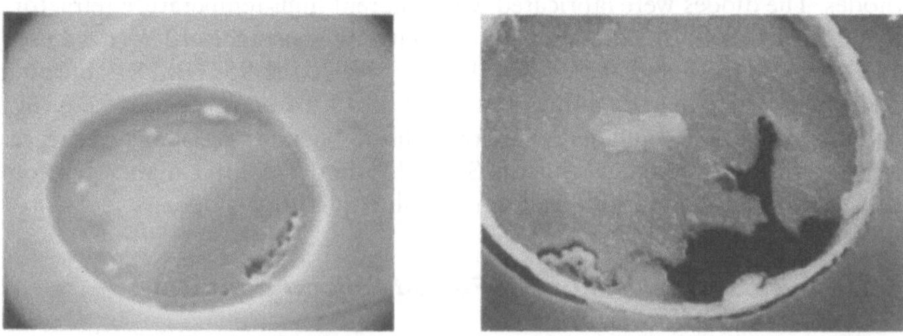

Figure 23. (Left) SEM picture (10,000 ×) of Ti–Mo–Au Schottky diode failed under 10-nsec rf pulses. (Right) SEM picture (13,900 ×) of Pt–Ti–Mo–Au Schottky diode failed under 3-nsec rf pulses.

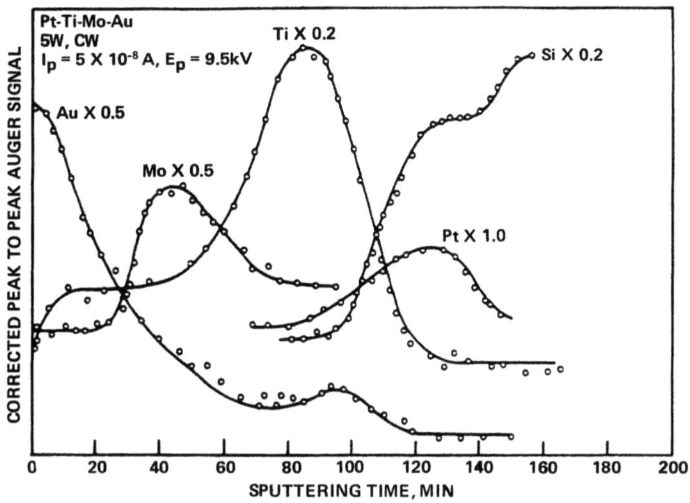

Figure 24. Sputter Auger profile of Pt–Ti–Mo–Au diode failed due to cw or microsecond pulses. (After Anand, Christou, and Dietrich. Reprinted with permission from *IEEE Trans. CHMT*, December 1978.)

SEM photograph of the degraded Pt–Ti–Mo–Au Schottky diode that failed at 100 W under 3-nsec pulses (1000 pps) is shown in Fig. 23b; failure points are again at the junction periphery. These data show that the high-temperature Pt– Si interface is capable of handling high rf power. A microspot sputter Auger profile of the Pt–Ti–Mo–Au diode is shown in Fig. 24. Significant interdiffusion between platinum and silicon is evident. Gold has diffused through the barrier metals to the silicon region. No such interdiffusion was observed for diodes which failed by edge burnout.

5.5. Physical Analysis of RF Pulsed Millimeter GaAs Schottky Barrier Failed Diodes

Recently, Anand and Christou[99,100] have investigated the failure mechanism of millimeter GaAs Schottky barrier diodes. The (TiW)–Au–GaAs, (Ti–Mo–Au)–GaAs, Pd–(Ti–Mo–Au) and Ni–Au diodes were fabricated and examined for rf burnout at 36 and 94 GHz.

Gradual degradation in noise figure as a function of rf power was observed in all types of GaAs Schottky diodes as shown in Fig. 25. The silicon Schottky diodes (all types) degrade instantaneously at certain rf power levels. This gradual degradation of GaAs Schottky diode is not well understood but may indicate the presence of two competing effects: metal–GaAs interdiffusion and substrate-edge breakdown.

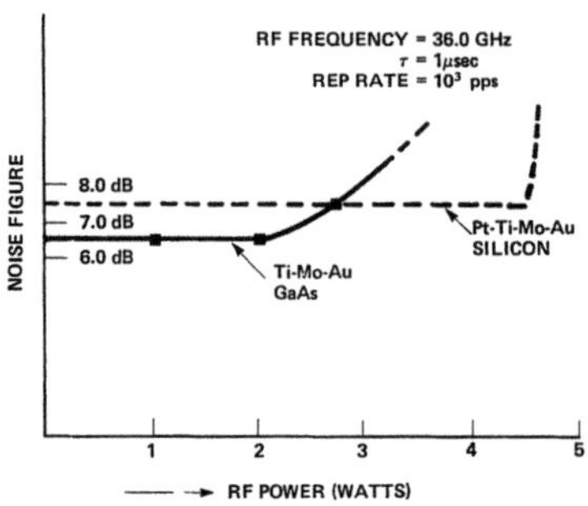

Figure 25. Comparison of noise figure vs. rf power of millimeter silicon and GaAs Schottky diode. (After Anand and Christou. Reprinted with permission from *IEEE Elect. Device Conf.* December 1980.)

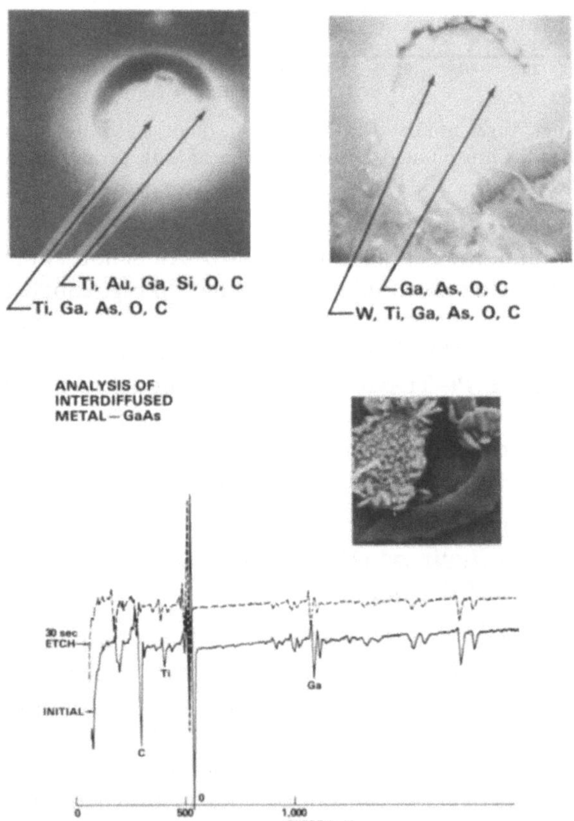

Figure 26. (Top) Comparison of cw burnout at 36 Ghz; (left) TiMoAu/GaAs, 1.5 W; (right) TiW–Au/GaAs, 2.5 W. (b) Analysis of interdiffused metal–GaAs. (After Anand and Christou, reprinted with permission from *IEEE Elect. Device Conf.* December 1980.)

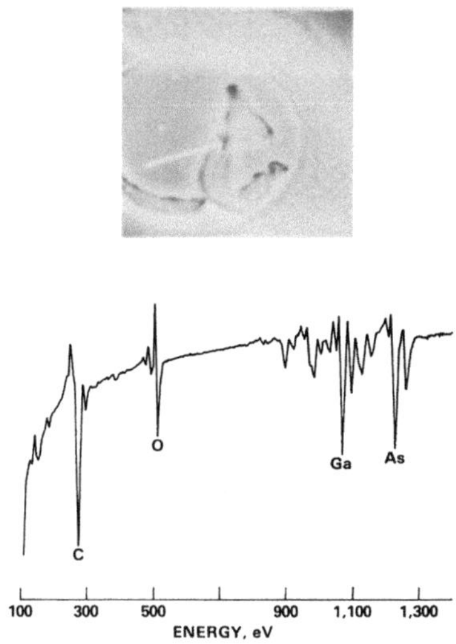

Figure 27. AES analysis of active GaAs area. (After Anand and Christou. Reprinted with permission from *IEEE Elect. Device Conf.* December 1980.)

SEM and microspot Auger electron spectroscopy (AES) analysis of the failed diodes indicates the presence of two different failure modes. Figure 26 shows a cw-burnout mode at 36 GHz indicating an interdiffusion region where the TiMoAu has completely reacted with the substrate. Adjacent to the interdiffusion action is the exposed GaAs active area. AES analysis did not detect traces of the metal barriers in the exposed region, indicating the possibility of substrate damage occurring at the onset of failure followed by rapid interdiffusion at adjacent areas. Another typical failure mode is shown in Fig. 27. The exposed active area indicates that substrate burnout has occurred. Adjacent to this area is the metal–GaAs interdiffusion region where PdGa, TiAs, TiGa, and AuGa compound were detected.

The diodes which failed as a result of the 5-nsec pulses also exhibited two distinct types of contributing failure modes: substrate burnout (edge burnout) and metal–GaAs interdiffusion. The failure analysis investigations are continuing in order to identify the Ga–metal compounds which form for both pulse durations.

5.6. Electrostatic Failure of Silicon Schottky Barrier Diodes

The various barrier silicon Schottky diodes were tested for accidental electrostatic discharge failure.[88] It was found that zero bias, low-barrier, medium Schottky diodes are quite sensitive but high-barrier Pt–Schottky

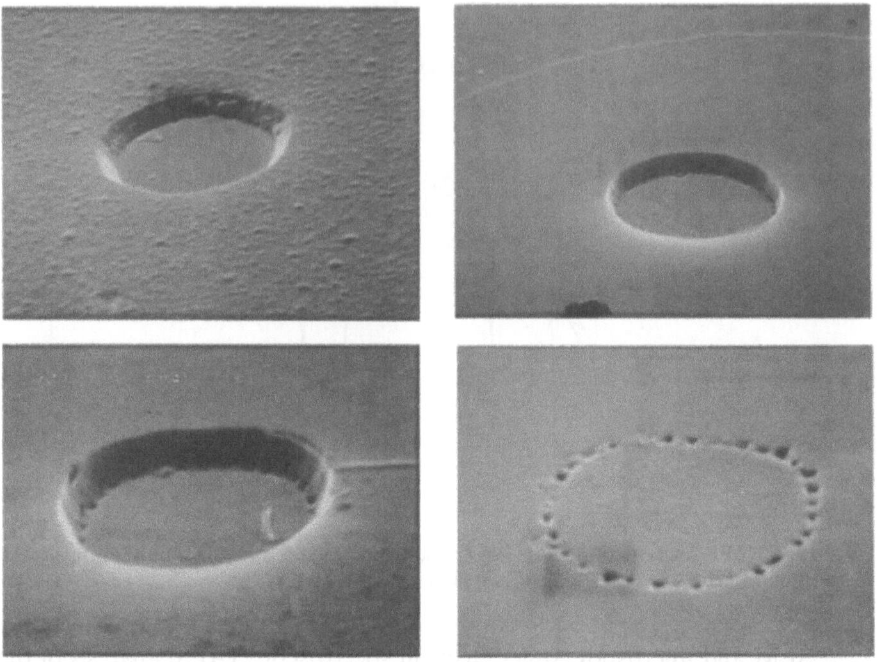

Figure 28. SEM pictures of burned-out Schottky diodes under accidental and simulated electrostatic discharge. (a) Ti–Mo–Au. (b) Pt–Ti–Mo–Au. (c) Pt–Ti–Mo–Au diode failed at 1600 V. (d) Same diode after SiO_2 removed.

was found to be least sensitive to electrostatic discharge. Electrostatically burned out diodes from both natural and circuit simulation techniques exhibited failure points at the periphery of the diode junction similar to that observed for nanosecond rf burnout. Figure 28a and 28b show accidental electrostatic (human body) failures of Ti–Mo–Au and Pt–Ti–Mo–Au Schottky barrier diodes. The Pt–Ti–Mo–Au Schottky diode which failed at 1600V using a static simulator is shown in Figs. 28c and 28d. The experimental results suggest that there is strong correlation between electrostatic and nanosecond rf burnout of various types of Schottky barrier diodes.

6. CONCLUSION

Schottky barrier diodes have made many rapid strides in the last two decades. At low frequencies, silicon Schottky diodes have surpassed point contact diodes in $1/f$ noise, overall noise figure, and rf burnout performance. They are presently being used in wide variety of rugged thermal compression bonded packages and beam lead configurations for waveguide and integrated circuit applications. GaAs Schottky diodes are preferred to silicon in application above X-band frequencies due to higher cutoff frequency. This results from the higher mobility of electrons in GaAs than in silicon. Cryogenically cooled, high-quality millimeter-wave GaAs Schottky mixer diodes exhibit noise performance approaching theoretical limit.

The recent advances in microfabrication techniques and availability of abrupt GaAs epitaxial layers grown on high-quality semi-insulating GaAs material will make low-cost monolithic microwave and millimeter-wave receivers a reality in the near future.

REFERENCES

1. R.J. Archer and M.M. Atalla, *Ann. NY Acad. Sci.* **101**, 697–709 (1963).
2. D. Kahng and M.P. Lepselter, Planar epitaxial silicon Schottky barrier diodes, *Bell Syst. Tech. J.* **44**, 1525–1528 (1965).
3. Y. Anand, X-band high burnout resistance Schottky barrier diodes, *IEEE Trans. Electron Devices* **ED-24**, 1330–1336 (1977).
4. Y. Anand and A. Christou, Millimeter high-burnout GaAs Schottky barrier diodes, International Electron Device Meeting, IEEE, Washington, D.C. (1979).
5. J.C. Irvin and D. Young, Millimeter frequency conversion using Au–n-type GaAs Schottky barrier epitaxy diode with a novel contacting technique, *Proc. IEEE* **53**, 2130–2131 (1965).
6. M.P. Lepselter, Beam-lead technology, *Bell Syst. Tech. J.* **45**, 233–253, (1966).
7. R.D. Dupuis *et al.*, *Inst. Phys. Conf.* **45**, 1–9 (1978).
8. A.Y. Cho and H.C. Casey, Properties of Schottky barriers and p–n junction prepared with GaAs and $Al_xGa_{1-x}As$ molecular beam epitaxy layers, *J. Appl. Phy.* **45**, 1258–1263 (1974).
9. D.N. Held and A.R. Kerr, Conversion loss and noise of microwave and millimeter wave mixers Part I and II, *IEEE Trans. Microwave Theory Tech.* **MIT-26**, 49–61 (1978).
10. A.R. Kerr, Low-noise room-temperature and cryogenic mixers for 80-120GHz, *IEEE Trans. Microwave Theory, Tech.* **MTT-23**, 781–787 (1975).
11. E.R. Carlson, M.V. Schneider, and T.F. McMaster, Subharmonically pumped millimeter wave mixers, *IEEE Trans. Microwave Theory Tech.* **MTT-26**, 706–715 (1978).
12. C. Baron, A simplified theory of crystal mixer, Royal Radar Development Establishment Report 378 (1958).
13. H.C. Torrey and Whitmer, *Crystal Rectifiers*, MIT Radiation Lab. Ser., Vol. 15, McGraw-Hill, New York (1948), pp. 111–173, 293–296.
14. Y. Anand, Characterization of microwave silicon mixer diodes, Ph.D. dissertation, Lehigh University, Bethlehem, Pennsylvania (1970).
15. M.R. Barber, Noise-figure and conversion loss of the Schottky barrier mixer diodes, *IEEE Trans. Microwave Theory Tech.* **MIT-15**, 629–635 (1967).
16. A.A.M. Saleh, Theory of resistive mixers, PH.D. dissertion, Massachusetts Institute of Technology, Cambridge, Massachusetts (1970).
17. S. Egami, Nonlinear analysis and computer aided design of resistive mixer diodes, *IEEE Trans. Microwave Theory Tech.* **MTT-22**, 270–275 (1975).
18. A.R. Kerr, Noise and loss in balanced and subharmonically pumped mixers Part I—Theory, *IEEE Trans. Microwave Theory Tech.* **MMT-27**, 135–140 (1979).
19. A.R. Kerr, Noise and loss in balanced and subharmonically pumped mixers Part II—application, *IEEE Trans. Microwave Theory Tech.* **MTT-27**, 938–950 (1979).

20. P.H. Siegel and A.R. Kerr, NASA Tech. Memo. NASA TM-X-80324, (1979).
21. N.J. Keen and R.A. Linke, Noise characterization for resistive mixers, Diode Mixers at Millimeter Wavelength Workshop, Max Planck Institute, April 26–28 (1977).
22. W. Baechtold, Noise behavior of GaAs field-effect transistors with short gates, *IEEE Trans. Electron Devices* **19**, 674–680 (1972).
23. A. Uhlir, Shot noise in *p–n* junction frequency converters, *Bell Syst. Tech. J.* **37**, 951–988 (1958).
24. A. Van der Ziel, Theory of shot noise in junction diodes and junction transistors, *Proc. IRE* **43**, 1639–1646 (1955).
25. C.T. Sah and F.H. Heilscher, Evidence of the surface origin of the $1/f$ noise, *Phys. Rev. Lett.* **17**, 956–958 (1966).
26. L.I. Schiff, Noise in crystal rectifiers, National Defense Research Committee Report 14–126, University of Pennsylvania (1963).
27. V.J. Weisskoph, On the theory of the noise in conductors, semiconductors and crystal rectifiers, National Defense Research Committee Report 14–133, University of Pennsylvania (1963).
28. O. Jantch, A. theory of $1/f$ noise at semiconductor surfaces, *Solid State Electron.* **11**, 267–272 (1968).
29. O. Jantch, Inversion behavior of silicon rectifier in moist gases, *Z. Naturforsch.* **15**, 141 (1960).
30. M.E. Sprinks, G.T.G. Robinson, and B.E. Bosch, The frequency dependence of noise temperature ratio in microwave crystals, *Brit. J. Appl. Phys.* **8**, 275 (1957).
31. P.H. Miller, Noise spectrum of crystal rectifiers, *Proc. IRE* **35**, 252–256 (1947).
32. H.T. Friis, Noise figures of radio receivers, *Proc. IRE* **32**, 419–422 (1944).
33. A.M. Cowley and H.O. Sorenson, Quantitative comparison of solid state microwave detectors, *IEEE Trans. Microwave Theory Tech.* **MTT-14**, 588–602 (1966).
34. E.R. Beringer, Crystal detectors and the crystal video receivers, MIT Radiation Laboratory Report 638, November 16 (1944).
35. A. Uhlir, Characterization of crystal diodes for low-level microwave detection, *Microwave J.* 59–67 (1963).
36. J.F. Reynolds and M.R. Rosenzweig, Learn the language of mixer specification, *Microwaves* **17**, 72–80 (1978).
37. W. Schottky *Naturwissenschaften* **26**, 843 (1939).
38. M.M. Atalla, Metal semiconductor Schottky barriers, devices and applications, in Proc. *Munich Symp. Microelectronics*, R. Oldenbourg-Verlag, Munich, pp. 123–157 (1966).
39. S.M. Sze, *Physics of Semiconductor Devices*, John Wiley and Sons, New York (1969).
40. H.A. Watson, *Microwave Semiconductor Devices and their Circuit Applications*, McGraw-Hill, New York (1969).
41. T.J. Viola, R.J. Mattauch, United theory of high frequency noise in Schottky barriers, *J. Appl. Phys.* **44**, 2805–2808 (1973).
42. M.V. Schoeider, Metal semiconductor junctions as frequency converters, in *Infrared and Millimeter Waves* (K.J. Bulton Ed.) Vol. 6, pp. 209–275, Academic Press (1982).
43. W.M. Kelley and G.T. Wrixon, Optimization of Schottky barrier diodes for low-noise, low-conversion operation at near millimeter-wave wavelength, in *Infrared and Millimeter Waves*, (K.J. Bulton, ed.), Vol. 3, pp. 77–110, Academic Press, New York (1980).
44. P. Kennedy, Spreading resistance in cylindrical semiconductor devices, *J. Appl. Phys.* **31**, 1490–1497 (1960).
45. L.E. Dickens, Spreading resistance as a function of frequency, *IEEE Trans. Microwave Theory Tech.* **MIT-15**, 101–109 (1967).
46. B.J. Clifton, W.T. Lindly, R.W. Chick, and R.A. Cohen, *Proceedings of Third Biennial Cornell Electrical Engineering Conference*, 1971, pp. 463–475.

47. M.L. Korwin–Pawlowski and E.L. Heasell, *Solid State Electron.* **18**, 849 (1975).
48. W.M. Kelly and G.T. Wrixon, Proc. Int. Conf. Submillimeter Waves Appl. 3rd Guildford (1978).
49. M. McColl and M.F. Millea, *J. Electron Mater.* **5**, 191, 208 (1976).
50. A. Christou, W.T. Anderson, and M.L. Bark, International Electron Device Meeting, IEEE, Washington, D.C., pp. 449–451 (1980)
51. F.A. Kroger, G. Dimer, and H.A. Klasens, Nature of an ohmic contact metal–semiconductor contact, *Phys. Rev.* **103**, 279 (1956).
52. A.Y.C. Yu, Electron tunneling and contact resistance of metal–silicon contact barriers, *Solid State Electon.* **13**, 239–247 (1970).
53. J. Bardeen, Surface states and rectification at a metal–semiconductor contact, *Phys. Rev.* **71**, 717–727 (1947).
54. S.M. Sze *et al.*, Photoelectric determination of the image force dielectric constant for hot electrons in Schottky barriers, *J. Appl. Phys.* **35**, 2534 (1964).
55. B.L. Smith and E.H. Rhoderick, Possible source of error in the deduction of semiconductor impurity concentration from Schottky barrier (*C*, *V*) characteristics, *J. Phys. D.* **2**, 465–467 (1969).
56. M.J. Turner and E.H. Rhoderick, Metal–silicon Schottky barriers, *Solid State Electron.* **II**, 291–300 (1968).
57. C.F. Genzabella and C. Howell, Gallium arsenide Schottky mixer diodes, *Proc. First Int. Symp. Gallium Arsenide* (Reading, England), 1966, Inst. Phys. Soc. Conf. Ser. 3, pp. 131–137.
58. W.J. Moroney and Y. Anand, Low barrier height Gallium arsenide microwave Schottky diodes using gold—germanium alloy, in *Proc. Third Int. Symp. Gallium Arsenide* (Reading, England), 1970, Inst. Phys. Soc. Conf. Ser. 3.
59. N. Braslave, J.B. Gunn, and J.L. Staples, Metal–semiconductor contacts for GaAs bulk effect devices, *Solid-State Electron.* **10**, 381 (1967).
60. Y. Anand, Low barrier height ion implanted GaAs mixer diodes, *Sixth Biennial Cornell University Conference, August* 16–18 (1977).
61. J.M. Shannon, Increasing the effective height of a Schottky barrier using low energy ion implantation, *Appl. Phys. Lett.* **25**(1), 75 (1974).
62. A. Christou, W. Anderson, J.E. Davey, and Y. Anand, A low barrier height Ge–GaAs millimeter wave mixer diode, *Seventh Biennial Cornell University Conference, August* 14–16 (1979).
63. Y. Anand, Low barrier height ion implanted silicon Schottky barrier diode, IEEE Elec. Device Conf., Washington, D.C., December, 1977.
64. Y. Anand and Steve Ellis, Manufacturing technology program for a high-burnout silicon Schottky barrier mixer diodes for Navy Avionics M/A-COM Silicon Products, Inc., Burlington, Massachusetts Final Report, Contract No. N00173-79-C-0107 (1982).
65. Y. Anand, Zero bias Schottky barrier detector diodes, U.S. Patent 3,968,272 (1976).
66. A.R. Kerr and Y. Anand, Schottky diode MM detectors with improved sensitivity and dynamic range, *Microwave J.* 67–71 (1981).
67. Y. Anand, *X*-band image enhancement mixer, *IEEE Solid State Circuits Conference*, Philadelphia, Pennsylvania (1968).
68. D. Steinbrecker, private communication (1982).
69. L.E. Dickens and D.W. Maki, A new "phased-type" image enhanced mixer, *IEEE Trans. Microwave Theory Tech.* **MTT-5**, 149–151 (1975).
70. Y. Utsumi, Analysis of image recovery down converter may be planar circuit mounted in waveguide, *IEEE Trans. Microwave Theory Tech.* Vol. **MTT-30**, 858–868 (1982).
71. J.B. Cahalan, J.E. Degenford, and M. Cohen, An integrated *X*-band, image and sum frequency enhanced mixer with 1 GHz IF, *IEEE MTTS Digest* 16–17 (1971).

72. L.T. Yuan, Low noise octave bandwidth mixer, *IEEE Conf. Digest MTTS*, San Diego, California, pp. 698–704 (1977).
73. R.A. Linke, M.V. Schneider, and A.Y. Cho, Cryogenic millimeter-wave receiver using molecular beam epitaxy diodes, *IEEE Trans. Microwave Theory Tech.* **MTT-26**, 935–938 (1978).
74. N.J. Keen, W.M. Kelly, and G.T. Wrixon, Pumped Schottky diodes with noise temperature of less than 100 K at 115 GHz, *Electron Lett.* **15**, 689–690 (1979).
75. E.R. Carlson and M.V. Schneider, Subharmonically pumped millimeter wave receivers, *Int. Conf. Infrared Millimeter Wave their Appl. Tech.* Dig. 4th, pp. 82–83 (1979).
76. S. Weinreb, Unpublished report (1981).
77. A.R. Kerr, R.J. Mattauch, and J.A. Grange, A new mixer design for 140–220 GHz, *IEEE Trans. Microwave Theory Tech.* **MTT-25**, 399–401 (1977).
78. S.T. Eng., Low noise properties of microwave backward diodes, *IRE Trans. Microwave Theory Tech.* **MTT-9**, 419–425 (1961).
79. Y. Anand, Low frequency noise in Schottky barrier diodes, *Proc. IEEE* (*Lett.*) **57**, 855–856 (1969).
80. Y. Anand, unpublished report (1982).
81. E. Kollberg and H. Zirath, On the optimization of cryogenic mixers, 12th European Microwave Conference Helsenki, Finland (1982).
82. J.A. Calviello and J.E. Wallace, Performance and reliability of an improved high-temperature GaAs Schottky junction and native oxide passivation, *IEEE Trans. Electron Devices* **ED-24**, 698–704 (1977).
83. Y. Anand and W.J. Moroney, Microwave mixer and detector diodes, *Proc. IEEE* **59**, 1182–1190 (1970).
84. T.F. McMaster *et al.*, Subharmonically pumped millimeter-wave mixer built with notch front and beam lead diodes, *IEEE Conf. Digest* **MTT-5**, San Diego, California, pp. 389–391 (1977).
85. G.T. Wrixon, Low-noise diodes and mixers 1–2 mm wavelength region, *IEEE Trans. Microwave Theory Tech.* **MTT-22**, 1159–1165 (1974).
86. R.J. Baur, M. Cohen, J.M. Cotton, and R.F. Packard, Millimeter wave semiconductor diode detectors, mixers and frequency multipliers, *Proc. IEEE* **54**, 595–605 (1966).
87. K. Tomiyasu, On spurious outputs from high-power pulsed microwave tubes and their control, *IRE Trans. Microwave Theory Tech.* **MTT-9**, 480–484 (1961).
88. Y. Anand, G. Morris, and V. Higgins, Electrostatic failure of *X*-band silicon Schottky barrier diodes conference, September 1979, Denver, Colorado, IEEE EOS/ESD.
89. G.E. Morris, Y. Anand, V. Higgins, C. Cook, and G. Hall, RF burnout of mixer diodes as induced under controlled laboratory conditions and correlation to simulated system performance, *MTT Symp. Rec.*, Palo Alto, pp. 182–183 (1975).
90. Y. Anand, *X*-band high-burnout silicon Schottky barrier diodes, *Microwave J.* 55–61 (1979).
91. M.P. Lepselter and S.M. Sze, Silicon Schottky barrier diodes with near ideal *I–V* characteristics, *Bell Syst. Tech. J.* **47**, 195–208 (1968).
92. H.M. Day and A.C. McPherson, Design and fabrication of high-burnout Schottky crystal video diodes, *Solid State Electron.* **15**, 409–416 (1972).
93. Y. Anand, RF burnout dependence on variation in barrier capacitance of mixer diodes, *Proc. IEEE* (*Lett.*) **61**, 247–248 (1973).
94. W.J. Moroney and Y. Anand, Reliability of microwave mixer diodes, presented at Reliability Symposium, Las Vegas, April 1972.
95. G.E. Morris, G.A. Hall, C.F. Cook, and V.J. Higgins, Investigation of RF induced burnout in microwave mixer diodes: A continuing study, presented at MTT Int. Symp., Atlanta, Georgia, June 12–14 (1974).

96. Y. Anand, High-burnout mixer diodes, presented at *IEEE Elec. Device Conf.*, Washington, D.C., December (1974).
97. W.H. Weisenberger, A. Christou, and Y. Anand, High spatial resolution scanning auger spectroscopy applied to analysis of *X*-band diode, *J. Vac. Sci. Technol.* **12**(6), 1365–1368 (1975).
98. Y. Anand and A. Christou, Millimeter high-burnout GaAs, Schottky Barrier Diodes, IEDM, Washington, D.C. 1979.
99. A. Christou and Y. Anand, GaAs mixer diode burnout mechanisms at 36–94 GHz, 1980 International Reliability Physics Symposium, April 8–10, Las Vegas, Nevada (1980).
100. Y. Anand, A. Christou, S. Ellis, L. Mang, and D. Bensen, 95 GHz High burnout GaAs Schottky barrier diodes, GOMAC Conference November 2, 1982, Orlando, Florida (1982).
101. Y. Anand and C. Howell, A burnout criterion for Schottky barrier mixer diode, *Proc. IEEE (Lett.)* **56**, 2098 (1968).
102. Y. Anand and C. Howell, The red real culprit in diode failure, *Microwave* **August** 1–3 (1970).
103. D. Rees, Wright–Patterson AFB Avionics Lab., private communications.
104. H. Cong, A.R. Kerr, and R.J. Mattauch, The low-noise 115 GHz receiver on the Columbia-Giss 4 ft. radio telescope, *IEEE Trans. Microwave Theory Tech.* **27**, 245–248 (1979).
105. F.J. Solmon, C.D. Berglund, R.W. Chick, and B.J. Clifton, Ka-band communication system of the Lincoln experimental satellites Les-8 and Les-9, *J. Space Craft Rockets* **16**, 181–186 (1979).

7

Metal–Semiconductor Field Effect Transistors

James A. Turner

1. INTRODUCTION

It is now over 15 years ago that the first gallium arsenide field effect transistors were fabricated. Since then major programs worldwide have enabled the device to be developed into a commercial production item opening up many new applications areas both in low-noise receivers and in transmitting circuitry.

The intense interest in field effect transistors in GaAs is due to two main reasons:

1. In gallium arsenide the electron mobility is some six times, and the peak drift velocity approximately twice, that of silicon (Fig. 1).
2. GaAs FETs are fabricated in layers grown on a semi-insulating substrate with a resistivity of approximately $10^7\,\Omega$cm.

These properties combine to give the GaAs FET a marked advantage in operating frequency over a comparable geometry device in silicon.

Early workers, Turner *et al.*,[1] fabricated the devices using a diffused gate technology. However, due to the anomalous behavior of *p*-type diffusants in GaAs, particularly the enhanced diffusion at the edge of the masking silicon dioxide film (Fig. 2), gate lengths were limited to a minimum of a few micrometers. The advent of the Schottky barrier gate FET process overcame these problems and is now universally used to produce all but the most specialized FET structures.

James A. Turner • Plessey Research (Caswell) Ltd., Allen Clark Research Centre, Caswell, Towcester, Northants NN12 8EQ, England.

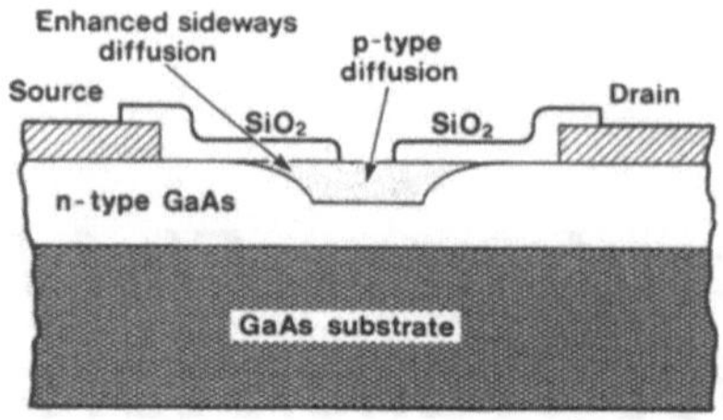

Figure 1. Diffused gate FET showing how enhanced sideways diffusion increases the effective gate length.

2. SMALL-SIGNAL FET THEORY

As gate lengths have become progressively shorter the original FET theory of William Shockley[2] has been modified to take into consideration the situation that carriers will reach their limiting velocity in the channel region of the device leading to the fact that the switching time of the FET will become dependent on the saturation velocity.

Small-Signal Equivalent Circuit of a Short Gate Length FET

It has been shown by Grove[3] that the channel current of a field effect transistor can be expressed as

$$I_{\text{ch}} = g_0 \left[V_i - \frac{2}{3} \frac{(V_i + V_{Bi} - V_g)^{3/2} - (V_{Bi} - V_g)^{3/2}}{V_p^{1/2}} \right] \tag{1}$$

where V_i is the voltage drop across the gate region, V_{Bi} is the barrier height voltage, and, V_g is the gate voltage;

$$g_0 = \frac{q\mu N_D Za}{L} \tag{2}$$

$$V_p = \frac{qN_D a}{2\epsilon_0 \epsilon} \tag{3}$$

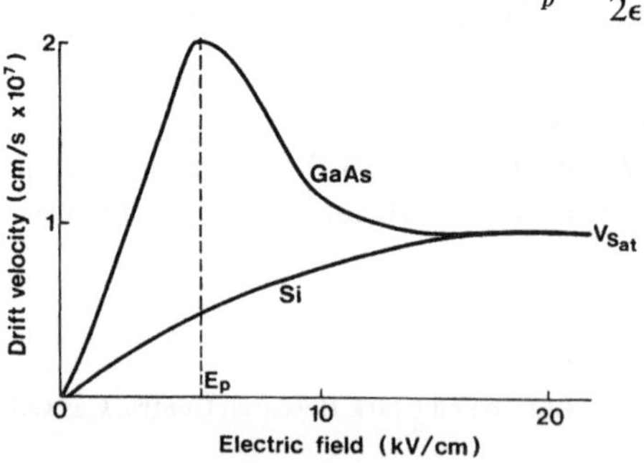

Figure 2. Velocity-field characteristics of GaAs and Si.

where V_p is the voltage to reduce the current through the device to essentially zero, $\epsilon_0\epsilon$ is the permittivity of the channel material, a is the channel thickness, L is the gate length, and Z is the gate width.

Equation (1) is valid when

$$V_i = V_s, \qquad \text{where } V_s = E_s L$$

For a short gate length FET where $L \simeq 1\ \mu\text{m}$

$$V_s \ll (V_{Bi} - V_g)$$

Thus

$$I_{\text{ch}} \simeq g_0\left(1 - \frac{a_0}{a}\right)V_i = g_d V_i \tag{4}$$

where

$$a_0 = \left[\frac{2\epsilon\epsilon_0(V_{Bi} - V_g)}{qN_d}\right]^{1/2} \tag{5}$$

$$= a\left[\frac{V_{Bi} - V_g}{V_p}\right]^{1/2} \tag{6}$$

and $g_d \simeq g_0(1 - a_0/a)$ is the drain conductance. The saturation current I_{sat} is equal to

$$I_{\text{sat}} = g_d V_s \tag{7}$$

The small signal transconductance of the device at the drain voltage at which current saturation occurs is given by

$$g_m \frac{\partial I_{\text{ch}}}{\partial V_g}\bigg|_{V_i = V_s} = g_0\left[\frac{(V_s + V_{Bi} - V_g)^{1/2} - (V_{Bi} - V_g)^{1/2}}{V_p^{1/2}}\right] \tag{8}$$

which for a short gate device can be simplified to

$$g_m = \left[\frac{qN_d\epsilon_0\epsilon}{2(V_{Bi} - V_g)}\right]^{1/2} V_s Z \tag{9}$$

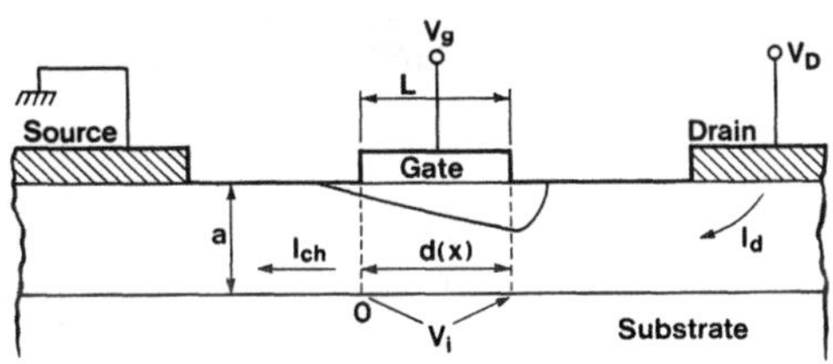

Figure 3. A cross section through the channel of a FET.

These expressions can be used to accurately predict the I_{sat} and g_m values of GaAs FETs with gate lengths around 1 μm.

Referring to Fig. 3, the total charge Q under the gate in the region where the voltage drop across the gate is small compared to V_s is given by

$$Q = qN_d Z \int_0^{Z_g} d(x)\, dx \tag{10}$$

where as $V_s = (V_{Bi} - V_g)$ simplifies to

$$Q = qN_g Z a_0 L \tag{11}$$

We can now derive expressions for the drain-to-gate and gate-to-source capacitances C_{dg} and C_{gs} in the equivalent circuit given in Fig. 4:

$$\begin{aligned} C_{\text{dg}} &= \left(\frac{\partial Q}{\partial V_i}\right)_{V_g = \text{const}} \\ &= \frac{2\sqrt{2}}{3} \frac{ZL(\epsilon_0 \epsilon q N_d)^{1/2}}{V_i^2} \times [\tfrac{3}{2} V_i (V_i + V_{Bi} - V_g)^{1/2} \\ &\quad - (V_i + V_{Bi} - V_g)^{3/2} + (V_{Bi} - V_g)^{3/2}] \end{aligned} \tag{12}$$

and

$$\begin{aligned} C_{\text{gs}} &= \left(\frac{\partial Q}{\partial V_g}\right)_{V_i - V_g = \text{const}} \\ &= \frac{2\sqrt{2}}{3} \frac{ZL(\epsilon_0 \epsilon q N_d)^{1/2}}{V_i^2} \times [(V_i + V_{Bi} - V_g)^{3/2} \\ &\quad - (V_{Bi} - V_g)^{3/2} - \tfrac{3}{2}(V_{Bi} - V_g)^{1/2} V_i] \end{aligned} \tag{13}$$

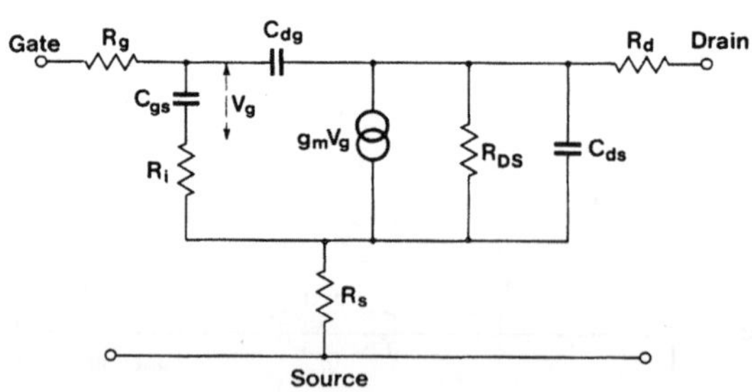

Figure 4. An equivalent circuit of a small signal FET.

For small values of V_i

$$C_{ds} = C_{gs} = \frac{1}{2}\frac{\epsilon_0 \epsilon ZL}{a_0} \tag{14}$$

From this expression can be derived a cutoff frequency

$$f_T \simeq \frac{1}{2\pi}\frac{g_m}{C_{gs}} \tag{15}$$

which for a FET with a gate length of 1 μm gives an f_T of $\sim$ 25 GHz.

The characteristic switching time of the FET is given by

$$\tau = \frac{Q(V_s)}{I_{sat}}$$

Using (11) and (4) we obtain

$$\tau = \frac{L}{V_s} \cdot \frac{a_0}{a - a_0} \tag{16}$$

This shows that the switching time is proportional to the transit time under the gate electrode of the FET which is determined by the saturation velocity of the carriers in the channel region of the device.

In this section, expressions for the vital elements of the small signal equivalent circuit have been given. These can be related to the physical structure of the FET and therefore enable the device designer to optimize gain and noise performance.

3. DESIGN PARAMETERS OF A LOW-NOISE DEVICE

Shown in Fig. 5 are the equivalent circuit parameters of the device as they are physically related to the device structure. The intrinsic elements of the device are represented by $C_{dg} + C_{gs}$, the total gate-to-channel capacitance, R_i the open channel resistance, and R_{DS} the output resistance. The transconductance g_m relates the current i_{ds} to the voltage across C_{gs}. Parasitic elements R_s, the source series resistance, R_d the drain series resistance, R_g the gate metallization resistance, and C_{ds} the source-to-drain substrate resistance all serve to reduce the effective microwave performance of the device. It is the object of the device technologist to reduce these values to an absolute

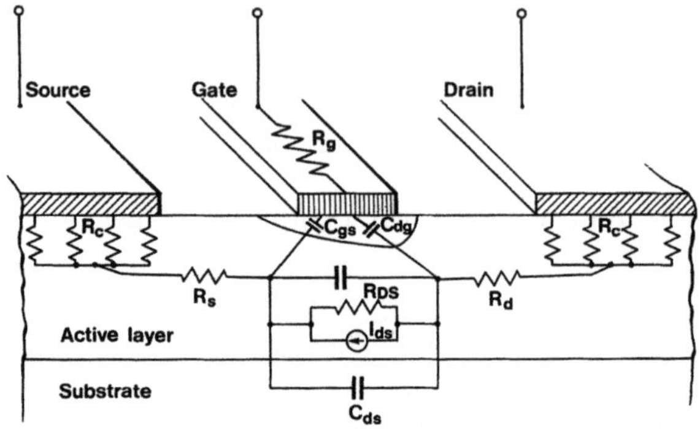

Figure 5. Physical origins of the circuit elements of the FET.

minimum. R_s and R_d can be reduced by placing the ohmic source and drain contacts as close to the ends of the channel as possible and by choosing a contact metal having a low specific contact resistance to the semiconductor. R_g can be reduced by the use of a gate metal of low electrical conductivity and by depositing as thick a metal layer as possible. C_{ds} depends on the quality of the semi-insulating GaAs substrate material.

The equivalent circuit of Fig. 4 together with the model of Fig. 5 can be used to derive expressions for the gain and noise figure of a FET in terms of the physical parameters of the structure.

Pengelly[4] has shown that the maximum available gain of a FET at a frequency f is given by

$$\text{MAG} = \left(\frac{f_T}{f}\right)^2 \left[\frac{4}{R_{\text{DS}}}\left(R_i + R_s + 2R_g + \frac{\omega_t L_s}{2}\right) + 2\omega_t C_{\text{dg}}(R_i + R_s + 2R_g + \omega_t L_s)\right]^{-1} \tag{17}$$

where L_s is the inductance in the common source lead of the transistor. From this expression it is clear that f_T must be maximized by fabricating a FET with a gate length as short as possible [equation (15)] and by minimizing R_g, R_s, C_{dg}, and the source lead inductance.

Pucel[5] and Fukui[6] have developed theories for the noise generated in the FET and have derived an expression relating the noise performance of the FET to its intrinsic and extrinsic elements. Figure 6 gives a more practical representation of the device showing the now commonly used recessed channel structure. The following expression for noise figure of the FET

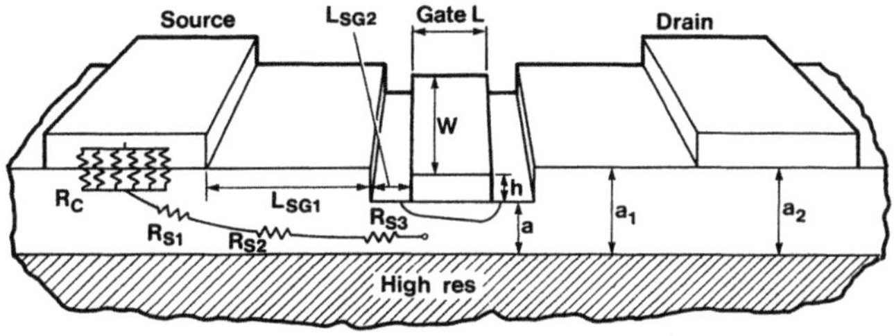

Figure 6. Schematic diagram of the FET showing the elements used in the noise theory.

incorporates all the elements shown in Fig. 6:

$$F = 1 + kfL^{5/6}\left[\left(\frac{N}{a}\right)^{1/6}\right]\left[\left(\frac{3.3W^2\rho}{hL} + 0.6W^2\frac{\rho f}{hL}\right.\right.$$
$$\left.\left.+\frac{1.8}{N}\left(\frac{L_{sg1}}{a_0} + \frac{L_{sg2}}{a}\right) + \frac{0.18R_c}{Na_2}\right]^{1/2} \tag{18}$$

where F is the noise factor, K is the material quality factor (0.333), f is the frequency in GHz, N is the carrier density of the channel in $10^{16}\,\text{cm}^{-3}$, W is the unit gate width in millimeters, h is the metalization thickness in micrometers, L is the gate length in micrometers, ρ is the gate metal resistivity in $10^6\Omega$cm, R_c is the specific contact resistance in $10^6\Omega\text{cm}^2$, and a's and L's are the dimensions in micrometers of the elements shown in Fig. 6.

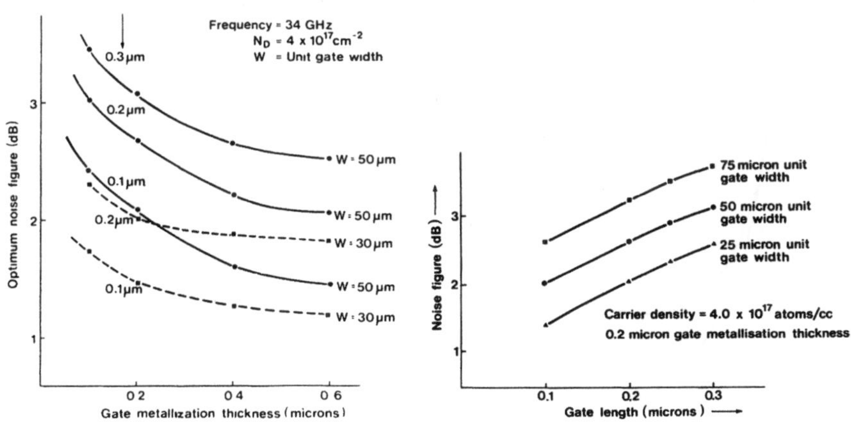

Figure 7. (a) Theoretical variation of optimum noise figure of an FET with gate metallization thickness. (b) Theoretical variation of optimum noise figure with gate length and unit gate width (at 35 GHz).

Terms 1 and 2 in Eq. (18) represent the noise contribution from gate metallization loss and depend on the gate length, the gate metallization thickness, the unit gate width, and the gate metal conductivity, Figure 7a–7c show the effect of gate dimensions on noise performance at 34 GHz. It can be seen how quite large improvements to noise performance can be achieved by increasing the gate metal thickness to 0.4 μm. A reduction in unit gate width has a similar effect.

The noise figure expression of Eq. (18) gives no indication of the source impedance of the device required to give the minimum noise figure. The device designer has to reach a compromise between the minimum noise figure he can obtain and the source impedance required to be presented to the device to achieve this noise figure. It is pointless for instance if the lowest noise figure occurs at a very high input impedance as the device will be difficult to match over even the narrowest of frequency ranges. The width of the gate of the transistor gives the device designer a certain degree of flexibility in this regard.

Term 3 in Eq. (18) relates the noise performance to the parasitic resistance between the source and gate electrodes (resistances R_{s2} and R_{s3} in Fig. 6). Figure 8 shows how these resistances vary with typical starting material thickness. It is clear that the undercut resistance R_{s3} is an important resistance in determining the final noise figure of the device. R_{s2} and R_{s3} can be minimized by choosing a suitable carrier density and thickness for the initial starting material.

The final term of Eq. (18) relates to the contact and spreading resistance Figure 9 shows how the contact resistance affects the noise figure. In modern low-noise devices a heavily doped (n^+) contact layer is grown on top of the active channel layer to assist in producing a metal contact with a low specific contact resistance.

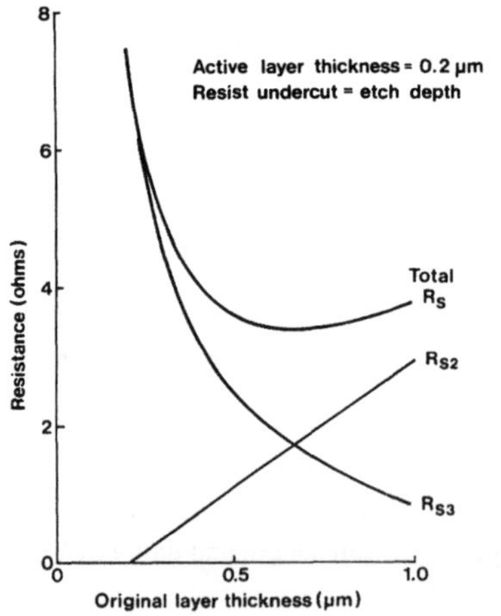

Figure 8. The variation of the elements of source resistance as a function of layer thickness.

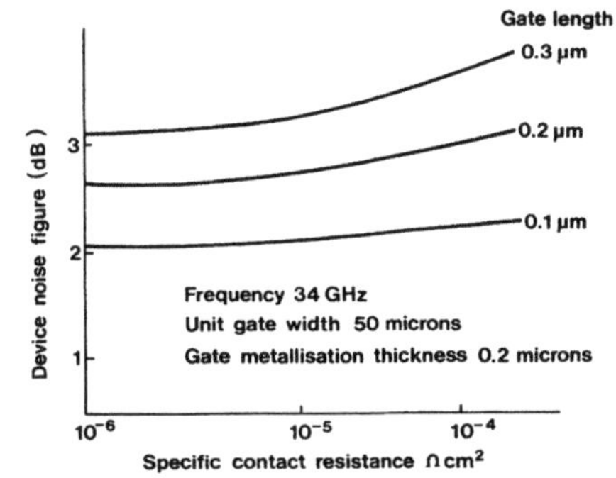

Figure 9. The effect on FET noise figure with changes in specific contact resistance.

4. PRACTICAL SMALL-SIGNAL FET FABRICATION TECHNIQUES

In previous sections (Sections 2 and 3), the physical factors affecting microwave performance of the FET have been highlighted. In this section fabrication technologies that can be used to achieve the chosen design are discussed.

4.1. Material Growth Techniques

Four major techniques for the production of material suitable for low noise FETs are currently being researched, these are

a. Vapor phase epitaxy,
b. Molecular beam epitaxy,
c. Organometallic epitaxy,
d. Ion implantation.

All four techniques have the ability to produce material with electrical properties such that state-of-the-art-noise performance transistors can be produced. The two techniques that are likely to become the predominant ones are (c) and (d) as they can be readily adapted for large-scale material production, and very large quantities of material will be required as GaAs monolithic circuits containing FETs become widely accepted. One major advantage of method (d), ion implantation, is the extremely high uniformity of material parameters that can be maintained across a whole slice of GaAs. Pinch-off voltages of 1.18 V with a standard deviation of only 60 mV can be achieved. This indicates a variation of less than 40 Å in the implantation depth at a particular electron concentration over the entire wafer surface.

4.2. FET Fabrication Technology

The defining of the gate electrode represents the most difficult fabrication task for the device technologist. Two main techniques are currently employed to perform this operation—a self-aligned technology and a recessed gate technology.

Figure 10 shows the critical steps taken during the processing of a self-aligned gate device. Firstly mesas are formed to isolate the active device areas, then the gate metal is deposited over the entire wafer. Photoresist areas defining the position of the gates are opened up over the wafer (these will eventually form the source and drain contact areas of the device). The gate metal is etched away so as to undercut this resist layer (Fig. 10b) until a thin strip of gate metal is left having a width equal to the finally required gate length. With the resist left in place (Fig. 10c) the ohmic contact metal is deposited. The whole wafer is immersed in a suitable solvent until the excess ohmic metal is removed from above the gate area. The gate metal is left positioned accurately between source and drain areas (Fig. 10d). This technique requires that the layer thickness of the material used is grown precisely to a value that allows devices of the required dc parameters to be obtained; no "tailoring" of these parameters can be made during device fabrication.

A technique that enables this disadvantage to be overcome is the recessed channel technology. Figure 11 shows the stages in the fabrication of this type of structure. The ohmic contacts and mesa isolation processes are first carried out. Photoresist is then spun over the whole of the slice and areas developed out where the gate electrode is to be defined. A suitable etch is then used to etch away a small amount of GaAs in the gate region such that the dc current flowing from a source contact to an adjacent drain contact reaches a predetermined value. The gate metal is then deposited with the resist still in

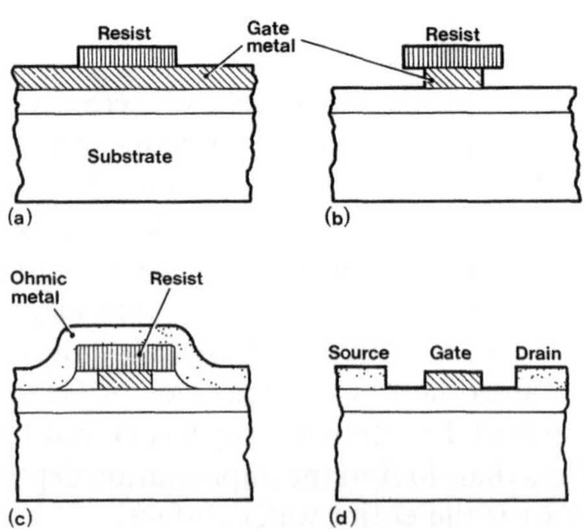

Figure 10. The process stages in a self-aligned gate fabrication technology.

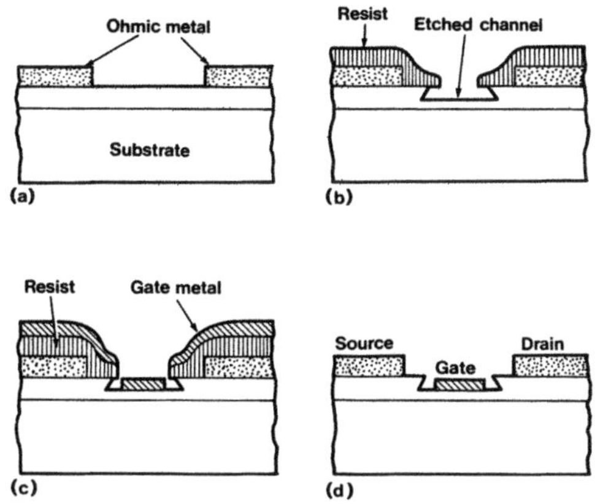

Figure 11. The process stages in the recessed channel fabrication technology.

place. Excess gate metal is floated off by immersion in a suitable solvent. Figure 12 shows a scanning electron microscope photograph of the channel region of a 0.3-μm FET.

The operating frequencies of the GaAs FET are continually being pushed up by improved fabrication technologies. Figure 13 shows an FET produced entirely by electron beam lithography techniques. At gate lengths of around 0.5 μm normal photolithographic (ultraviolet processes) become impracticable and yields rapidly deteriorate. The use of electron beam lithography[7] overcomes this problem and devices with gate lengths of 0.3 μm and less can be produced with high working yields. The gate length of the device in Fig. 13 is 0.3 μm. Its microwave performance is shown in Fig. 14.

Figure 12. A scanning electron microscope picture of the gate region of a 0.3 μm gate device.

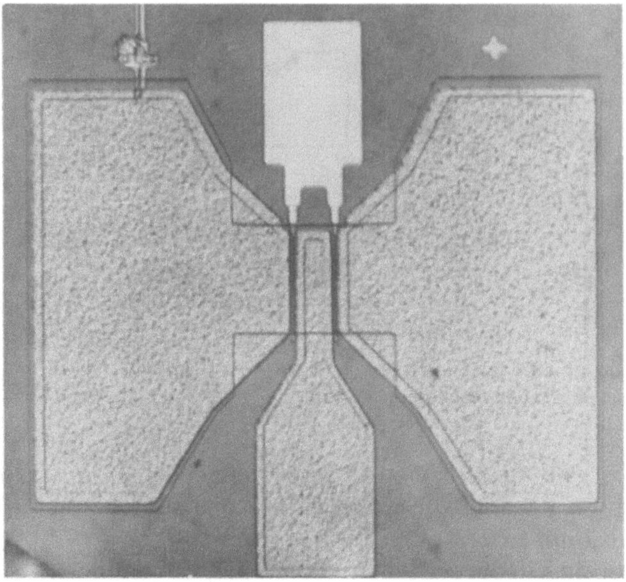

Figure 13. A photograph of a 0.3 μm gate length FET.

5. GaAs Power Field Effect Transistors

Up to this point all the considerations have been towards the small signal device. However, the FET is now also emerging as potentially a very good device for power generation. In the past few years powers in excess of 10 W at X-band frequencies and 25 W at 3 GHz have been obtained by combining within a transistor package the power from two or more chips.

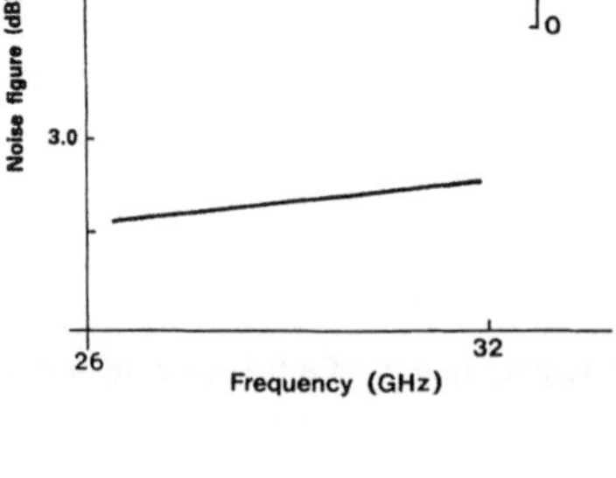

Figure 14. The noise figure and associated gain of a 0.3 μm gate length FET as a function of frequency.

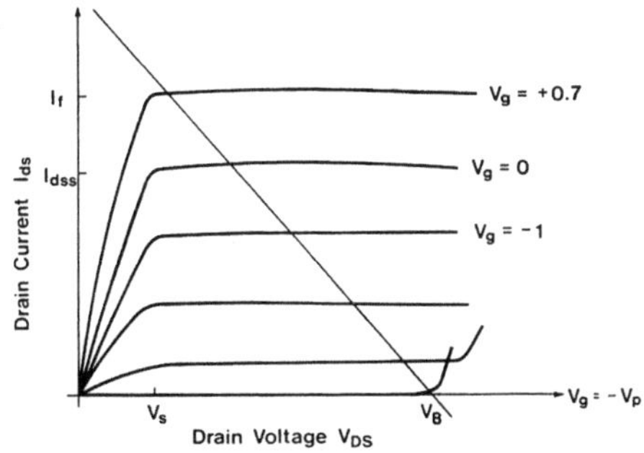

Figure 15. The output characteristics of a power FET.

5.1. Principle of Power FET Operation

Figure 15 shows the drain characteristics of a power FET from which the maximum output power obtainable under class A conditions can be determined. The maximum output voltage and current swings give an output power for a sinusoidal waveform of

$$P_{\text{out}} = \frac{I_f(V_B - V_s)}{8} \tag{19}$$

where I_f is the value of I_{ds} corresponding to a forward gate bias of about 0.7 V for a typical Schottky barrier diode to GaAs ($I_f \simeq I_{dss} + 0.7g_m$), V_B is the source drain voltage at which breakdown occurs in the (just) pinched-off region of the characteristic, and V_s is the saturation or "knee" voltage of about 2 V for the GaAs FET.

The optimum ac load resistance is

$$R_L = \frac{V_B - V_s}{I_f} \tag{20}$$

Large values of I_f are obtained for power devices by paralleling up a number of FET cells to give a large total gate width. Since R_L decreases with I_f the limitation to the total gate width is determined by the minimum load resistance which can be presented to the device. I_f per unit gate width is usually determined by the value for optimum gain. The breakdown voltage V_B is determined principally by avalanche breakdown of the reverse biased gate to drain diode giving the relationship

$$V_B = V_A - V_P \tag{21}$$

where V_A is the avalanche breakdown voltage for the semiconductor Schottky

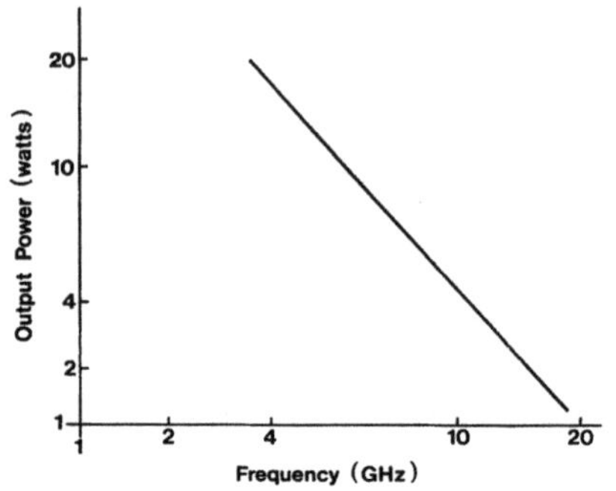

Figure 16. The power–frequency performance for state-of-the-art power FETs in Japan, the United States, and Europe.

barrier junction which increases with decreasing carrier density. In choosing an optimum carrier density a compromise must be reached with gain since the device transconductance decreases with decreasing carrier density. An additional breakdown mechanism which is important at high current levels is current crowding at the drain contacts[8] which limits the bias capability of the device and may override the limitation due to V_B. The use of n^+ ohmic contact regions are believed to overcome this effect[8,9] as well as reduce the parasitic contact resistance.

Figure 16 shows the output power versus frequency performance of modern day power FETs. The line represents the present limit in output power from a number of laboratories from Japan, the US, and Europe. Current programes are aimed towards 50 W at 4 GHz to 2–4 W at 20 GHz.

5.2. Thermal Impedance

The rise in temperature of the channel region of an FET above ambient due to dissipated dc power is given by the expression

$$\Delta T = R_{\mathrm{Th}} P\left(1 - \frac{1}{G}\right)\left(\frac{100}{\eta} - 1\right) \qquad (22)$$

where R_{Th} is the thermal impedance of the device, P is the output power, G the power gain, and η the power added efficiency (the ability of the device to convert dc power to rf power). If ΔT is to be minimized, which is obviously necessary for high reliability considerations, then R_{Th} must also be kept to a minimum. This can be achieved by taking extreme care to ensure that the heat is properly removed from the device chip (Section 5.3 describes practical schemes to do this). An example of the effect of gate-to-gate spacing on thermal impedance is given in Fig. 17. It can be seen that the thermal impedance is a strong function of chip thickness and for thermal impedances of less than

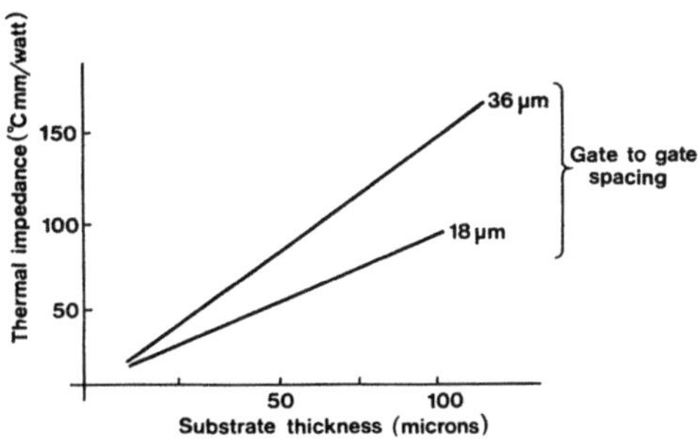

Figure 17. The effect on thermal impedance of substrate thickness and gate-to-gate spacing of a power FET.

100°C/W the chip thickness is such that the chip has to be supported on an integral heat sink to enable it to be handled without breaking. Figure 17 also shows how the thermal impedance can be reduced by modifying the layout of the metallized areas of the FET design. By reducing the gate-to-gate spacing a factor of 2 reduction in thermal impedance is seen to be possible for a chip thickness of 50 μm.

5.3. Power FET Technology

The microwave power performance of the FET is dominated by four factors

1. Gate width,
2. Minimization of source inductance,
3. Reduction of thermal impedance,
4. The ability of structure to sustain high source-drain voltages.

The output power of a FET is directly proportional to its gate width. Powers per millimeter of gate width being obtained currently are between 0.5 and 1 W. An interconnection scheme to allow up to 10 mm of gate width to be used is therefore required if powers in excess of 5 W are to be obtained. Four approaches to this problem are currently being examined in detail with each manufacturer having its own favorite technique. The techniques being pursued are

1. Wire bonding,
2. Air bridging/dielectric crossovers,
3. Via hole connections,
4. Flip chip bonding.

An examination of a typical power FET structure, Fig. 18, will show the extent of the problem. For satisfactory microwave operation the source pads

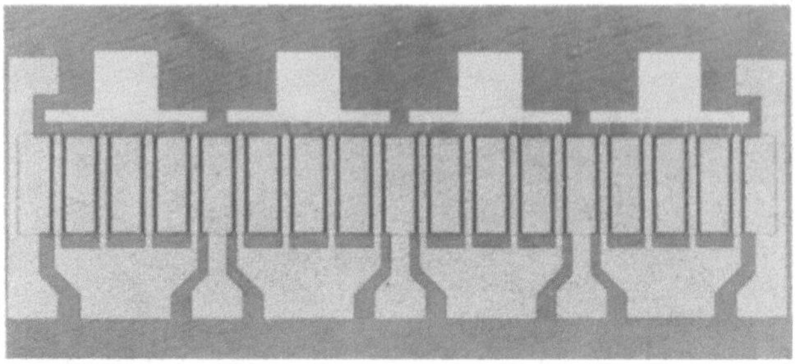

Figure 18. A photograph of a typical power FET structure.

on this design must be interconnected and then attached to the ground of the package through low-inductance connections.

The method used by most workers during the early days was to wire bond to the source pads (Fig. 19). This proved successful electrically to 8 GHz or so and for relatively small devices (power output < 1 W), but at higher frequencies the inductance of the wire bonds degraded the microwave performance and at higher powers the size of the device (the bonding pads have to be sufficiently large to facilitate the bonding process) becomes prohibitive. So alternative techniques for contacting to the source pads are being developed.

The first of these is flip chip bonding—here the source pads are plated up so that they can be ultrasonically attached to a heat sink and rf grounding plate. Figure 20 shows this configuration. This technique was developed by I. Drukier[10] and has been used successfully for output powers of 5 W at X-band

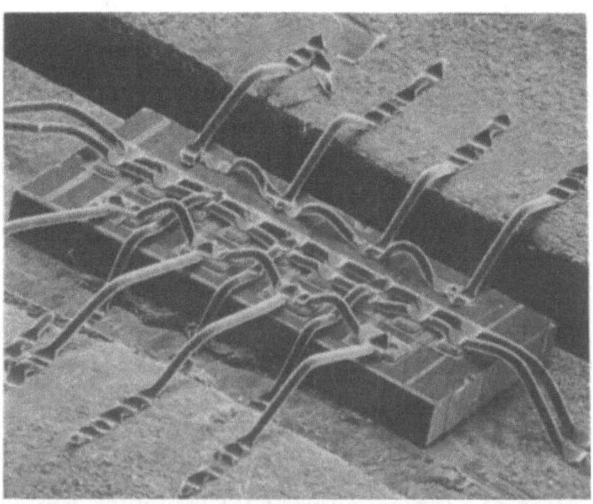

Figure 19. A bonded power FET showing the complex nature of the wire bonding operation.

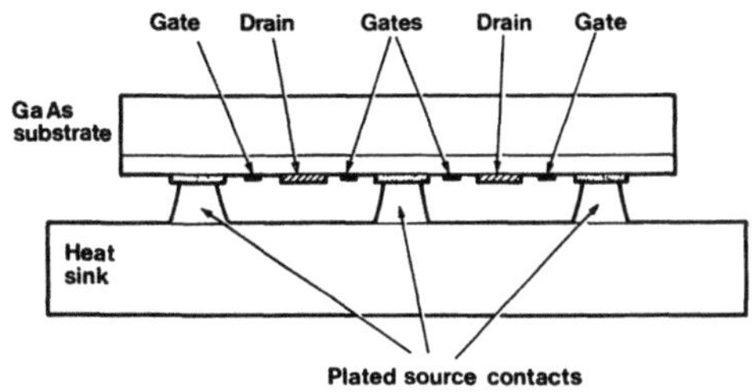

Figure 20. Flip chip mounting technology showing the plated up source contact connecting to the heat sink.

frequencies. It has not been universally adopted, however, due to two main reasons: (1) It is difficult to attach a very large device chip in this way and ensure that all source pads are connecting the heat sink. (2) As it is a flip-chip process the active device face is not visible after mounting and precap visual inspection becomes impossible.

It is a particularly good mounting technique from a thermal standpoint as the heat is being extracted at the site of generation—the surface of the chip.

A second popular interconnection scheme is to use dielectric crossovers. In this scheme connection between source pads is made by metal bridges over a suitable low-loss dielectric layer. Silicon dioxide, silicon nitride, or air is used for this dielectric. The photograph in Fig. 21 shows an air bridge technology. This technique has many advantages over the wire bond interconnection approach: (1) The bridges are formed whilst the devices are in slice form by a photolithographic and gold plating process and so it is therefore potentially a "production" method. (2) The broad metal stripes form low-inductance

Figure 21. A photograph showing an air bridge crossing the channel region of a power FET.

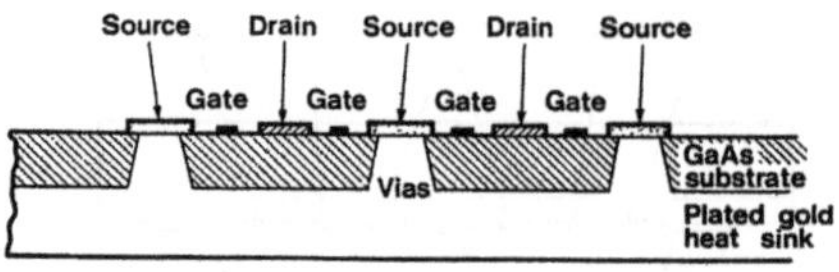

Figure 22. Via hole plating technique showing the via metal contacting the back face of the source metallization.

connections to the device which enable high-frequency performance to be obtained.

The via hole connection process possibly holds the most promise in terms of ultimate microwave performance and thermal efficiency. A schematic of this technique is shown in Fig. 22. Holes are etched from the back of the wafer to contact to the source pad metallization. A plating process allows electrical contact to be made to this pad area so that mounting the device face upwards on to a heat sink–ground plate automatically contacts all the sources to the plate. This technique was developed first by D'Azaro[11] and is becoming more fully utilized not only for power devices but also as a means of grounding specific contact areas in monolithic GaAs integrated circuits.

A combination of these technologies may well be the one finally chosen by power FET manufacturers to obtain the power–frequency performance potential of the device. However, one overriding problem remains—that of device reliability. Power devices run at channel temperatures of typically 150–200°C, these high temperatures put particular stress on the metallizations currently used, and MTTFs of 10^4 at 100°C are being reported. The failure mode of the devices is either electron migration of the gate electrode or a degradation of the channel properties. Refractory metals are being examined to prevent the movement of metal in the gate electrode which at present forms voids and prevents the voltage on the gate modulating the channel resistance. The degradation of the conductivity of the channel is a longer-term problem that requires a greater understanding of the material growth technologies.

6. CONCLUSIONS

This chapter has dealt with the operation and fabrication of small signal and power GaAs MESFETs. It is clear that the full potential of the device has not yet been reached and we can expect further advances in high-frequency performance of both device types in the next few years. One exciting area that has not been covered in this chapter is the use of these devices in monolithic GaAs integrated circuits. Workers throughout the world now are participating in major programs both in linear and digital circuitry exploiting this new area of application. The successful exploitation of this technology will lead to large volume circuit requirements for phased array radar systems, miniature gain blocks, fast computing elements, and for consumer products such as microwave receivers for the reception of TV signals from satellites.

REFERENCES

1. J.A. Turner, 1967 *Symp. GaAs and Related Compounds*, Inst. Phys. Conf. Series 3, London (1967).
2. W. Shockley, A unipolar field effect transistor, *Proc. IRE*, **40**, 1365–1367 (November 1952).
3. A.S. Grove, *Physics and Technology of Semiconductor Devices*, John Wiley and Sons, New York (1967).
4. R.S. Pengelly, *Microwave Field Effect Transistors—Theory, Design, and Applications*, Research Studies Press, John Wiley and Sons, New York (1982).
5. R. Pucel, H. Haus, and H. Statz, Signal and noise properties of GaAs microwave field effect transistors, *Adv. Electron. Electron Phys.*, **38**, 195–265 (1975).
6. H. Fukui, Optimal noise figure of microwave GaAs MESFETs, *IEEE Trans. Electron Devices*, **ED-26** (7) 1032–1037 (July 1979).
7. J.A. Turner, A.J. Waller, R. Bennett, and D. Parker, An electron beam fabricated GaAs microwave field effect transistor, 1970 *Symp. GaAs and Related Compounds*, Inst. Phys. Conf. Series No. 9, London (1971).
8. M. Fukuta, K. Suyama, H. Suzuki, Y. Nakayama, and H. Ishimawa, *IEEE Trans. Microwave Theory Tech.*, **MTT-24**, 312 (1976).
9. W.C. Niehaus, H. Cox, B. Hewitt, S.H. Wemple, J.V. Di Lorenzo, W.O. Schlosser, and F.M. Magelhae, *Proc. 6th Intl. Symp. on GaAs*, St. Louis, Inst. Phys. Ser. No. 33b, London (1976), pp. 271.
10. I. Drukier, R.L. Camisa, S.T. Jolly, H.C. Huang, and S.Y. Narayan, *Electron. Lett.*, **2**, 104 (1975).
11. L.A. D'Asaro, J.V. Di Lorenzo, and H. Fukui, Int. Electron Devices Meeting Technical Digest, pp. 370–371, December (1977).

8

Schottky Barrier Gate Charge-Coupled Devices

Dieter K. Schroder

1. INTRODUCTION

A charge-coupled device (CCD) is a shift register in an integrated, compact format. The idea of making a shift register for analog signals and using it as a delay line, dates to the early fifties.[1] The basic principle is quite simple. Sampled values of the analog signal are stored in the form of charges on a series of capacitors. A switch between each capacitor transfers the charge from one capacitor to the next, following a command from a clock pulse.

A first implementation of such a shift register, using bipolar transistors as the switch, was reported in 1965[2] and an MOS version in 1969.[3] However, a really practical realization of this concept had to await the development of the integrated circuit where both the switch and the capacitor could be implemented on the same semiconductor chip. A bipolar version in which bipolar transistors were used as switches and the transistors' collector-base capacitances as the storage elements, was demonstrated in 1969.[4] Shortly thereafter an MOS transistor version was implemented.[5] This latter device is known as a bucket-brigade device (BBD) and is shown schematically in Fig. 1.

The cross section of Fig. 1a shows that the device consists of a *p*-type substrate with *n*-diffusions. Overlying this is an insulator and gates. The detail of Fig. 1b shows how the electrically floating diffusions act simultaneously as the drain of one transistor and the source of the next. In other words they act merely as regions through which electrons flow from channel to channel and are therefore sometimes forward and sometimes reverse biased. This is an important point, because as we shall see later on, diffusions and Schottky

Dieter K. Schroder • Department of Electrical and Computer Engineering, Arizona State University, Tempe, Arizona 85287.

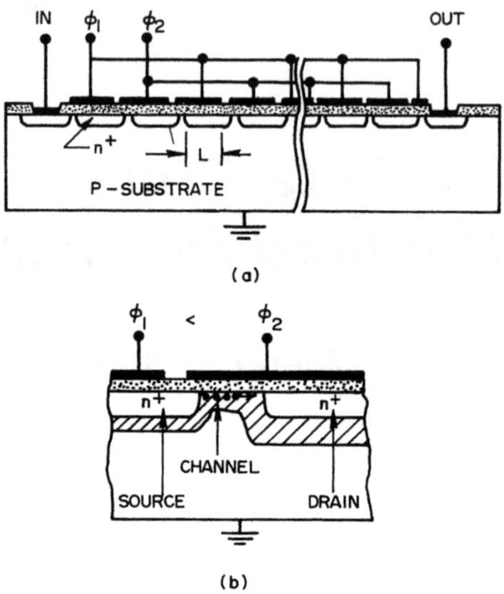

Figure 1. The MOS bucket brigade device. (a) Charge is stored in diffused diodes, shown by L. Diagram (b) shows the gate overlap and the diffusions acting as source and drain.

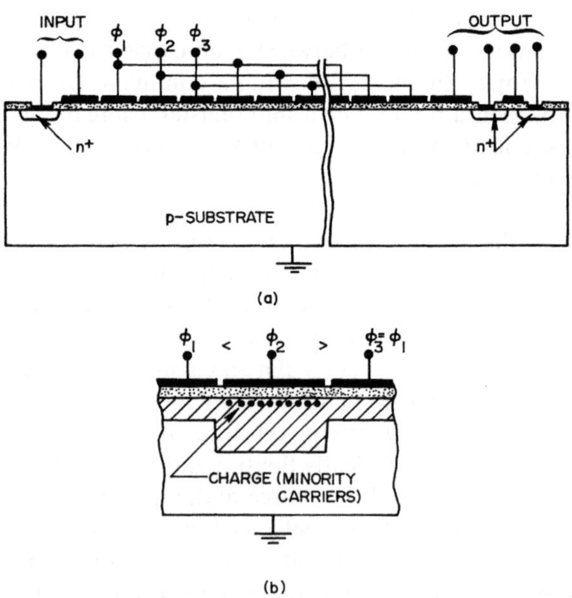

Figure 2. The surface channel charge-coupled device. Diagram (b) shows the minority carrier charge stored at the insulator–semiconductor interface and the space–charge region surrounding the charge.

contacts can be used as gates. Then, however, they are no longer electrically floating, but have an electrical contact and must always be reverse-biased.

A major disadvantage of BBDs is the feedback from the drain to the source. It modulates the threshold voltage of the MOS transistor and limits the charge transfer efficiency—a parameter describing the completeness of charge transfer from one element to the next—to approximately 0.999 per transfer. Although this value appears quite close to unity it is too low for high-quality transfer in a device consisting of many transfers. A typical device might have 1000 or more transfer stages.

A concept that eliminates this feedback effect is one incorporating *complete* charge transfer. This requires that there be no diffused regions in the transfer path and the charge-coupled device came into being in 1970.[6] It is shown as a three-phase version in Fig. 2. The original concept was very similar to the BBD, with the exception of the absence of diffused regions. This feature allowed all the charge in a given charge packet to be transferred, something that was not possible with the BBD since the diffused regions could never be completely depleted.

The charge in CCDs is stored in potential wells created by voltages

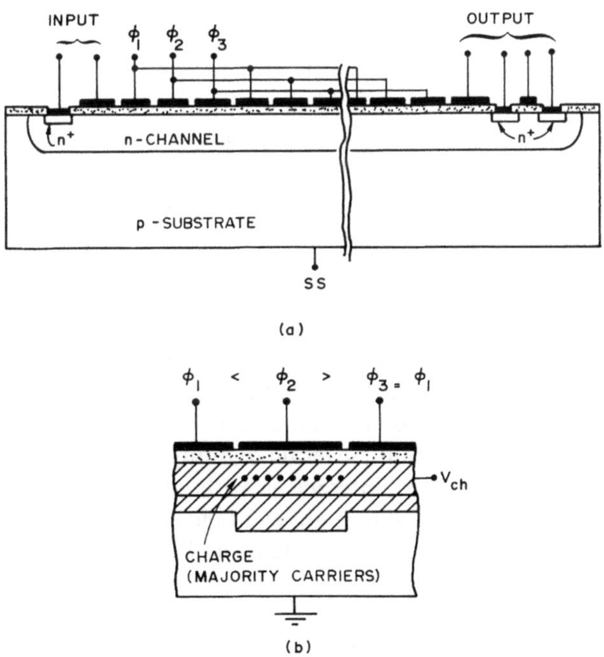

Figure 3. The bulk channel charge-coupled device. Diagram (b) shows the majority carrier charge stored in the n-channel surrounded by a depleted space–charge region.

applied to electrodes located on an insulating layer. By periodically varying the electrode voltages, potential wells, and hence the charge packets, are shifted along the semiconductor. By varying the frequency of the applied voltages, also called clock voltages, the delay time can be varied electronically over a wide range—a very desirable feature.

The surface channel CCD (SCCD) has two shortcomings that are the result of the charge location right at the semiconductor–insulator interface: (i) Interactions with interface states cause charge to be captured and emitted resulting in reduced transfer efficiency and noise, and (ii) drift fields induced by the externally applied voltages are low, limiting the frequency of operation to typically a few MHz.

These limitations were overcome with the advent of the bulk channel CCD (BCCD) in 1972[7] shown in Fig. 3 Its chief advantage is the location of the charge packet in the interior of the semiconductor, away from the semiconductor–insulator interface. The price to be paid for this feature is a more complex structure, because a layer of one type of conductivity is required on top of a substrate of the opposite type. However, the benefits far outweigh this added complexity. Because the charge resides in the bulk it no longer interacts with interface states but only with bulk states or recombination centers. Since the density of bulk states is easier to control than interface states and can be reduced to lower values, both the transfer efficiency and noise performance are improved. Additionally, because the location of the charge is further removed from the gates, higher drift fields give higher operational frequencies. An operational penalty is a smaller charge handling capacity.

Two slightly different types of bulk channel CCD have been discussed. The buried channel device[7] is generally fabricated by forming a thin upper region of one conductivity by ion implantation—its thickness is typically less than 1 μm—while the upper region in the peristaltic CCD[8] is a thicker epitaxial layer.

If the epitaxial layer is lightly doped, the charge packet, although it resides in the bulk, extends all the way to the interface and again interacts with interface states similar to surface channel devices. By incorporating a nonuniformly doped epitaxial layer in which the doping near the surface is high and decreases into the bulk, several advantages are attained. The charge handling capacity is increased, interaction with interface states is eliminated, and the high-speed feature due to high drift fields is maintained. In fact, with this type of device operational frequencies as high as 180 MHz have been achieved using a Si substrate.[9]

All the devices discussed so far utilize an insulating layer on the semiconductor on which the gate electrodes are formed. While this makes for an excellent device on Si where a high quality SiO_2 layer can be formed by thermal oxidation with a low density of interface states at the Si–SiO_2

interface, it becomes much more difficult for any other semiconductor. Attempts to make CCDs with other semiconductor materials have only been reasonably successful. For example, for infrared applications narrow bandgap materials are required and CCD performance has been shown in Ge,[10] InSb,[11] GaInSb,[12] and HgCdTe.[13] While the performance has not equaled that of Si devices, HgCdTe devices especially have shown very respectable results, considering the relatively small effort that has gone into it.

An entirely different approach was suggested by Schuermeyer *et al.*[14] in which *p–n* junctions or Schottky barriers are utilized for the gate structure instead of the oxide barrier. A schematic diagram is shown in Fig. 4. The junctions are reverse-biased and in conjunction with the *p–n* junction in the body of the semiconductor, form potential wells for majority carriers very much like those found in bulk channel devices. The first implementation occurred in 1977 in GaAs[15] by a group from Siemens, followed by a similar device from the Rockwell group in 1978.[16] Schottky barrier electrodes on epitaxial *n*-type GaAs layers grown on semi-insulating substrates as well as ion-implanted channels were used in these devices.

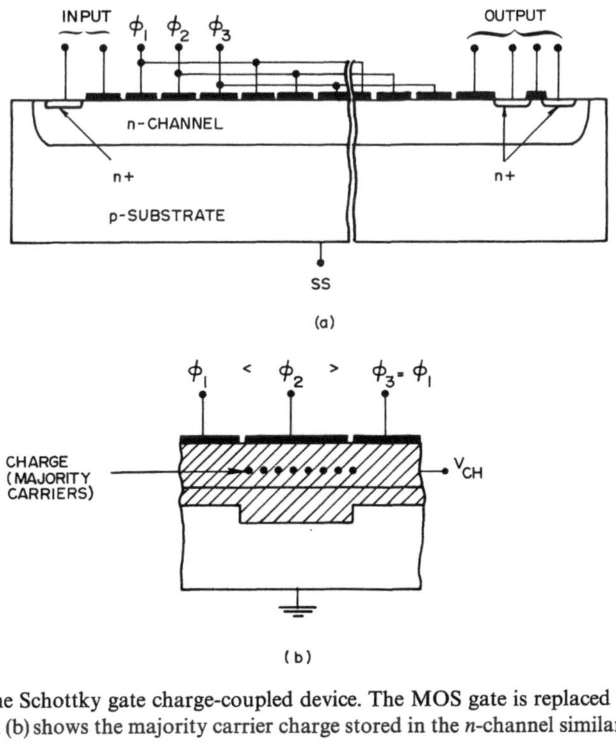

Figure 4. The Schottky gate charge-coupled device. The MOS gate is replaced by a Schottky gate. Diagram (b) shows the majority carrier charge stored in the *n*-channel similar to that in the bulk channel CCD.

The Schottky-barrier approach can be implemented on semiconductor materials in which it is difficult, if not impossible, to form metal–insulator–semiconductor structures. One reason to use GaAs as the active CCD material is the higher electron mobility compared to Si. This, however, is not the main reason, for even in Si, the ultimate speed determined by intrinsic charge transfer is in the gigahertz range. In real devices, the operational frequency is limited by the on-chip output amplifier to frequencies around 100–200 MHz. In GaAs, by contrast, Schottky-gate MESFET amplifiers have been operated to frequencies of 20 GHz.[17] These devices utilize the same technology as CCDs, and can be integrated on the same chip. They have given CCDs operating at frequencies up to 1 GHz.[18]

Such high-speed CCDs can perform specialized analog signal-processing functions at substantial gains in speed and reduction in volume and power consumption compared to an all-digital approach.[19] This is especially true when the analog device can operate at clock frequencies around 1 GHz.

2. SCHOTTKY GATE CCDs

The workhorse among CCDs is the MOS CCD in both the surface and bulk channel versions, as a result of the advanced state of development and high perfection of the silicon material and the $Si–SiO_2$ interface. Other materials are only considered if they promise significant performance enhancements over Si. GaAs is such a material because it has higher electron mobility. But it is only considered if the higher mobility is important for device applications. This is clearly the case for high-frequency applications because (i) the intrinsic propagation delay is less and (ii) the output amplifier has a higher bandwidth than silicon. For example, in order to operate a CCD at sample rates of 1 GHz, the bandwidth of the on-chip amplifier must be several GHz. However, the bandwidth of Si *n*-channel MOSFETs is typically around 0.5–1 GHz; values of several gigahertz are very difficult to achieve, while for GaAs MESFETs bandwidths of up to 20 GHz have been demonstrated.

Given the frequency advantage of GaAs over Si, and existing applications for CCDs with GHz frequency capabilities, it then becomes a technology problem to design and fabricate such structures. The metal–insulator–semiconductor type of device is very difficult if not impossible to make in GaAs because the GaAs–insulator interface has a high density of interface states that appear to be a fundamental property independent of the type of insulator used.[20] It is, however, quite easy to make Schottky barrier contacts to GaAs. This feature allows Schottky-gate CCDs to be realized. The same technology, furthermore, is used for the on-chip output MESFET amplifier.

GaAs, when doped with chromium or EL2 donors, is highly resistive. The resistivity of such semi-insulating substrates is typically 10^8–$10^9\Omega$cm. The CCD channel is formed by ion-implantation or epitaxial growth. The clock lines and other peripheral circuitry run largely over the semi-insulating substrate and because the capacitance is less, there will be reduced power dissipation on the chip. This is an important consideration at high frequencies because the power dissipation due to charging and discharging of a capacitor C is given by

$$P = CV^2 f_c \tag{1}$$

where V is the voltage swing of the driving voltage and f_c its frequency. This applies to a square wave driver and corresponds to 30 mW for $C = 1$ pF, $V = 5$ V and $f_c = 1$ GHz. It has been pointed out(21) that the power dissipation can be substantially reduced for tuned sinusoidal drivers to

$$P = (\pi/Q)CV^2 f_c \tag{2}$$

where Q is the quality factor of the circuit. For example, for a Q of 30 the tuned sinusoidal driver dissipates only 10% of the power of a square wave driver. This is a considerable reduction.

The Schottky-gate structure has a built-in antiblooming feature. When in imaging applications an unusually bright light spot is incident on the device, it can generate a sufficient density of carriers to overflow the potential well. In an MOS-CCD, the carriers cannot flow into the gate because of the insulator potential barrier. They will instead flow laterally into adjacent wells. The result is an enlargement of the original spot into a much larger area, called blooming. In the extreme it can encompass the entire imaging device. In the Schottky gate approach, once a potential well is full, additional charge will flow into the gate associated with that well and be removed as gate current, because the barrier to current flow will have been eliminated. Lateral charge flow and therefore blooming is eliminated. It is a built-in feature of this type of structure, and no additional antiblooming features need be incorporated into the design.

Schottky-gate CCDs are more radiation-tolerant than MOS devices because there is no insulator or insulator–semiconductor interface to be degraded by the energetic radiation. For example, in Si–SiO_2 devices oxide charge increases upon radiation as does the interface state density causing threshold voltage shifts, transfer efficiency degradation, and dark current and noise increases. High-energy radiation has a long penetration depth in semiconductors, creating many electron–hole pairs during its passage through the material. In long diffusion length semiconductors, such as Si, a good fraction of these extraneous carriers will be collected in potential wells. In GaAs and other III–V materials, this is a decidedly less serious problem because the minority carrier diffusion length is much smaller.

3. POTENTIAL–CHARGE RELATIONSHIPS

An important quantity in the design of CCDs is the potential at that point of the structure where the charge resides. For a surface channel CCD, this is the surface potential, while for both bulk MOS CCDs and Schottky CCDs it is the minimum potential. Any charge introduced into the device will reside at the point of minimum potential.

3.1. Surface Channel CCD

The surface potential is[22]

$$\phi_s = V_G' + V_0 + Q/C_{ox} - [2(V_G' + Q/C_{ox})V_0 + V_0^2]^{1/2} \qquad (3)$$

where $V_G' = V_G - V_{FB}(V)$ is the applied gate voltage minus the flat band voltage, Q is the charge density in the potential well (C/cm^2), C_{ox} is the oxide

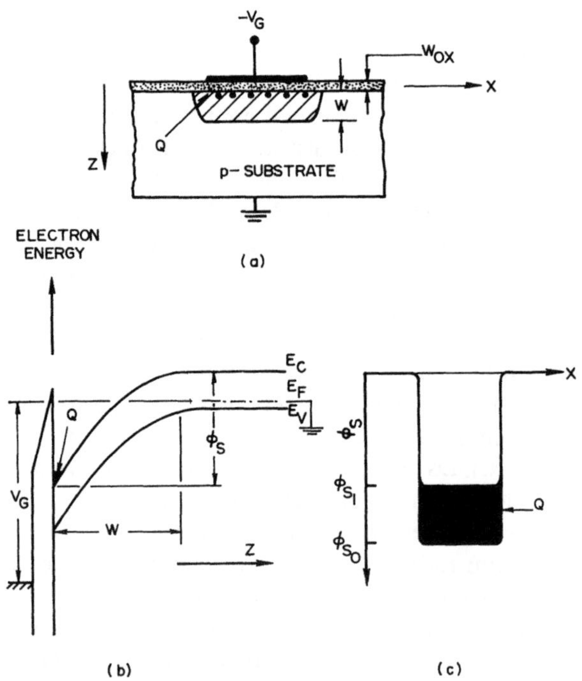

Figure 5. The surface channel CCD. (a) A single gate with charge stored in the inversion layer. (b) The potential diagram from the gate into the semiconductor. The charge Q is shown as minority carriers at the interface. (c) The surface potential along the semiconductor surface for the case of zero charge (ϕ_{s0}) and charge $Q(\phi_{s1})$.

capacitance/unit area (F/cm^2), V_0 is $qK_s\epsilon_0 N_A/C_{ox}^2(V)$, N_A is the substrate doping concentration (cm^{-3}). The other symbols have their usual meaning.

A graphical presentation of Eq. (3) is shown in Fig. 5. In Fig. 5a a cross section through the CCD shows one gate, the charge Q in the inversion layer at the semiconductor–insulator interface, and a space-charge region surrounding this charge. In Fig. 5b, the energy band diagram shows the band bending in the oxide and the semiconductor, the location of Q, and the space-charge region characterized by its width W. In Fig. 5c the surface potential ϕ_s is plotted along the semiconductor surface. The value ϕ_{s0} corresponds to an empty well, $Q = 0$, while ϕ_{s1} is the value for the case of Fig. 5b, i.e., charge Q in the potential well.

From Eq. (3) it is obvious that the surface potential depends on the oxide thickness through $C_{ox} = K_{ox}\epsilon_0/W_{ox}$ and the substrate doping concentrations N_A, for a given V_G and Q. The device designer has these two parameters at his disposal. Once the device is designed, the surface potential can be externally controlled through the applied gate voltage. It changes internally with the amount of charge in the potential well, which can vary from zero to its maximum value

$$Q_{max} = C_{ox}(V_G' - 2\phi_B - V_B)$$
$$= C_{ox}(V_G - V_T) \quad (4)$$

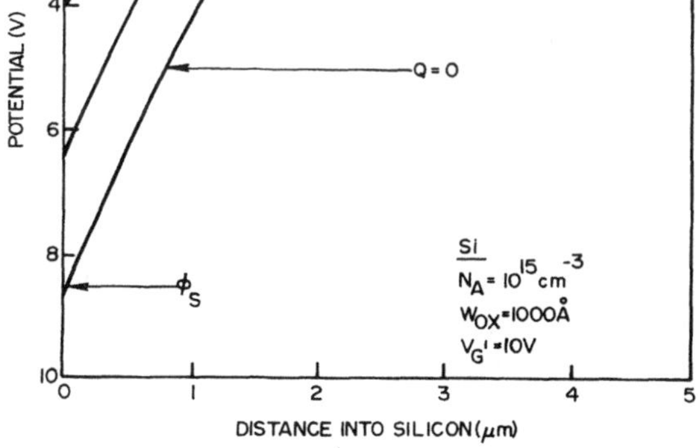

Figure 6. The potential from the semiconductor surface into the semiconductor as a function of charge Q.

where $\phi_B = (kT/q)\ln(N_A/n_i)$ and $V_B = (4qK_s\epsilon_0 N_A \phi_B / C_{ox}^2)^{1/2}$. $2\phi_B$ is the surface potential at strong inversion and V_T is the threshold voltage, i.e., the gate voltage at which strong inversion is established. Typically $2\phi_B$ and V_B are much smaller than V_G' and can be neglected. Equation (4) then simplifies to

$$Q_{\max} \simeq C_{ox} V_G' \tag{5}$$

For an oxide thickness of 1000Å and $V_G' = 10\,\text{V}$ this gives $Q_{\max} = 3.4 \times 10^{-7}\,\text{C/cm}^2$ or 2×10^{12} electrons/cm^2.

The potential distribution in the semiconductor is plotted in Fig. 6 as a function of the charge in the potential well. The potential at the semiconductor surface (zero distance) corresponds to the surface potential. These curves show that the potential is always at a minimum at the surface and the charge will always reside there. The drop in surface potential with increasing Q is accompanied by an increased oxide voltage drop for a constant gate voltage.

3.2. Bulk Channel CCD

The potential-charge calculations are more complex for a bulk channel device, because it is a three-terminal structure and has, in addition to the oxide thickness and substrate doping, the additional thickness and doping of the channel to be considered. It turns out that the substrate doping has the smallest effect, leaving the other three as the major design parameters. Instead of only considering the surface potential, as in Fig. 5, for a bulk channel device the minimum potential, defined in Fig. 7, is of prime importance although the surface potential is also important.

The charge resides at the minimum potential point, given by[21]

$$\phi_{\min} = \{V_G' + V_1 + Q/C_{\text{eff}} + V_{01} - [2(V_G' + V_1 + Q/C_{\text{eff}}) \times V_{01} + V_{01}^2]^{1/2}\}(N_D + N_A)/N_D \tag{6}$$

where

$$V_G' = V_G - V_{FB} \qquad \text{(V)}$$

$$V_1 = qN_D t(W_{ox}/K_{ox}\epsilon_0 + t/2K_s\epsilon_0) \qquad \text{(V)}$$

is the voltage that must be applied across the p–n junction to deplete the n-channel completely, Q is the charge density in the potential well, which is negative for the n-channel devices

$$C_{\text{eff}}^{-1} = W_{ox}/K_{ox}\epsilon_0 + (t - W_n + Q/2qN_D)/K_s\epsilon_0 \qquad (\text{cm}^2/\text{F})$$

$$W_n = \{2K_s\epsilon_0 N_A \phi_{\min}/[qN_D/(N_D + N_A)]\}^{1/2}$$

$$= N_A W_p / N_D \qquad \text{(cm)}$$

$$V_{01} = qK_s\epsilon_0 N_A/C_1^2 \quad \text{(V)}$$
$$C_1^{-1} = W_{ox}/K_{ox}\epsilon_0 + t/K_s\epsilon_0 \quad (\text{cm}^2/\text{F})$$

Equation (6) reduces to Eq. (3) when $t \to 0$ and $N_D \to \infty$, i.e., when the bulk device becomes a surface device, because then $V_1 \to 0$, $C_{eff} \to C_{ox}$, $C_1 \to C_{ox}$, and $V_{01} \to V_0$.

The surface potential is important because it specifies the potential barrier that keeps the charge, Q, away from the interface. For the device to operate properly, the charge must be held in the channel and not contact the interface. The barrier $\Delta\phi_s$, that keeps it away from the interface is given by

$$\Delta\phi_s = (qN_D/2K_s\epsilon_0)(t + Q/qN_D - W_n)^2 \tag{7}$$

Figures 7a and 7b show a cross section of the bulk-channel device. It

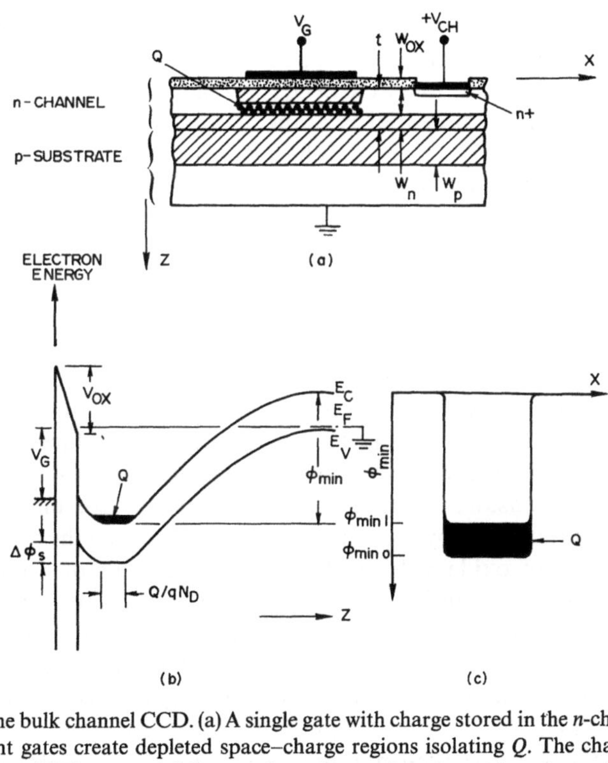

Figure 7. The bulk channel CCD. (a) A single gate with charge stored in the n-channel. In a real device adjacent gates create depleted space–charge regions isolating Q. The charge consists of majority carriers. (b) The potential diagram from the gate into the semiconductor. The charge Q is shown as majority carriers in the n-channel. It is a portion of the original electrons in the n-channel surrounded by depleted space–charge regions. (c) The minimum potential along the n-channel for zero charge (ϕ_{min0}) and charge $Q(\phi_{min1})$.

consists of a substrate that is lightly doped, p-type in this example. A thin, more heavily doped n-channel layer is formed on top of this substrate. If this channel is formed by ion implantation, it is thin and heavily doped and is referred to as a buried-channel device.[7] If it is formed by epitaxial growth, it is not as heavily doped and thicker and referred to as a peristaltic CCD.[8] The semiconductor is covered with an insulator and gate electrodes.

Application of the voltage V_{ch} causes two space-charge regions to form, one originating at the semiconductor–insulator interface, the other at the buried p–n junction. These two space-charge regions form, even if $V_G = 0$. They gradually deplete the n channel and when they touch, the channel is totally depleted, i.e., $Q = 0$. A charge Q corresponds to part of the channel not depleted and is merely part of the original *majority* carrier concentration. This contrasts with a surface-channel device in which the charge, Q, corresponds to *minority* carriers at the interface.

A more positive gate voltage on the device in Fig. 7 will increase Q_{min} and electrons in the potential well will flow towards that potential. The minimum potential is achieved when $Q = 0$ and is shown as ϕ_{min0} in Fig. 7c. As in SCCDs, there is a maximum charge that can be stored in a BCD. The limitation, however, is very different. In a SCCD Q_{max} is reached when $\phi_s \simeq 2\phi_B$. Beyond that point, the surface potential is pinned. In a BCD, the quantity $\Delta\phi_s$ in Fig. 7b decreases as Q increases. When $\Delta\phi_s = 0$, the charge is no longer confined to the n-channel bulk, but spreads out and touches the interface. It interacts with interface states and the device loses the advantages of a bulk channel device. This condition is reached when in Eq. (7) $\phi_s = 0$, or

$$Q_{max} = -qN_D(t - W_n) \tag{8}$$

where for the n-channel device considered here, Q is negative. Q_{max} represents the charge in the channel when the space-charge region is due to V_{ch} only, i.e., when the space-charge region due to the gate has been reduced to zero. That the charge in the channel will move towards the interface instead of the bulk is obvious if one considers that the voltage between the channel and the substrate is held constant by V_{ch}, while the voltage between the channel and the gate is the sum of oxide voltage, V_{ox}, and semiconductor voltage, $\Delta\phi_s$. A decrease in $\Delta\phi_s$ causes an increase in V_{ox}, holding the sum constant.

The potential distribution into the semiconductor is shown in Fig. 8 as a function of charge in the potential well for one value of gate voltage. The most obvious change from the potential plots in Fig. 6 is the location of the potential minima. They are now located within the n channel and as the charge is increased, the widths of these minima increase and the potential barrier, $\Delta\phi_s$, diminishes until finally the charge extends all the way to the interface.

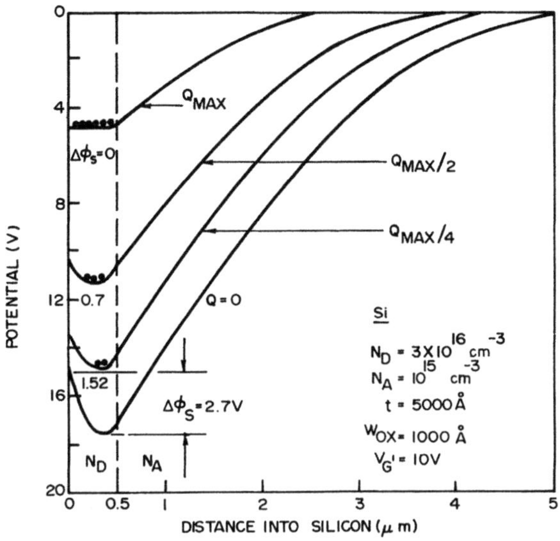

Figure 8. The potential from the semiconductor surface into the semiconductor as a function of charge Q. The potential minimum in the n-channel is clearly seen and the barrier to the surface ($\Delta\phi_s$) decreases as Q increases.

3.3. Schottky Gate CCD

The Schottky gate CCD is very similar to a bulk channel CCD in terms of the potential–charge relationship. Equation (6) remains unchanged

$$\phi_{\min} = \{V_G' + V_1 + Q/C_{\text{eff}} + V_{01} - [2(V_G' + V_1 + Q/C_{\text{eff}}) \times V_{01} + V_{01}^2]^{1/2}\}(N_D + N_A)/N_D \tag{9}$$

where now, however,

$$V_G' = V_G + V_{\text{bi}} \qquad \text{(V)}$$

with V_{bi} the built-in potential at the Schottky barrier junction; $V_1 = qN_Dt^2/2K_s\epsilon_0 (V)$ is the voltage that must be applied across the p–n junction to deplete the n channel completely

$$C_{\text{eff}}^{-1} = (t - W_n + Q/2qN_D)/K_s\epsilon_0 \qquad (\text{cm}^2/\text{F})$$

$$W_n = \{2K_s\epsilon_0 N_A \phi_{\min}/[qN_D(N_D + N_A)]\}^{1/2} = N_A W_p/N_D \qquad (\text{cm})$$

$$C_1^{-1} = t/K_s\epsilon_0 \qquad (\text{cm}^2/\text{F})$$

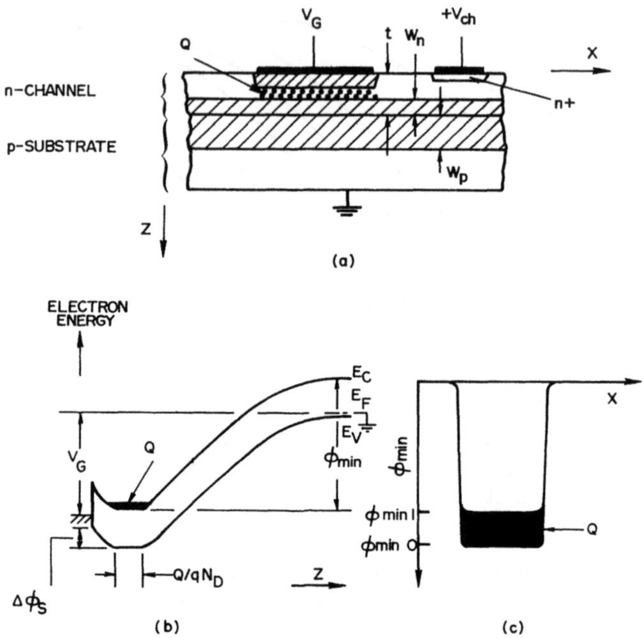

Figure 9. The Schottky gate CCD. (a) A single gate with charge stored in the *n*-channel. (b) The potential diagram from the gate into the semiconductor. The charge Q is shown as majority carriers in the *n*-channel. It is part of the original electrons in the *n*-channel. (c) The minimum potential along the *n*-channel for zero charge ($\phi_{\min 0}$) and charge Q ($\phi_{\min 1}$). The diagrams are similar to those of the bulk channel CCD with the exception that the insulator is absent.

The other terms remain unchanged from Eq. (6). The barrier $\Delta\phi_s$ is given by $V_G - \phi_{\min}$.

The main change from the bulk-channel device is the absence of the insulator, shown in Fig. 9. The gate voltage now appears directly across the semiconductor. This has the beneficial effect that if the charge in the channel exceeds its maximum value, it flows into the gate and does not spread laterally into the device, as discussed earlier. This provides for a built-in antiblooming feature. A detrimental effect is that the gate voltage must be more carefully controlled to prevent any forward bias which would inject charge into the potential well.

The diagrams of Fig. 9 look very similar to those of Fig. 7. It is obvious from Fig. 9b that a change in V_G towards more positive values results in a downward shift of the potential minimum. Recall that on an energy band diagram of this type positive *potential* is downward while positive *energy* is plotted up. Electrons flow down on this diagram so that an adjacent gate with

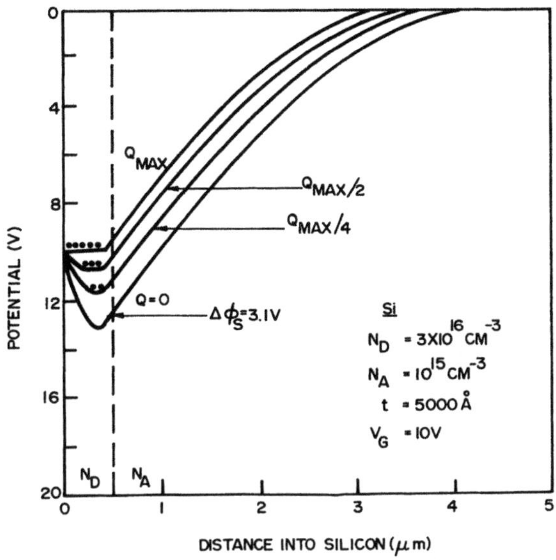

Figure 10. The potential from the semiconductor surface into the semiconductor as a function of charge Q. The surface potential is given by the gate voltage and is independent of Q.

more positive gate voltage causes the electron charge to flow to that potential minimum. This is the mechanism of charge transfer and in principle is identical to that of both SCCDs and BCCDs.

The potential distribution in Fig. 10 is similar to that of the BCCD in Fig. 8. The one difference between these two is the potential at the semiconductor surface. In Fig. 8 this value changes because the channel voltage is divided between semiconductor and oxide potentials. In the Schottky CCD there is no oxide and therefore the potential at the surface is fixed by the gate voltage. The minimum value, $\phi_{\min}$, changes with channel charge.

The band diagram of Fig. 9 and the potential profiles of Fig. 10 assumed an n channel on a p-substrate. Usually the p-doping is less than that of the n channel and the space-charge region extends further into the substrate than into the channel. However, the substrate doping is sufficiently high that the extent of the space-charge region into it is typically a few micrometers. One of the virtues of GaAs technology is the availability of semi-insulating substrate material. As pointed out earlier, this reduces wiring and bonding pad capacitances, making high-frequency devices possible.

High resistivities can be achieved by one of two methods, (i) reduce N_A to very low values so that the resistivity, given by

$$\rho = [q(\mu_n n + \mu_p p)]^{-1} \tag{10}$$

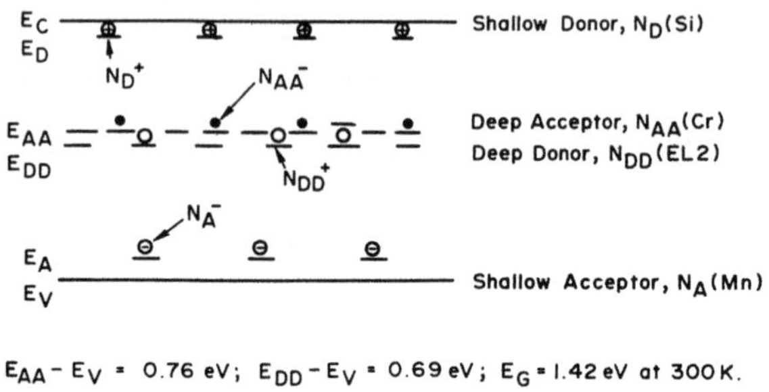

Figure 11. The band diagram of semi-insulating GaAs. This is a simplified picture consisting of shallow donors, shallow acceptors, deep acceptors (Cr), and deep donors (EL2). The values of the energy of the deep levels are from Ref. 23.

is determined by $p = N_A \approx n_i$, and (ii) introduce deep-lying impurities that increase ρ by reducing the mobile carrier concentration. The first method is impractical, because N_A would have to be reduced to n_i, which is around 10^6 cm^{-3} in GaAs. This is not possible. The second technique is actually used.

Semi-insulating GaAs usually has two deep levels in addition to shallow level donors and acceptors. The deep levels have been identified[23] as an electron trap and a hole trap. The electron trap, designated EL2, is a deep donor level and appears to be stoichiometry related.[24] The hole trap is due to Cr. The energy band diagram with all four levels is shown in Fig. 11. Knowledge of the energy levels and concentrations of the four impurities allows the Fermi level to be determined, which in turn is used to calculate n and p and the resistivity, ρ.

Charge neutrality considerations give

$$n + N_A^- + N_{AA}^- = p + N_D^+ + N_{DD}^+ \tag{11}$$

where

$$n = N_c \exp(E_F - E_c)/kT, \qquad p = N_v \exp(E_v - E_F)/kT$$

$$N_A^- = N_A/[1 + g_A \exp(E_A - E_F)/kT]$$

$$N_{AA}^- = N_{AA}/[1 + g_{AA} \exp(E_{AA} - E_F)/kT]$$

$$N_D^+ = N_D/[1 + g_D \exp(E_F - E_D)/kT]$$

$$N_{DD}^+ = N_{DD}/[1 + g_{DD} \exp(E_F - E_{DD})/kT]$$

N_c and N_v are the effective densities of states in the conduction and valence bands and g_A, g_{AA}, g_D, and g_{DD} are the degeneracy factors of the impurities.[25]

Equation (11) was solved numerically for E_F allowing n, p, and ρ to be

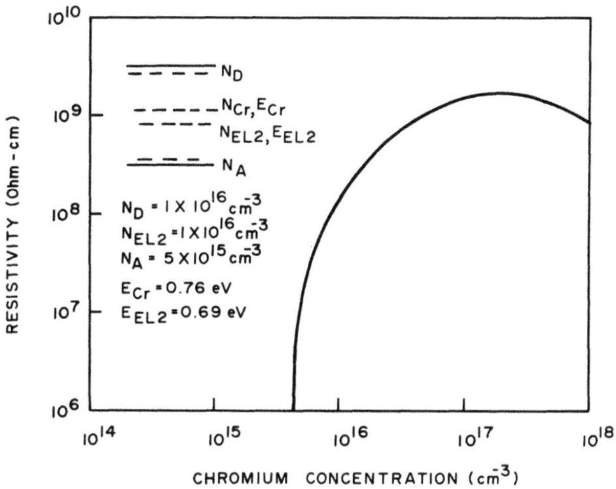

Figure 12. The resistivity of GaAs as a function of the deep C_r acceptor level. For the concentrations shown on the figure the resistivity is low at low C_r densities. It reaches a maximum when the Fermi level is near the middle of the band gap, then ceases as E_F moves down. In actual materials the situation is more complicated because there are usually additional impurities and they have more than one charge state.

calculated using the values from Figs. 11 and 12. The resulting resistivity is plotted in Fig. 12 as a function of N_{AA}, the chromium concentration. The results, as one would expect, show a sharp increase in resistivity when $N_{AA} \simeq N_D - N_A$. The maximum resistivity is given by

$$\rho_{\max} = [2qn_i(\mu_n\mu_p)^{1/2}]^{-1} \tag{12}$$

and is reached at $E_F = E_i - (kT/2)\ln(\mu_n/\mu_p)$. Equation (11) can also be solved graphically using the Shockley graph approach.[26] This is discussed by Zucca[27] and Martin *et al.*[23]

The two methods of growing semi-insulating substrates mentioned earlier give similar values of resistivity because both reduce the mobile carrier concentration. The question that needs to be answered is whether they behave the same way under reverse bias conditions as far as the width of the space-charge region is concerned. The space-charge region width, W_p, is given by

$$W_p = W_n N_D / N_A \tag{13}$$

Typically $N_D \simeq 10^{16}\,\text{cm}^{-3}$ and $W_n \lesssim 1\,\mu\text{m}$. It is clear that if $N_A \to 0$ then $W_p \to \infty$, as would occur if all impurities could be reduced to very low levels. For example, for $N_A = 10^6\,\text{cm}^{-3}$, required for semi-insulating materials, $W_p =$

$10^{10}\,\mu$m. This says that if the required high resistivity could be achieved by reducing all impurities to extremely low values, the space-charge region width would extend through the entire substrate. The situation is more complicated for the compensated case. For $N_D = 10^{16}\,\text{cm}^{-3}$, $N_A = 5 \times 10^{15}\,\text{cm}^{-3}$, $N_{DD} = 10^{16}\,\text{cm}^{-3}$, and $N_{AA} = 5 \times 10^{16}\,\text{cm}^{-3}$, the charge state of these four impurities is shown in Fig. 13. It shows that the deep donor level is essentially filled with electrons, making it neutral, while the deep acceptor level is only partially filled with electrons. In fact, the electrons residing on this level come chiefly from the shallow donor level since it is located above E_{AA}.

Now consider an n layer on top of the semi-insulating substrate, doped with N_{D1} donors/cm³. The energy band diagram of Fig. 13 shows the n layer, the substrate, and the zero-bias space-charge region. At the Schottky barrier surface, the space-charge region width, W_s, is determined by the barrier potential, ϕ_B, and the channel doping N_{D1}. W_n and W_p are determined by N_{D1} and the effective doping in the semi-insulating substrate. The net doping

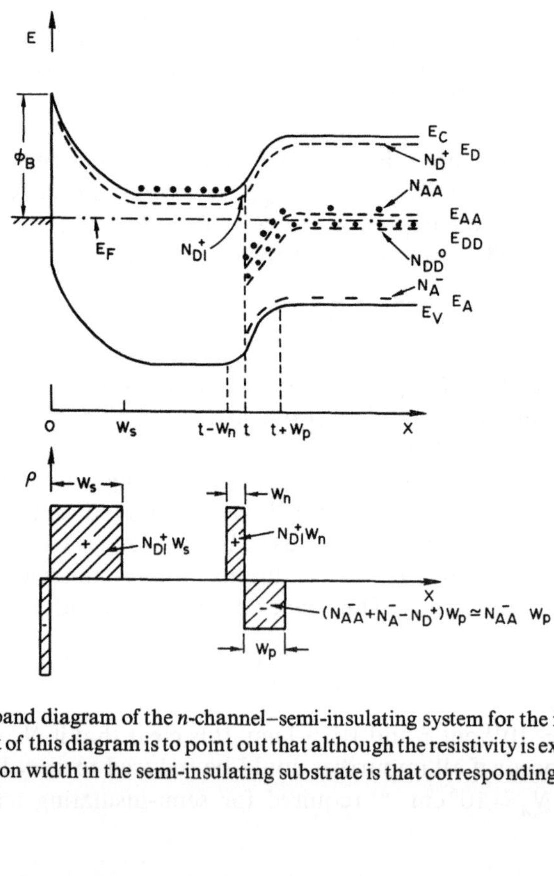

Figure 13. The band diagram of the n-channel–semi-insulating system for the impurities of Fig. 12. The main point of this diagram is to point out that although the resistivity is extremely high, the space–charge region width in the semi-insulating substrate is that corresponding to a high doping concentration.

concentration in the bulk substrate is effectively zero, but in the region W_p near the n channel the deep Cr acceptor level dips below the Fermi level and is therefore filled with electrons. The deep EL2 donor level is filled with electrons everywhere, but, being a donor, is in the neutral charge state. Hence the net charge density in the W_p portion of the substrate is the sum of all the charge species, which is $q(N_{AA}^- + N_A^- - N_D^+)W_p \simeq qN_{AA}^- W_p$, assuming $N_D, N_A < N_{AA}$. These considerations imply that the semi-insulating substrate can be viewed as a p^+ semiconductor of width W_p on top of a high-resistivity substrate. A similar argument for a more simplified system has been advanced earlier.[28]

When a positive voltage is applied to the n channel to deplete it, the channel–substrate junction becomes reverse-biased. The Fermi level now splits into electron and hole quasi-Fermi levels and the occupancy of the deep levels is no longer well specified. The net charge in the substrate near the channel side is $\rho = q(N_D^+ + N_{DD}^+ - N_{AA}^- - N_A^-)$. Using $N_D^+ = N_D$ and $N_A^- = N_A$, we get $\rho = q(N_D - N_A + N_{DD}^+ - N_{AA}^-)$. N_{DD}^+ and N_{AA}^- depend on the emission coefficients of the deep-lying donors and acceptors and the applied voltage and are in general very difficult to determine. It is obvious, however, that N_{AA}^- will no longer be equal to N_{AA}, nor will all the N_{DD} be in the neutral state. If N_{AA} is the dominant impurity and if, for simplicity, we assume $\rho \simeq qN_{AA}^- < qN_{AA}$, then we see that with applied bias, W_p will widen not only because the voltage increase, but also because N_{AA}^- changes and the space-charge region width can become quite wide. In addition, it has been found that the Cr[29] and the EL2.[24] impurities can exist in several charge states, which complicates the picture further. An exact analysis requires a detailed numerical analysis and is beyond the scope of this chapter. Qualitatively, however, we see that the potential plots of Fig. 10 are correct, even though they vary in detail for a Schottky gate CCD on an insulating substrate. The important point is that the potential minimum lies in the n channel. For a significant extension of W_p into the substrate there is the danger that the CCD charge interacts with the deep impurities there causing poor transfer efficiency.

For an epitaxial layer on a Cr-doped, semi-insulating substrate it has been shown[30] by voltage probe, photoresponse, and deep trap measurement that there exists a thin layer of localized deep traps at the interface. They cause an energy barrier there, complicating the understanding further.

4. CHARGE STORAGE CAPACITY

4.1. Surface Channel CCD

The maximum charge that can be stored in a SCCD is

$$Q_{\max} = C_{\mathrm{ox}}(V_G - V_T) \tag{14}$$

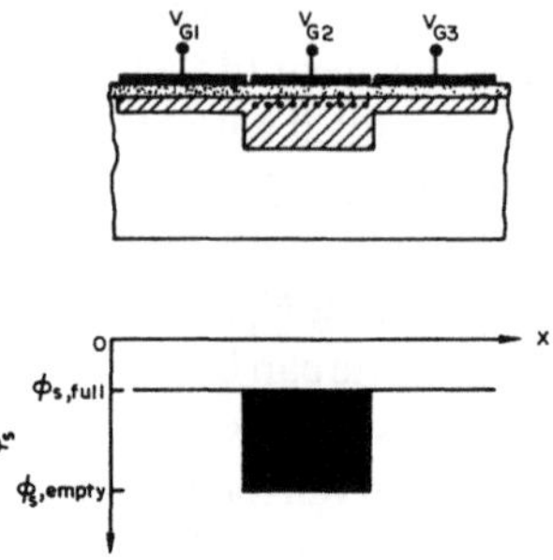

Figure 14. A cross section through a SCCD. The surface potential plot shows the surface potential values that determine the maximum charge that can be stored.

Aside from the threshold voltage, it is primarily determined by oxide capacitance, because the charge is stored at the semiconductor surface. Equation (14) represents the absolute maximum. In an actual CCD, Q_{max} is less than this, because it is determined by the surface potential of the blocking gates, indicated by $Q_{s,full}$ in Fig. 14. The full well charge is

$$Q_{full} = C_{ox}(V_{G2} - V_{G1}) \tag{15}$$

and represents the operational maximum charge this three-phase CCD can handle. A two-phase CCD can store less charge than this because the barrier to lateral charge movement is built into the device and not controlled by external voltages.[21]

4.2. Bulk Channel CCD

Intuitively it seems that the maximum charge stored in a BCCD should be less than in a SCCD because the charge is located not at the interface but within the n channel. If one defines an effective capacitance as the capacitance of a parallel plate capacitor with the spacing equal to the distance between the gate electrode and the centroid of the charge packet,[21] then it can be written as

$$C_{eff} = C_{s1}/(1 + C_{s1}/C_{ox}) \tag{16}$$

with $C_{s1} = K_s\epsilon_0/(t - W_n + Q/2qN_D)$. The C_{eff} expression is the same as that in Eq. (6) The maximum charge is then

$$Q_{max} = C_{eff}(V_G - V_T) \tag{17}$$

The location of the charge centroid should be as close to the surface as possible for highest Q_{max}. This implies that either the channel should be thin, as it is in ion-implanted buried-channel structures, or the n channel should be profiled so that the charge is located near the interface. This is exploited in the profiled peristaltic CCD, which consists of a lightly doped, thick channel with a thin, highly doped surface layer. This structure forces the charge towards the interface, but the highly doped surface layer prevents contact with the actual interface.

4.3. Schottky Gate CCD

Although the Schottky gate device is similar in potential distribution to a BCCD, it has a higher charge storage capacity because there is no insulator and the gate is in direct contact with the semiconductor. This reduces the distance between the gate and the charge centroid and the effective capacitance becomes

$$C_{\text{eff}} = C_{s1} = K_s\epsilon_0/(t - W_n + Q/2qN_D) \tag{18}$$

and

$$\begin{aligned} Q_{\max} &= C_{\text{eff}}\,(V_G - V_{\text{bi}}) \\ &= C_{s1}\,(V_G - V_{\text{bi}}) \end{aligned} \tag{19}$$

To get a better idea of the charge storage capacity of these three structures, consider Figs. 5–10. We see that when $Q \to Q_{\max}$ the potential in the n layer flattens out and $t = W_n + Q/qN_D$, or $C_{s1} = 2K_s\epsilon_0/(t - W_n)$. Assuming $W_n \ll t$

$$C_{s1} \simeq 2K_s\epsilon_0/t \tag{20}$$

This leads to

$$Q_{\max}(\text{SCCD})/Q_{\max}(\text{BCCD}) \simeq 1 + K_{\text{ox}}t/2K_sW_{\text{ox}} = 1.8 \tag{21}$$

for $t = 5000$ Å and $W_{\text{ox}} = 1000$ Å, the examples of Figs. 5–10. Similarly

$$Q_{\max}(\text{SCCD})/Q_{\max}(\text{Schottky}) \simeq K_{\text{ox}}t/2K_sW_{\text{ox}} = 0.82 \tag{22}$$

and

$$Q_{\max}(\text{Schottky})/Q_{\max}(\text{BCCD}) \simeq 1 + 2K_sW_{\text{ox}}/K_{\text{ox}}t = 2.22 \tag{23}$$

These ratios bring out the surprising fact that the Schottky gate CCD has the highest storage capacity of all three devices, if the n-channel layer is thin. The reason for this is that the effective capacitance is that of the semiconductor and the dielectric constant of a semiconductor is several times larger than that of an oxide. It applies only to thin channel devices. Its Q_{max} equals that of the SCCD when $C_{ox} = C_{s1}$ or $t = 2K_s W_{ox}/K_{ox} \lesssim 6W_{ox}$ for Si. So for $t \lesssim 6W_{ox}$ the Schottky gate device can store more charge than the surface channel device. This is generally the case for shallow, ion-implanted structures. Schottky gate devices made in GaAs have generally had thin n layers so that their storage capacity is very high.

5. CHARGE TRANSFER

5.1. Charge Transfer Efficiency

A functional relationship of fundamental interest is the CCD transfer efficiency as a function of clock frequency, the objective being to attain the highest possible frequency at a transfer efficiency consistent with the desired device performance. A plot that characterizes this behavior is shown in Fig. 15, where, however, the *transfer inefficiency*, ϵ, is plotted against frequency. ϵ is defined as follows[22]: as a charge packet is transferred from one well to the next, a fraction ϵ of charge is temporarily left behind and a fraction $\eta = (1 - \epsilon)$ flows into the next well. η is the transfer efficiency and ϵ the transfer inefficiency. The net result is that at the output the charge packet is reduced in size and the charge temporarily left behind follows as trailing charge packets. Transfer inefficiency is a temporary loss. It is reasonably constant up to some frequency, f_1, and then increases rapidly. For frequencies below f_1, where $\epsilon = \epsilon_1$, the transfer inefficiency is determined by extrinsic mechanisms like interface states, bulk recombination centers, interelectrode gaps, and clock voltages, while for $f > f_1$, it is determined by intrinsic mechanisms.

The transfer inefficiency degrades the CCD performance by reducing the gain and causing a phase shift in CCD delay lines according to[31]

$$G = \exp\{-n\epsilon[1 - \cos(2\pi f/f_c)]\} \tag{24}$$

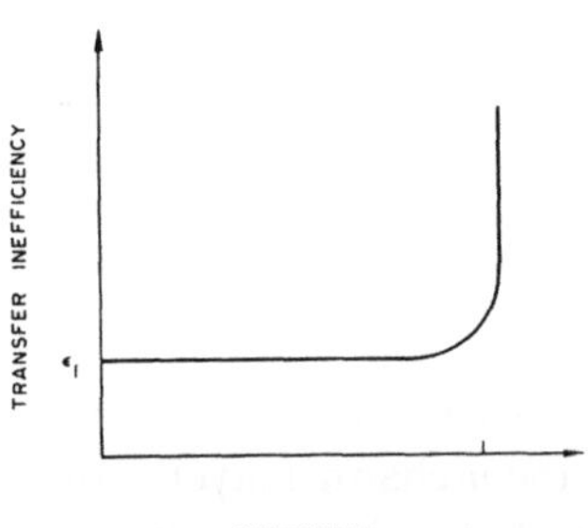

Figure 15. The charge transfer inefficiency as a function of the frequency. The low-frequency value ϵ_1, is determined by extrinsic charge transfer mechanisms. Above the frequency f, the transfer efficiency is limited by intrinsic mechanisms.

and

$$\Delta\Theta = -n\epsilon[2\pi f/f_c - \sin(2\pi f/f_c)] \quad (25)$$

where n is the total number of transfers and $n\epsilon$ the transfer inefficiency product, f the frequency of a sinusoidal signal and f_c the clock frequency. In imaging applications, transfer inefficiency gives rise to image smearing or modulation transfer function degradation.

Transfer inefficiencies for SCCDs are typically 10^{-3}, but they can be improved to 10^{-4} by introducing a background charge, sometimes called "fat zero." Bulk channel CCDs have enhanced transfer efficiencies through elimination of interaction of the charge packets with interface states. However, trapping by bulk centers, whose densities can be reduced to $\sim 10^{11}\,\text{cm}^{-3}$, limits ϵ to around 10^{-5}(32) for good quality Si devices. A small circulating background charge is still required for optimum performance.

5.2. Charge Transfer Mechanisms

The ultimate limit in $\epsilon - f$ performance is set by *intrinsic* transfer mechanisms. These are explained with the aid of Fig. 16. Consider the potential diagram of Fig. 16a, in which the potential represents the surface potential in SCCDs or the minimum potential in BCCDs and Schottky-gate devices. The charge resides initially under the ϕ_2 gate and is of uniform density so no charge gradients exist along the device. When ϕ_3 is pulsed into the "high" state, charge begins to flow into the higher potential well setting up large charge gradients in the ϕ_2 well. Now both ϕ_s for the SCCD and $\phi_{\min}$ for the bulk devices depend fairly linearly on Q. This indicates that as the charge becomes less uniformly distributed, a potential is established along the charge transfer direction to provide a self-induced electric field which provides the main driving force for charge transfer during the early stages of transfer.

A mathematical formulation, taking all three mechanisms into account, will now be discussed. It follows essentially the treatment by Esser and Sangster.[9] The current density due to drift and diffusion is

$$J(x,t) = Q(x,t)\,\mu\varepsilon_{SI}(x,t) + D\partial Q(x,t)/\partial x + Q(x,t)\mu\varepsilon_F(x,t) \quad (26)$$

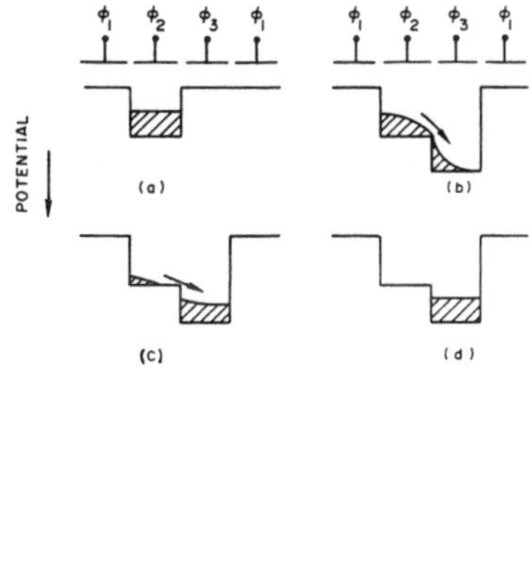

Figure 16. Schematic of the charge transfer mechanism. (a) The original isolated charge. (b) Initial charge transfer by self-induced drift in which charge gradients set up a self-induced electric field that aids charge transfer. (c) Final charge transfer when the remaining charge is a small fraction of the initial charge. (d) Final transferred charge.

where $Q(x, t)$ is the charge density, μ the mobility, $\varepsilon_{SI}(x,t)$ the self-induced electric field, D the diffusion coefficient, and $\varepsilon_F(x, t)$ the fringing or externally induced electric field. Using the continuity equation

$$\partial Q(x,t)/\partial t = -\partial J/\partial x \tag{27}$$

we derive the charge transfer equation

$$\partial Q(x,t)/\partial t = -\partial[D_{\text{eff}}\partial Q(x,t)/\partial x]/\partial x - \mu\partial[Q(x,t)\varepsilon_F(x,t)]/\partial x \tag{28}$$

where $D_{\text{eff}} = D + \mu Q(x,t)/C_{\text{eff}}$ with C_{eff} defined in Eq. (6). D_{eff} is the effective diffusion coefficient taking into account both self-induced drift and thermal diffusion.

The charge transfer inefficiency is defined by

$$\epsilon = Q_r(t)/Q_i \tag{29}$$

where

$$Q_r(t) = (1/L)\int_0^L Q(x,t)\,dx \tag{30}$$

is the remaining charge density integrated over the gate length L. Q_i is the initial charge density

$$Q_i = Q_r(0) \tag{31}$$

The time for charge transfer and therefore the frequency response will now be evaluated for the three cases.

5.2.1. Surface Channel CCD

The effective capacitance becomes the oxide capacitance and

$$D_{\text{eff}} = \mu[Q(x,t)/C_{\text{ox}} + kT/q] \tag{32}$$

For self-induced charge transfer, the term kT/q in Eq. (32) is negligible and $D_{\text{eff}} = \mu Q(x,t)/C_{\text{ox}}$ so that

$$\partial Q(x,t)/\partial t = (\mu/2C_{\text{ox}})\,\partial^2 Q^2(x,t)/\partial x^2 \tag{33}$$

It has been shown[33] that a short time after initiation of charge transfer, the shape of the charge concentration remains constant in time. This allows separation of variables and a solution of Eq. (33) gives

$$\epsilon = (1 + t/T_{\mathrm{SI}})^{-1} \tag{34}$$

with

$$T_{\mathrm{SI}} = 2L^2 C_{\mathrm{ox}}/\pi\mu Q_i \tag{35}$$

being the self-induced transfer time constant.

For thermal diffusion charge transfer the second term in Eq. (32) is dominant, being valid when $Q(x, t)$ has decayed to sufficiently low values and substitution into Eq. (28) gives

$$\partial Q(x, t)/\partial t = (\mu kT/q)\partial^2 Q(x,t)/\partial x^2 \tag{36}$$

The solution is

$$\epsilon = \exp(-t/T_d) \tag{37}$$

with

$$T_d = 4qL^2/\pi^2 \mu kT \tag{38}$$

being the diffusion transfer time constant.

For fringing field induced charge transfer

$$\partial Q(x, t)/\partial t = -\mu\partial[Q(x, t)\varepsilon_F(x, t)]/\partial x \tag{39}$$

Assuming a constant electric field for $0 < x < L$, zero charge density at $x = L$ and zero charge gradient at $x = 0$, then Eq. (39) can be solved[34] to give

$$\epsilon = \exp(-t/T_F) \tag{40}$$

with

$$T_F = 4L/\pi^2 \mu\varepsilon_F \tag{41}$$

being the fringing field transfer time constant.

The initial charge transfer, given by Eq. (35), is very fast. The last remnant of charge is transferred by either diffusion or fringing field. Which of these two mechanisms is dominant is very important for high-speed transfer. Since diffusion is an inherently slow process, the device should be designed for the fringing field mechanism to be dominant during the last moments of transfer. In fact, a comparison of Eqs. (38) and (41) shows that it amounts to a comparison of ε_F to $\varepsilon_{th} = kT/qL$, ε_F being the externally induced fringing field and ϵ_{th} the thermally induced field. Generally $\epsilon_F > \epsilon_{th}$ and fringing field charge transfer is dominant; this is especially true in bulk and Schottky gate CCDs.

5.2.2. Bulk Channel CCD

The derivations are the same as for the SCCD, the only exception being that the oxide capacitance is replaced by the effective capacitance. The three time constants become

$$T_{SI} = 2L^2 C_{eff}/\pi\mu Q_i \tag{42}$$

$$T_d = 4qL^2/\pi^2 \mu kT \tag{43}$$

$$T_F = 4L/\pi^2 \mu \varepsilon_F \tag{44}$$

with $C_{eff} = [C_{ox}^{-1} + (t - W_n + Q/2qN_D)/K_s\epsilon_0]^{-1}$, as defined in Eq. (6).

5.2.3. Schottky Gate CCD

The three time constants are identical to those in Eqs. (42)–(44) except that $C_{eff} = [(t - W_n + Q/2qN_D)/K_s\epsilon_0]^{-1}$. The charge is located deeper into the body of the device for the latter two devices and one would expect the externally induced fringing field to be stronger than for the SCCD.

To determine the maximum frequency of operation, Eqs. (34), (37), and (40) are solved for t. The initial self-induced transfer time is very short and the total time is determined predominantly by either the diffusion time or the fringing field induced time. A good estimate of the frequency is obtained by

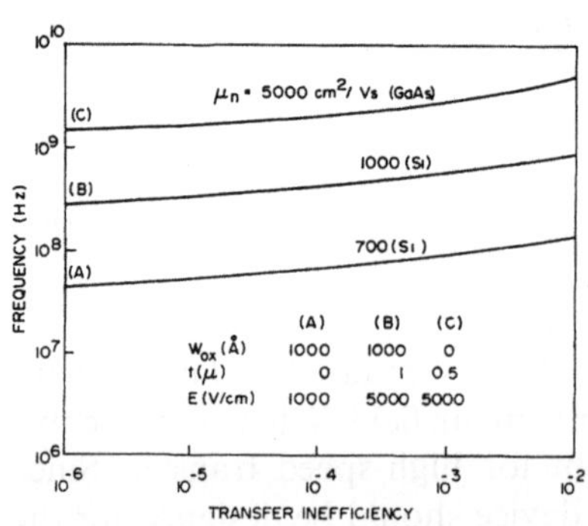

Figure 17. Maximum frequency of (A) SiSCCD, (B) SiBCCD, (C) GaAs Schottky gate CCD as a function of the transfer inefficiency. Gate length is 5 μm.

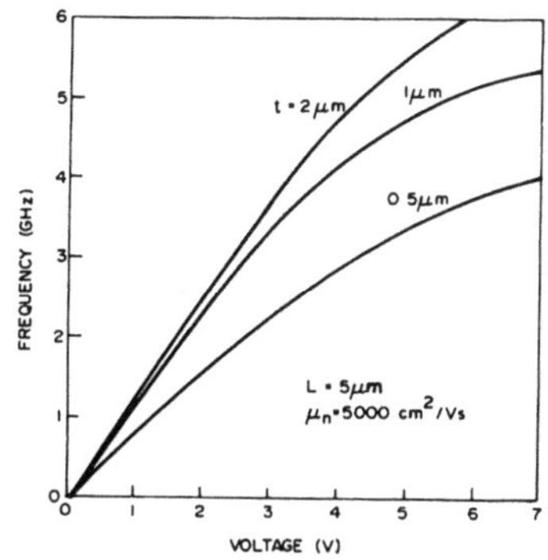

Figure 18. Maximum clock frequency for a four-phase GaAs Schottky gate CCD as a function of the clock voltage for various channel thickness (from Ref. 36).

adding the time as

$$t = 1/(1/t_{SI} + 1/t_d + 1/t_F) \tag{45}$$

and then calculating f from

$$f = 1/2\pi t \tag{46}$$

It is this frequency that is plotted in Fig. 17 as a function of transfer inefficiency. The frequency depends on ϵ, but not very strongly. The lower curves are for Si devices, (a) for a SCCD and (b) for a BCCD. For the 5-μm gate length chosen here, the predictions are several tens of megahertz for the SCCD and several hundreds of megahertz for the BCCD. The higher mobility of GaAs gives these devices a decided frequency advantage over Si devices, even though in these calculations a slightly shallower channel layer was considered.

The results are strongly dependent on the fringing field. This would imply that raising the gate voltage to increase the fringing field would increasingly give better frequency responses. This is only partially true, because, once the field reaches a critical field, the carrier velocity will saturate and no further frequency enhancement is possible.[35] For GaAs, it is more complicated due to the velocity–electric field behavior peaking and then decreasing. These effects have been considered in Fig. 18,[36] where the drift transit time limited maximum clock frequency is shown as a function of the peak–peak clock voltage. These curves indicate that CCDs operating up to several GHz appear feasible.

6. INPUT–OUTPUT CIRCUITS

There are a number of input and output circuits for CCDs and their use is judged by the ease of use, the linearity between input voltage and charge, signal

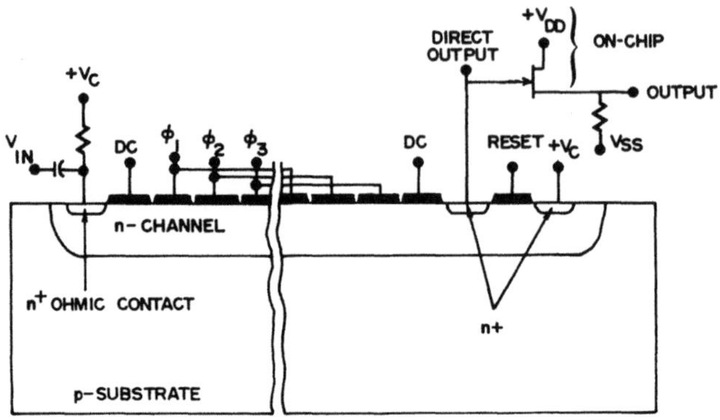

Figure 19. A GaAs Schottky gate CCD showing the input and output circuits used for the measurement of the 1 GHz device. At the output the "direct output" was used.

distortion, and noise performance. A nice discussion of the various trade-offs is given by Esser and Sangster.[9] The input technique that gives the best overall performance is the "fill and spill" method.[37,39] It is, however, difficult to implement at high frequencies. For this reason, the simpler circuit of Fig. 19 was used for the 1-GHz Schottky-gate device. The input charge packet is injected into the CCD by a pulse generator capacitively coupled to the input ohmic contact. The input pulse was several clock pulses wide, so that the input sampling is done by the clock waveform and charge is injected into several consecutive potential wells.

The best output circuit utilizes a floating diffusion connected to an on-chip electrometer amplifier.[39] The floating diffusion is reset every clock period to a reference voltage, V_c in Fig. 19. The charge transferred into the floating diffusion gives a voltage change proportional to the charge which is reset by the on-chip MESFET.

The circuit requires periodic resetting, and it is for this reason that it is difficult to use at high frequencies. A simpler circuit is that shown at "direct output" in Fig. 19. A load resistor and dc voltage is connected to that node and instead of the charge packet causing a voltage change in the on-chip amplifier, a current spike is sensed in the RC circuit. An off-chip source follower amplifier transforms the signal from its high impedance to low impedance, better suited for wide-bandwidth signals.

7. SCHOTTKY-GATE HETEROJUNCTION CCDs

The use of heteroepitaxial growth techniques with low interface state densities has made possible the low threshold current solid-state laser.[40] The same techniques have led to the design and fabrication of CCD imaging

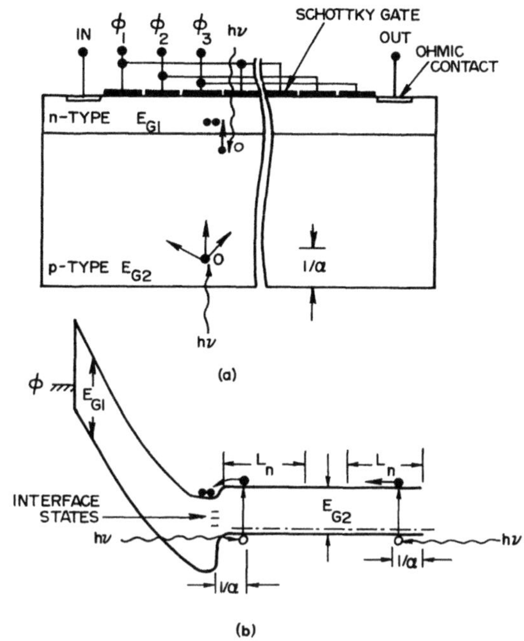

Figure 20. A. heterojunction Schottky gate CCD. Both top and bottom irradiation are shown. $1/\alpha$ is the optical absorption depth and L_n the minority carrier diffusion length in the narrow band gap substrate.

structures in which the light absorption and charge transfer functions have been separated, the goal being to optimize both independently. High quantum efficiencies are achieved by using a direct bandgap material, such as GaAs, for the absorber region. Low dark currents are attained by selecting a wider bandgap semiconductor, such as AlGaAs, that is well lattice-matched, for the charge transfer channel.

A Schottky-gate heterojunction CCD is shown in Fig. 20. It consists of a wide bandgap upper layer which is typically *n*-type to take advantage of the higher electron mobility for high-speed charge transfer. The substrate is a narrow bandgap *p*-type semiconductor. The bandgap of the upper layer should be as wide as possible for the lowest possible dark current while the bandgap of the substrate is adjusted for optimum optical absorption characteristics.

From the band diagram of Fig. 20b it is obvious that several requirements must be met for this device to function properly.

- The lattice match between the two semiconductors must be good for low interface state densities. This is necessary for minority electrons in the *p* substrate to diffuse into the *n* layer; it also ensures that interface-state generated dark current is low.

- There should not be a notch at the interface to impede the electron flow into the n layer.
- The Schottky barrier contact should have a sufficiently high barrier height to ensure low barrier-emitted dark current.

The radiation can be incident from either top or bottom in Fig. 20a. For top irradiation, the Schottky gates must be transparent for the light to reach the absorbing substrate. It passes unabsorbed through the upper layer, called the optical window. Such a transparent gate material is difficult to achieve because good Schottky barriers are usually achieved with metals, which must be very thin to be optically transparent. The situation is not much better for bottom irradiation. The light is absorbed in a thin layer near the surface since the absorption coefficient, α, is very high in direct bandgap semiconductors and the absorption depth, $1/\alpha$, is therefore quite small. Minority carriers generated at this depth must diffuse to the n layer, but the diffusion length is very shot, typically $1-10\,\mu$m, and very few carriers reach the CCD charge-transfer layer, because the substrate must be sufficiently thick for mechanical strength. An additional performance degrading mechanism is the lateral diffusive spread of the minority electrons, indicated by arrows in Fig. 20a, leading to a loss of resolution or modulation transfer function (MTF) degradation.[41] This effect is generally more important in Si than in III–V materials, because the diffusion lengths in Si are much higher. Nevertheless, it should be considered even in III–V devices when the gate lengths become very small and material improvements lead to longer diffusion lengths.

Structures of the type shown in Fig. 20 have been fabricated and tested. The first implementation was an n-type AlGaAs layer grown on a semi-insulating GaAs substrate.[42] The Al content of the CCD layer was varied between 8% and 22%. It should be kept below 35%, the direct–indirect bandgap crossover point, because beyond this value the electron mobility drops sharply. A second implementation consisted of an n-GaAlAsSb/p-GaSb structure.[43] The CCD layer has a bandgap of 1.1 eV while the light-absorbing substrate has a 0.72-eV bandgap, suitable for infrared imaging to 1.7-μm wavelengths.

The limitations of this heterojunction structure can be overcome by the three-layer device of Fig. 21. Here the light is incident from the bottom, eliminating the problems of top irradiation. The wide bandgap substrate is transparent and light absorption takes place in the narrow bandgap central layer. It can be made thick enough for good optical absorption, yet be sufficiently thin not to degrade the MTF. The additional energy discontinuity, ΔE, acts as a barrier to minority carriers. This is equivalent to a low surface recombination velocity interface.[40] However, since the middle layer is of the order of a diffusion length thick for high quantum efficiency, electrons

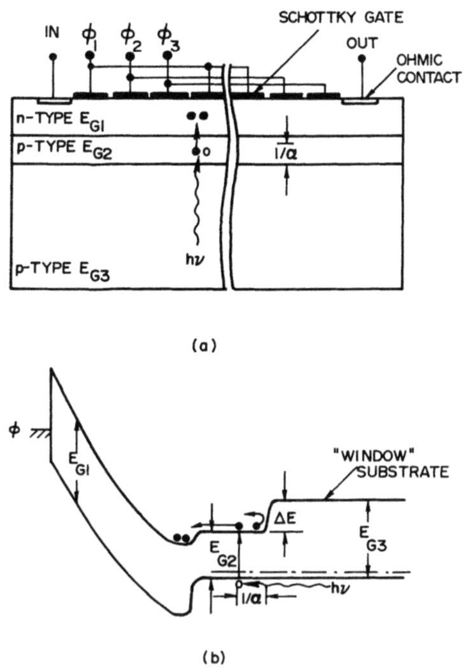

Figure 21. The device of Fig. 20 with the substrate being a wide band gap "window" substrate. The light-absorbing layer is sandwiched between the wide band gap substrate and the read-out layer.

thermally generated by interface states at the heterojunction interface will contribute to dark current. These same interface states can also act as recombination centers for optically generated minority carriers. High performance devices require low interface states because of these two requirements, and such high perfection interfaces can be made in the AlGaAs/GaAs system.

Back-illuminated devices using the concept of Fig. 21 have been designed and made. The first was an n-$Al_{0,3}Ga_{0,7}As$ ($1.5\mu m$ thick, $N_D = 1.5 \times 10^{16} cm^{-3}$)/$p$-GaAs ($1.0\mu m$ thick, $N_A = 5 \times 10^{17} cm^{-3}$)/$p$-$Al_{0,6}Ga_{0,4}As$ ($0.3\mu m$ thick, $N_A = 10^{17} cm^{-3}$) structure.(44) The wide bandgap substrate in this case was thin and the entire device was bonded to a thick glass substrate. The tansfer efficiency of this first device was 0.97 and the dark current density $\sim 10 nA/cm^2$. A more advanced version(24) using similar semiconductor layers bonded to a glass substrate with Au–Ge for ohmic contacts and Cr–Au for the Schottky gates had a dark current density of $\sim 1\ nA/cm^2$.

The dark current of such a double heterojunction device is due to several sources, as shown in Fig. 22. They are as follows:

(1) Schottky barrier thermionic emission current, J_{Sch}, is(46)

$$J_{Sch} = A^{**} T^2 \exp(-q\phi_B/kT) \exp[q(\Delta\phi + V)/kT] \tag{47}$$

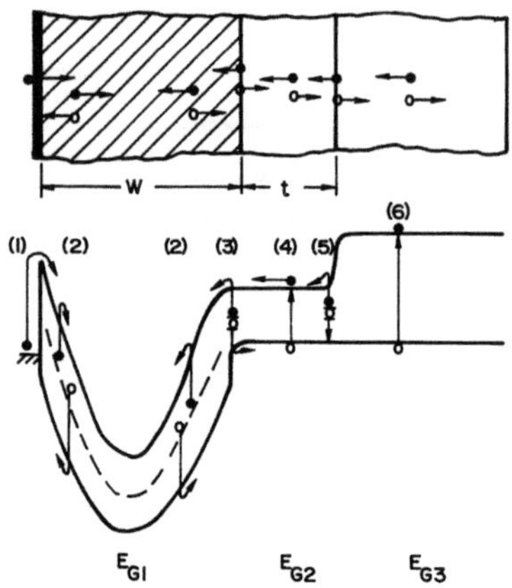

Figure 22. Dark current components in the device of Fig. 21 under operating bias conditions.

where A^{**} is the effective Richardson constant, ϕ_B is the zero field barrier height, $\Delta\phi$ is the Schottky barrier lowering potential, V is the applied voltage, and the other terms have their usual meaning.

(2) Space-charge region generation current is[46]

$$J_{\text{scr}} = qn_{i1}W_1/\tau_{g1} \tag{48}$$

where W_1 is the width of the depleted upper CCD layer, τ_{g1} is the generation lifetime in that layer, and n_{i1} is the intrinsic carrier concentration

$$n_i = (N_C N_V)^{1/2} \exp(-E_{G1}/2kT) \tag{49}$$

It is this quantity that is the chief reason that a wide bandgap semiconductor has low dark current, because of its exponential dependence on E_G.

(3) Generation current due to interface states at the E_{G1}/E_{G2} heterojunction interface

$$J_{is1} = qn_{i2}s_{01} \tag{50}$$

where s_{01} is the interface generation velocity.

(4) Diffusion current from the neutral narrow bandgap region

$$J_{\text{dif}2} = qD_{n2}n_{i2}^2/N_{A2}L_{n2} \tag{51}$$

where D_{n2}, L_{n2} are the diffusion constant and length and N_{A2} is the doping in this layer. This current can be made small by making N_{A2} large.

(5) Generation current due to interface states at the E_{G2}/E_{G3} heterojunction interface, modified by the fact that it is located a distance t from the

edge of the reverse-biased space-charge region (assuming the space-charge region extends mainly into the upper film, which is assured by doping the central layer sufficiently high):

$$J_{is2} = qn_{i3}s_{02}\exp(-t/L_n) \tag{52}$$

(6) Diffusion current from the neutral wide bandgap region, modified by its distance from the edge of the space-charge region:

$$J_{\mathrm{dif}\,3} = (qD_{n3}n_{i3}^2/N_{A3}L_{n3})\exp(-t/L_n) \tag{53}$$

Experimetnal measurements on $Al_{0.3}Ga_{0.7}As/GaAs$ n–p^+ heterojunction CCDs have been performed.[47] The results show that current components (47) and (48) are dominant and a plot of the dark current as a function of temperature is shown in Fig. 23. Also shown for comparison are the lowest reported dark currents of GaAs homojunctions[48] and high-quality Si

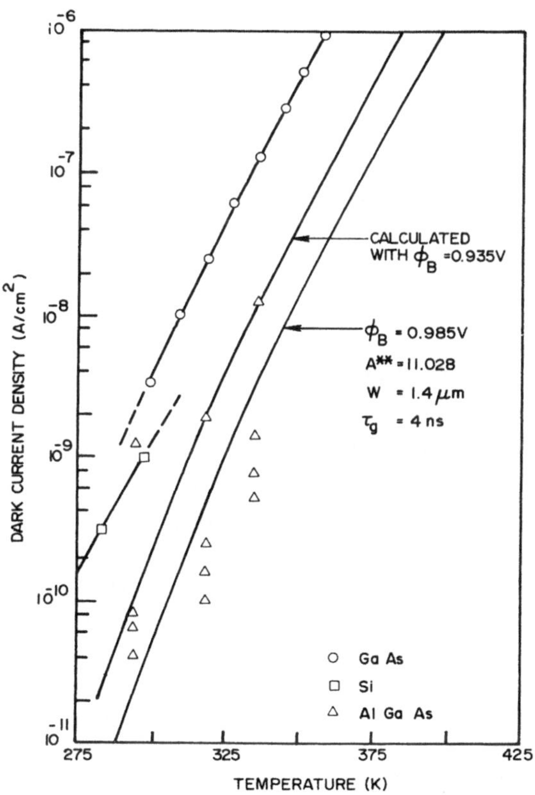

Figure 23. Dark current density as a function of temperature for GaAs, Si, and AlGaAs heterojunction devices. The calculated curves are taken from Ref. 47.

CCDs.[49] These curves show that GaAs alone has a dark current higher than Si, even though n_i considerations imply that it should be much lower. The heterojunction device current is lower than Si, but nowhere near the reduction in n_i expected. This is because the generation lifetime is so much lower than in Si. Nevertheless, the trend is in the right direction and further work should reduce the dark current to values significantly below those of Si.

This concept of heterojunction CCDs can also be used for very narrow band-gap semiconductors used for imaging. A fundamental limitation to dark current in such semiconductors is band-to-band tunneling[50] determined mainly by the small band gap. By utilizing the heterojunction approach, the light absorbing layer could be HgCdTe ($E_G = 0.1$ eV), for example, while the CCD read-out takes place in a lattice matched CdTe film ($E_G = 1.56$ eV). Substantial performance increases should be possible once the heterojunction interface is optimized. A HgCdTe/CdTe CCD has already been described,[51] but was designed for the "window" effect in which the CdTe substrate is transparent to the infrared irradiation, while CCD read-out takes place in the narrow band-gap HgCdTe. The next step would be to add an additional CdTe layer on top of the HgCdTe to reduce the dark current as shown in Fig. 21.

Whenever heterojunction devices are used in image sensing applications, it is important that energy barriers that are the result of the two band gaps with differing band gaps not prevent the collection of minority carriers. The AlGaAs/GaAs system has such an energy barrier in the conduction band and it has been shown[52] that it does not degrade the collection efficiency to any considerable extent. The reason is that the barrier is lowered by interface states to approximately 0.14 eV although the ideal interface should have a 0.31-eV barrier.

8. EXPERIMENTAL RESULTS

8.1. High-Frequency Devices

Although this chapter is concerned with Schottky gate CCDs, both surface and bulk channel MOS devices have been considered in some of the derivations to bring out the sequence in these three device types. The performance of MOS CCDs has been summarized by Esser and Sangster,[9] who show that Si peristaltic devices have been operated at frequencies of 180 MHz.

The early GaAs Schottky gate CCDs had transfer efficiencies of 0.99[15] and 0.999,[16] limited mainly by the large gap of 1.5–2 μm between gates. These efficiencies were obtained at low frequencies. A later version,[53] consisting of a 131-gate, 32-bit, 4-phase device with 5-μm gate length and 1-μm gaps exhibited a transfer efficiency higher than 0.999 at 150 MHz and was operated at frequencies up to 500 MHz. The *n* channel was a 1-μm epitaxial layer, doped to $5 \times 10^{15}\,cm^{-3}$ and grown on a Cr-doped semi-insulating substrate. The active part of the device was isolated by mesa etching.

An advanced version of this device with 259 gates (64 bits) was operated up to 1 GHz with a transfer efficiency of 0.994. At the lower frequency of 1 MHz it was 0.9999, indicating that material quality is quite good to allow devices with such high values to be fabricated. The intrinsic frequency response was estimated to be higher than 1 GHz, and was limited by measurement difficulties. For example, fast rise time, sufficiently high voltage clock pulses were not available.

These early results indicate that gigahertz response devices can be made. So far, neither the devices nor the measurement techniques have been optimized and the performance is capable of further enhancements.

The mesa-etched structures achieve a channel stop by running the gates from active channel region over the etched slope onto the semi-insulating substrate. The charge is confined to the interior of the *n* layer because the *n* region near the edge of the mesa is always operated in a punch-through condition. The epitaxial layer approach has the advantage over the ion-implanted approach, that the layer can be a high-quality, low defect density film. For ion-implanted layers, the channel is formed in a region of very high deep-level impurity concentrations. These impurities act as trapping centres thereby affecting device performance adversely. The depth of these impurities, being near midgap, causes electrons captured on them to have very long emission times. This implies that high frequency performance is not degraded, other than a slight decrease in mobility. A transfer efficiency of 0.9999 indicates a high level of perfection. It has previously been shown[(48)] that epitaxial layers with very low impurity concentrations can be grown.

Planar structures, with the advantage of no steps and no metal step coverage problems, have been made by using a Schottky gate channel stop.[(54)] In this approach metal crossovers must be provided for. These devices have not yet been characterized at high frequencies. Ion-implanted, planar structures have been operated up to 100 MHz.[(55)] The *n* channels were formed by sulfur implants at 100 keV at a dose of $10^{13}\,cm^{-2}$. The 0.99 transfer efficiency was believed to be caused by 2-μm gaps between the 5-μm gates. These gaps are much larger than the thickness of the implanted layer (0.2 μm). For good performances, these two parameters should be about the same. This implies that an overlapping gate technology is required. The ion-implant approach, however, is compatible with high-speed GaAs integrated circuits.

Considering the modest investment that has been made in GaAs CCDs, the results to date have been very encouraging. They show that GHz frequency devices can be built and theory predicts that frequencies of 5 GHz are feasible.

8.2. Heterojunction Devices

III–V materials have the unique property of low interface state heterojunction formation between compounds of different band gaps. The concept, extensively used for light-emitting diodes and solid-state lasers, is equally applicable for imaging devices in which the optical absorption layer can be of a different band gap than the CCD read-out layer. The former can be tailored for optimum photoresponse while the latter can be chosen for lowest dark current, for example. The concept has been successfully demonstrated using a GaAs optical layer and an AlGaAs read-out layer. A further extension of the concept is possible by utilizing two heterojunctions and using an "optical window" substrate.

This concept is useful for imaging applications. At present, however, the advanced state of Si CCD imaging arrays makes it difficult for a new technology to compete in a market that is not yet well established. The substantial dark current reductions expected on the basis of the wide band gaps of AlGaAs have yet to be verified. This is an important experimental gap since the concept rests largely on this premise.

9. APPLICATIONS

Schottky barrier gate CCDs have a few inherent advantages over their MOS counterparts. One of these is that they are more radiation resistant, because there is no gate insulator to degrade in a radiation environment. Another is their built-in antiblooming feature, because excess charge can flow into the gate. This is useful for imaging devices. A third, and perhaps the most important, is that they can be made in materials in which it is difficult, if not impossible, to make MOS devices. The formation of good semiconductor–insulator interfaces is difficult in materials other than Si.

When Schottky gate CCDs are made in GaAs or related III–V compounds, additional benefits accrue. The radiation resistance improves further because the minority carrier lifetime in direct bandgap semiconductors is low and cannot be degraded significantly. For example, the best radiation-hard values for Si NMOS and bipolar transistors are 10^5 and 5×10^6 rads, while for Si and GaAs MESFETs the corresponding values are 5×10^6 and 10^8 rads. Similar values as those for GaAs MESFETs are expected to apply to GaAs CCDs.

The main advantage that GaAs Schottky gate CCDs have, however, is their high-frequency performance. This is where their real promise lies. Another consideration is the ability to make heterojunction devices that are attractive for certain imaging applications. If it were not for these attributes,

GaAs devices would not be sufficiently attractive, compared to Si MOS CCDs, to be seriously considered.

Given these advantages, where might they find application? Their first use will almost certainly be in signal processing, where their high-frequency performance is the deciding factor. To gain a better perspective for this, it is necessary to look at signal processing trends. The rate of advancement of electronic systems performance is very high today, higher than ever before. This is largely the result of improvements in information processing techniques, and the overwhelming trend recently has been the increased speed and reduced cost of digital components, driven largely by the immense growth of the computer industry. It is becoming increasingly more difficult for analog signal processing techniques to progress from the laboratory to the marketplace. The attraction of digital systems is their flexibility, software programmability, and accuracy. Analog signal processing technologies have carved out special niches because of their enhanced computational speed, lower cost, smaller size, and reduced power requirements. Their weak points have traditionally been precision and dynamic range.

The main analog technologies to have shown good performance or sufficient promise to be considered are CCDs, acoustic wave devices,[19] magnetostatic wave devices,[56] and acousto-optic devices.[57] Of these, only CCDs and acoustic wave devices have found their way into specialized systems applications. For example, CCDs are used as comb filters in television receivers[58] and as time-axis converters in oscilloscopes. Acoustic wave devices have found applications as filters in television receivers, dispersive delay lines in pulse compression radars, and in correlators, to mention a few examples. GaAs CCDs operate in the same frequency range as surface acoustic waves devices but have the advantage that their frequency of operation can be easily altered by merely changing the clock frequency.

Linear signal processing functions such as filtering, correlation, and convolution at high sample rates of 10^9 samples/sec or more and high input resolution (>8 bits) are well suited for GaAs CCDs in the analog sampled data mode. Transversal filters, especially programmable, are very attractive. For example, a 250 cell CCD transversal filter operating at a 4-GHz clock rate is performing 10^{12} shift–multiply–add operations per second.[59] Frequency conversion or time-axis conversion delay lines is which data are acquired at gigahertz rates and subsequently read out at much lower megahertz rates are very useful for signal processing because digital postprocessors are incapable of handling the high data rates. One approach to implement a similar function today is to use a Si electron beam tube. A fast transient signal is written by a writing electron beam on the back of a thin Si target. The electron–hole pairs generated by the energetic incident electrons discharge diodes on the other side of the target. The charge on the diodes is subsequently read out by a second low-energy reading electron beam on the other side of the tube at a much slower rate and the information is displayed on an oscilloscope. A solid-state, nonvacuum implementation would have significant power consumption and reliability advantages. GaAs CCDs can further be combined with the more advanced GaAs integrated circuit technology to implement unique analog–digital logic combinations.

The full potential of the narrow bandgap–wide bandgap heterojunction imaging CCD has not yet been realized, but feasibility has been established. Here the chief contribution is the potentially low dark current. This would allow long integration time devices operable at room temperature. Today, if long integration times are required, Si imaging devices must be cooled to reduce dark current. Such an approach was recently used,[60] in which a Si CCD was cooled so that the integration time, which is typically around 1 sec, was extended to 10 min. This allowed near-infrared imaging of very faint sources in the galactic center. The heterojunction has the potential of being used at room temperature with long integration times and might open up a whole new series of imaging experiments, astronomical and others.

It is evident from these few examples that there are a number of applications unique to GaAs Schottky gate CCDs. They are, however, still early in their development and where they will eventually be used depends on the effort being put into their development, the progress of competing technologies, and finally the most severe critic—the marketplace. Schottky gate CCDs on other semiconductor materials have not yet been reported.

BIBLIOGRAPHY

The following are review articles and books devoted to charge-coupled devices:

D.F. Barbe (ed.), *Charge-Coupled Devices*, Springer-Verlag, New York (1980).

D.F. Barbe and S.B. Campana, in *Advances in Image Pickup and Display* (B. Kazan, ed.), Vol. 3, pp. 171–296, Academic Press, New York (1977).

J.D.E. Beynon and D.R. Lamb (ed.), *Charge-Coupled Devices and Their Applications*, McGraw-Hill, London (1980).

L.J.M. Esser and F.L.J. Sangster, in *Handbook on Semiconductors* (C. Hilsum, ed.), Vol. 4, pp. 335–421, North-Holland Publishing Co., Amsterdam (1981).

G.S. Hobson, *Charge Transfer Devices*, John Wiley and Sons, New York (1978).

R. Melen and D. Buss, *Charge-Coupled Devices: Technology and Applications*, IEEE Press, New York (1977).

M.J. Howes and D.V. Morgan (eds.), *Charge-Coupled Devices and Systems*, Wiley-Interscience, Chichester (1979).

C.H. Sequin and M.F. Tompsett, *Charge Transfer Devices*, Academic Press, New York (1975).

REFERENCES

1. J.M.L. Janssen, Discontinuous low frequency delay line with continuously variable delay, *Nature* **169**, 148–149 (1952).

2. W.J. Hannan, J.F. Schanne, and D.J. Woywood, Automatic correction of timing errors in magnetic tape recorders, *IEEE Trans.* **MIL-9**, 246–254 (1965).
3. R.A. Mao, K.R. Keller, and R.W. Ahrons, Integrated MOS analog delay line, IEEE Int. Solid State Circuits Conf. Philadelphia, pp. 164–165 (1969).
4. F.L.J. Sangster and K. Teer, Bucket brigade electronics, *IEEE J. Solid State Circuits* **SC-4**, 131–136 (1969).
5. F.L.J. Sangster, The bucket brigade delay line, A shift register for analogue signals, *Phil. Tech. Rev.* **31**, 97–110 (1970).
6. W.S. Boyle and G.E. Smith, Charge-coupled semiconductor devices, *Bell Syst. Tech. J.* **49** 587–593 (1970).
7. R.H. Walden, R.H. Krambeck, R.J. Strain, J. McKenna, N.L. Schryer, and G.E. Smith, The buried channel charge coupled device, *Bell Syst. Tech. J.* **51**, 1635–1640 (1972); C.K. Kim, J.M. Early, and G.F. Amelio, Buried channel charge-coupled devices, NEREM, Boston, pp. 161–164 (1972).
8. L.J.M. Esser, Peristaltic charge coupled device: A new type of charge transfer device, *Electr. Lett.* **8**, 620–621 (1972).
9. L.J.M. Esser and F.L.J. Sangster, in *Handbook on Semiconductors* (C. Hilsum, ed.), Vol. 4, pp. 335–421, North-Holland Publishing Co., Amsterdam (1981).
10. D.K. Schroder, A two-phase germanium charge-coupled device, *Appl. Phys. Lett.* **25**, 747–749 (1974).
11. R.D. Thom, T.L. Koch, J.D. Langan, and W.J. Parrish, A fully monolithic InSb infrared CCD array, *IEEE Trans. Electron Devices* **ED-27**, 160–170 (1980).
12. E.E. Barrowcliff, L.O. Bubelac, D.T. Cheung, A.M. Andrews, J.D. Blackwell, F. Cox, E.R. Gertner, W.E. Tenant, J.J. Ludowise, and L.E. Wood, Planar GaInSb CCDs, CCD Applic. Conf., San Diego, pp. 2–15 (1978).
13. R.A. Chapman, S.R. Borrello, A. Simmons, J.D. Beck, A.J. Lewis, M.A. Kinch, J. Hynecek, and C.G. Roberts, Monolithic HgCdTe charge transfer device infrared imaging arrays, *IEEE Trans. Electron Devices* **ED-27**, 134–145 (1980).
14. F.L. Schuermeyer, R.A. Belt, C.R. Young, and J.M. Blasingame, New structures for charge-coupled devices, *Proc. IEEE* **60**, 1444–1445 (1972).
15. W. Kellner, H. Bierhenke, and H. Kniepkamp, A. Schottky-barrier CCD on GaAs, Int. Electron Device Meet., Washington, p. 599 (1977).
16. I. Deyhimy, J.S. Harris, Jr., R.C. Eden, D.D. Edwall, S.J. Anderson, and L.O. Bubelac, GaAs charge-coupled devices, *Appl. Phys. Lett.* **32**, 383–385 (1978).
17. R.C. Eden, Comparison of GaAs device approaches for ultrahigh-speed VLSI, *Proc. IEEE* **70**, 5–12 (1982).
18. I. Deyhimy, W.A. Hill, and R.J. Anderson, Continuously clocked 1 GHz GaAs CCD, *IEEE Electr. Dev. Lett.* **EDL-2**, 70–72 (1981).
19. H. Gautier and P. Tournois, Signal processing using surface-acoustic-wave and digital components, *IEE Proc.* **127F**, 92–98 (1980).
20. H.H. Wieder, Problems and prospects of compound semiconductor field-effect transistors, *J. Vac. Sci. Technol.* **17**, 1009–1018 (1980).
21. D.F. Barbe and S.B. Campana, in *Advances in Image Pickup and Display*, (B. Kazan, ed.), Vol. 3, pp. 171–296, Academic Press, New York (1977).
22. J.D. Beynon and D.R. Lamb (eds.), *Charge-Coupled Devices and Their Applications*, McGraw-Hill, London (1980).
23. G.M. Martin, J.P. Farges, G. Jacob, J.P. Hallais, and G. Poiblaud, Compensation mechanisms in GaAs, *J. Appl. Phys.* **51**, 2840–2852 (1980).
24. E.J. Johnson, J. Kafalas, R.W. Davies, and W.A. Dyes, Deep center EL2 and anti-Stokes luminescence in semi-insulating GaAs, *Appl. Phys. Lett.* **40**, 993–995 (1982).
25. J.S. Blakemore, *Semiconductor Statistics*, Pergamon Press, New York (1962).

26. W. Shockley, *Electrons and Holes in Semiconductors*, Van Nostrand, Amsterdam (1950).
27. R. Zucca, Electrical compensation in semi-insulating GaAs, *J. Appl. Phys.* **48**, 1987–1994 (1977).
28. Y.M. Houng and G.L. Peason, Deep trapping effects at the GaAs–GaAs: Cr interface in GaAs FET structures, *J. Appl. Phys.* **49**, 3348–3352 (1978).
29. L. Eaves and P.J. Williams, Decay of the deep-level extrinsic photoconductivity response of *n*-GaAs (Cr, Si) at liquid helium temperature, *J. Phys. C: Solid State Phys.* **12**, L725–728 (1979).
30. N. Yokoyama, A. Shibatomi, S. Ohkawa, M. Fukuta, and H. Ishikawa, Electrical properties of the interface between an *n*-GaAs epitaxial layer and a Cr-doped substrate, *Inst. Phys. Conf. Ser.* **33b**, 201–209 (1977).
31. M.F. Tompsett, Charge transfer devices, *J. Vac. Sci. Technol.* **9**, 1166–1181 (1972).
32. L.J.M. Esser, in *Solid-State Imaging* (P.G. Jespers, F.v.d. Wiele, and M.H. White, eds.), pp. 343–425, Noordhoff International Publishers, Leyden (1976).
33. J.E. Carnes, W.F. Kosonocky, and E.G. Ramberg, Free charge transfer in charge-coupled devices, *IEEE Trans. Electron Devices* **ED-19**, 798–808 (1972).
34. G.F. Amelio, Computer modeling of charge-coupled device characteristics, *Bell Syst. Techn. J.* **51**, 705–730 (1972).
35. K. Hess and C.T. Sah, The ultimate limits of CCD performance imposed by hot electron effects, *Solid-State Electronics* **22**, 1025–1033 (1979).
36. I. Deyhimy, R.C. Eden, and J.S. Harris, Jr., GaAs and related heterojunction charge-coupled devices, *IEEE Trans. Electron Devices* **ED-27**, 1172–1180 (1980).
37. M.F. Tompsett and E.J. Zimany, Use of charge-coupled devices for delaying analog signals, *IEEE J. Solid-State Circuits* **SC-8**, 151–157 (1973).
38. J.E. Carnes, W.F. Kosonocky, and P.A. Levine, Measurements of noise in charge-coupled devices, *RCA Rev.* **34**, 553–565 (1973).
39. W.F. Kosonocky, Charge-coupled digital circuits, *IEEE J. Solid-State Circuits* **SC-6**, 314–322 (1971).
40. H. Kressel and J.K. Butler, *Semiconductor Lasers and Heterojunction LEDs*, Academic Press, New York (1977).
41. M.M. Blouke and D.A. Robinson, A method for improving the spatial resolution of frontside-illuminated CCDs, *IEEE Trans. Electron Devices* **ED-28**, 251–256 (1981).
42. Y.Z. Liu, I. Deyhimy, R.J. Anderson, J.S. Harris, Jr., and L.R. Tomasetta, GaAlAs/GaAs heterojunction Schottky barrier gate CCD, Int. Electron Device Meet., Washington, pp. 622–624 (1979).
43. Y.Z. Liu, I. Deyhimy, J.S. Harris, Jr., R.J. Anderson, J. Appelbaum, and J.H. Polland, Observation of charge storage and charge transfer in a GaAlAsSb/GaSb charge-coupled device, *Appl. Phys. Lett.* **36**, 458–461 (1980).
44. Y.Z. Liu, I. Deyhimy, R.J. Anderson, R.A. Milano, M.J. Cohen, J.S. Harris, Jr., and L.R. Tomasetta, A backside-illuminated imaging AlGaAs/GaAs charge-coupled device, *Appl. Phys. Lett.* **37**, 803–805 (1980).
45. Y.Z. Liu, R.A. Milano, R.J. Anderson, I. Deyhimy, and M.J. Cohen, Low dark current glass bonded AlGaAs/GaAs Schottky gate imaging CCD, Int. Electron Device Meet., Washington, pp. 338–341 (1980).
46. S.M. Sze, *Physics of Semiconductor Devices*, 2nd ed., Wiley-Interscience, New York (1981).
47. R.A. Milano, Y.Z. Liu, R.J. Anderson, and M.J. Cohen, Very-low dark-current hetrojunction CCDs, *IEEE Trans. Electron. Device* **ED-29**, 1294–1301 (1982).
48. K.W. Loh, D.K. Schroder, R.C. Clarke, A. Rohatgi, and G.W. Eldridge, Low leakage current GaAs diodes, *IEEE Trans. Electron Devices* **ED-28**, 796–800 (1981).
49. G.A. Antcliffe, L.J. Hornbeck, W.W. Chan, J.W. Walker, W.C. Rhines, and D.R. Collins, A backside illuminated 400 × 400 charge-coupled device imager, *IEEE Trans. Electron Devices* **ED-23**, 1225–1232 (1976).

50. W.W. Anderson, Tunnel current limitations of narrow bandgap infrared charge coupled devices, *Infrared Phys.* **17**, 147–164 (1977).
51. M.E. Kim, Y. Taur, S.H. Shin, G. Bostrup, J.C. Kim, and D.T. Cheung, CCDs in epitaxial HgCdTe/CdTe heterostructure, *Appl. Phys. Lett.* **39**, 336–338 (1981).
52. Y.Z. Liu, R.J. Anderson, R.A. Milano, and M.J. Cohen, Effect of heterojunction spike on the quantum efficiency of an AlGaAs/GaAs heterojunction charge coupled device, *Appl. Phys. Lett.* **40**, 967–969 (1982).
53. I. Deyhimy, R.C. Eden, R.J. Anderson, and J.S. Harris, Jr., A 500-MHz GaAs charge-coupled device, *Appl. Phys. Lett.* **36**, 151–153 (1980).
54. M.D. Clark, C.L. Anderson, R.A. Jullens, and G.S. Kamath, Planar sealed-channel GaAs Schottky-barrier charge-coupled device, *IEEE Trans. Electron Devices* **ED-27**, 1183–1188 (1980).
55. U. Ablassmeier, W. Kellner, H. Herbst, and H. Kniepkamp, Three-phase GaAs Schottky-barrier CCD operated up to 100-MHz clock frequency, *IEEE Trans. Electron Devices* **ED-27**, 1181–1183 (1981).
56. M.R. Stiglitz and J.C. Sethares, Magnetostatic waves take over where SAWs leave off, *Microwave J.* **25**(2), 18–111 (1982).
57. D.B. Anderson, Integrated optical spectrum analyzer: an imminent chip, *IEEE Spectrum* **15**(12), 22–29 (1978).
58. D.H. Pritchard, A CCD comb filter for color TV receiver picture enhancement, *RCA Rev.* **41** 3–28 (1980).
59. R.C. Eden and I. Deyhimy, Applications of GaAs integrated circuits and charge-coupled devices for high-speed signal processing, *Soc. Photo-Instrum. Eng.* **214**, 39–47 (1980).
60. J.W.V. Storey, J.O. Straede, P.R. Jorden, D.J. Thorne, and J.V. Wall, A CCD image of the galactic centre, *Nature* **296**, 333–334 (1982).

9

Schottky Barriers on Amorphous Si and Their Applications

R.J. Nemanich and M.J. Thompson

1. INTRODUCTION

While amorphous Si films had been produced for several years, it was in the 1960s that significant efforts were undertaken to produce pure *a*-Si which might be used for thin film devices. Researchers proceeded with diligence, but the norm of the time was that very few transport measurements exhibited consistent results between different samples. Furthermore, even brief exposure to air produced changes which could not be reversed by standard etching procedures. The discovery that changed all this was the method of producing *a*-Si by rf-plasma decomposition of silane (SiH_4) gas.[1] While it was initially thought that the material was pure (i.e., devoid of H), it was later shown that at least 3 at.% and as much as 50 at.% H could be incorporated into the film. It was found that stable and reproducible films could be produced in which the Fermi energy could be shifted by "substitutional" doping. This material thus represented a true semiconductor and has nucleated a whole field of research into its properties and applications.

The field is relatively new, thus there have been few in-depth studies into the properties of the Schottky barrier on *a*-Si:H, but because of the device potential there have been several attempts to qualitatively understand the basic diode properties. The unique character of the semiconductor and the desire for rapid understanding has led to several novel experiments, but it

R.J. Nemanich and M.J. Thompson • Xerox Palo Alto Research Center, 3333 Coyote Hill Road, Palo Alto, California 94304.

has also been the case that results have been misinterpreted because complete measurements of structural and electrical properties were not obtained.

This paper reviews the efforts to date starting with the first report of *a*-Si:H Schottky diodes in 1973 by Carlson *et al.*[2] To set the stage the properties of the unique amorphous semiconductor *a*-Si:H are described in Section 2. In Section 3 a qualitative description of the properties of *a*-Si:H Schottky barriers is first presented and then refined in detail in the sections on electrical measurements. The all-important (but least studied) interface kinetics are described in Section 4. The applications of the Schottky barrier configuration benefits both research and applied science and this is described in Section 5. Lastly, the current status is summarized with speculations for both future research and applications in Section 6.

2. PROPERTIES OF AMORPHOUS Si

Before the Schottky barrier properties of metal–*a*-Si:H interfaces can be addressed, it is necessary to have an understanding of the properties of the amorphous semiconductor. Since *a*-Si:H is amorphous, it therefore does not have a unit cell which describes its structure. The properties of the material are then dependent on the preparation method and as will be noted not all *a*-Si:H is alike. We will, therefore, begin by reviewing the most commonly used methods of producing the *a*-Si:H. The details of the structural properties are then addressed with particular attention to vibrational spectroscopy. The techniques which yield the vibrational spectra have proven to be the standard structural analysis. In addition to the atomic bonding configurations, microstructural anisotropy has been observed or deduced from several techniques.

Once the basic structural properties have been determined, the electrical properties can be explored. Intrinsic effects due to disorder play a major role in the transport properties of *a*-Si:H. Lastly of particular importance to Schottky barriers is the nature of the free surface. Unfortunately there have been few studies in this area.

2.1. Deposition Methods

The most significant break from traditional deposition methods occurred in 1969 when Chittick *et al.*[1] produced amorphous silicon by plasma decomposition of silane gas in contrast to the more traditional sputtering or evaporation. These latter techniques had been pushed to the state of the art, but still the material exhibited weak photoconductivity and high defect densities in the valence to conduction band gap. In contrast even the earliest efforts at producing amorphous silicon by rf plasma decomposition yielded high photoconductivity and a two orders of magnitude decrease in the density of states in the gap.[3] Coincident with the development of the plasma deposition (or glow discharge) technique was the development of reactive sputtering of Si in a hydrogen atmosphere.[4] This technique produced *a*-Si:H films with properties similar to those produced by plasma decomposition.

While it was quickly established that the reactivity sputtered films contained H bonded to the Si, it was several years before it was established that the plasma deposited films also contained H. With this revelation it became clear that the H served as a terminator to intrinsic defects in amorphous silicon which adversely affect the electrical properties. This particular point is somewhat ironic because in the years before plasma deposition was developed, techniques were optimized to produce the purest films.

Several configurations were initially utilized to produce *a*-Si:H, but the diode configuration developed by J. Knights[5] has become widely used. A schematic of the diode deposition system is shown in Fig. 1. In this configuration a mixture of gases flows into the chamber at a pressure of 10 to 100 mtorr. The plasma is excited by applying a 13.56-MHz rf field between the cathode and ground. Here the anode on which the substrates are placed is usually held at ground potential, but experiments have been conducted with varying substrates-bias potential.[5,6] Typically, 1 to 50 W of rf power are used to excite the plasma. The silane is decomposed into various radicals but this complicated reaction depends on gas phase (or plasma) constituents and on reactions occurring at the substrate surface. The rf power and gas phase constituents clearly affect the plasma; in addition the silane gas can be diluted by inert noble gases or replaced by higher-order silanes such as disilane. A

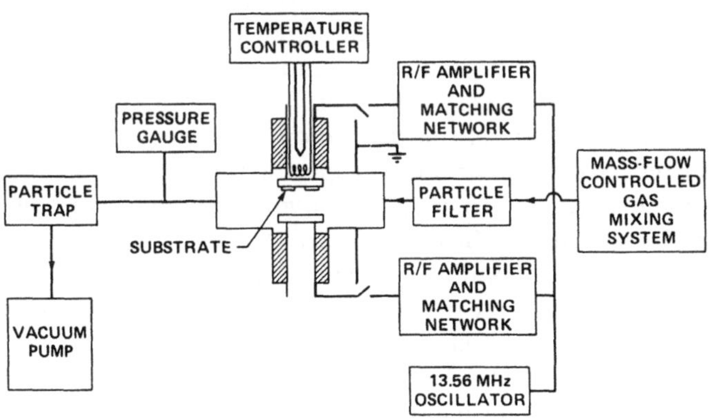

Figure 1. Schematic of a diode configuration plasma-deposition system used for producing *a*-Si:H (from Knights.[5])

parameter that has proven to be very important to the substrate interactions is the substrate temperature. Unfortunately as the plasma constituents clearly affect the surface reactions, it is difficult to study the effect and details of specific parameters. Generally experimentors have empirically varied the deposition parameters to optimize for lowest gap state defect density material. While the optimum conditions can vary from system to system, it is generally found that 1 to 2 W of rf power exciting a plasma from pure silane (or disilane) gas and depositing on substrates held at between 200 and 350°C produces optimal material. Typically, deposition rates of 1 to 5 Å/sec are achieved.

While for several years it was not clear whether rf SiH_4 plasma deposition or sputtering produced the most optimal *a*-Si:H, it seems that most investigators are now favoring the rf plasma decomposition technique as there is considerably more experience in producing high-quality material by this technique. However, developments continue for both techniques and it is difficult to state definitively that one method will produce *a*-Si:H suitable for all electronic applications. Furthermore totally new schemes are also being considered. One of these called Homo CVD[(7)] seems to be especially useful for producing *a*-Si:H with high H concentrations and a large band gap while retaining a low gap state defect density.

2.2. Structural Properties

The structural properties of the disordered film range from understanding the atomic bonding configurations to examining microstructure, anisotropy, and macroscopic properties. In describing the structure we shall begin on the atomic scale.

The vibrational spectra obtained from Raman scattering and infrared absorption were proven initially controversial, but now provide one of the most useful techniques characterizing differences in the atomic properties of the amorphous films.[8–11] It should be reemphasized here that *a*-Si:H does not have a single structure as a crystal might. Instead, both the disorder and the atomic constituents vary as the deposition conditions are changed. These changes are reflected in the vibrational spectra. Of the two techniques, Raman scattering is more sensitive to vibrations involving homopolar bonded atoms; thus the properties of the Si network are displayed while ir absorption is sensitive to heteropolar bonding and will be reflective of the H environments. Representative Raman and ir spectra are shown in Fig. 2. In the Raman spectra the features at $\omega < 600\,\mathrm{cm}^{-1}$ are reminiscent of the vibrational density of states of crystalline Si. The characteristic continuum spectra clearly indicate an amorphous rather than microcrystalline structure. The similarity of the vibrational density of states of crystalline Si and *a*-Si:H indicates that the Si bonding is similar in both cases. Diffraction measurements also indicate a

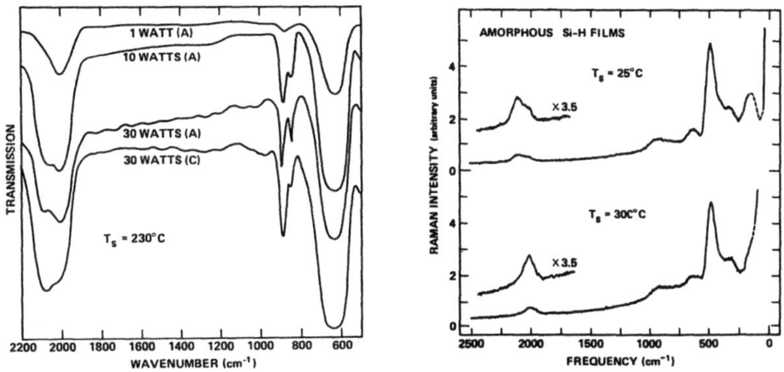

Figure 2. (a) The infrared transmission of a series of *a*-Si:H films prepared with different rf power and (b) the Raman spectra of films prepared at different substrate temperatures (from Lucovsky *et al.*[10] and Knights *et al.*[9])

bond length nearly identical to that in crystalline Si.[12] It has recently been demonstrated that the variations in the network disorder are reflected in the broadening of the Raman spectra and this is most likely due to bond angle fluctuations.[13]

While slight differences have been observed in the Si network vibrations of *a*-Si:H prepared under different conditions,[13] larger changes were observed in the H environments.[8–10] The H vibrations are split into three groups of bands which are evident in the ir spectra and the results are summarized in Fig. 3. The high-frequency band at 2000 to 2150 cm^{-1} is due to H motion along the Si–H bond. In contrast the bands at 600 cm^{-1} and the group of lines from 800 to 1000 cm^{-1} represent H motion transverse to the bond. The lower-frequency band at 600 cm^{-1} is due to all H atoms (one, two, or three) bonded to a single Si moving in phase, while the band at 800 at 900 cm^{-1} is due to complicated displacements due to multiple bonded H (two or three) to a single Si atom. Analysis of the spectra has led to the suggestion that the H can be incorporated in several environments.[8–10] Configurations of one, two, or three H bonded to a single Si are identified. Furthermore vibrational coupling of these modes [due to polysilanelike chains—$(SiH_2)_n$] leads to identifiable changes in the vibrational spectra. It has been found empirically that material with a minimum of SiH_2 and SiH_3 units but still containing 3 to 8 at. % H was the most optimum for electrical properties.[14] Thus the ir signature of an optimal film will show single modes at 2000 and 640 cm^{-1} and very little absorption at 800 to 950 cm^{-1}.

Once it is realized that a significant quantity of H is incorporated in the films, it should be determined whether the material is homogeneously distributed. For material containing large quantities of H ($>$ 10 at. %)

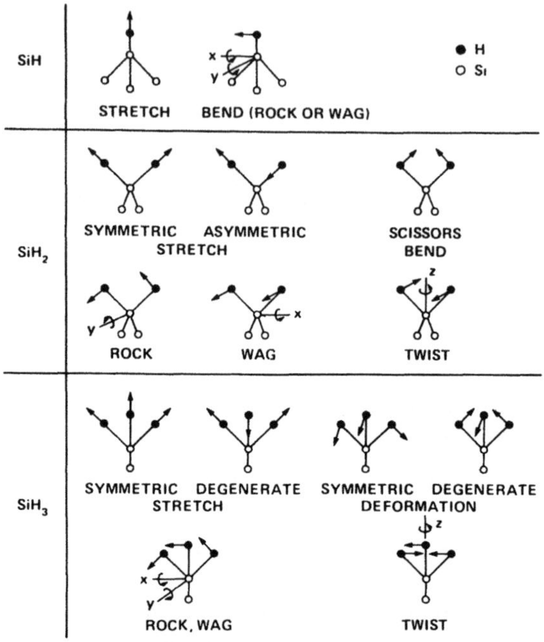

Figure 3. The vibrational displacement of the Si–H groups incorporated into *a*-Si:H films (from Lucovsky *et al.*[10])

columnar morphology is often evident.[15] Two scales of microstructure are observable. For thin films TEM has indicated ~100 Å diameter islands while SEM of thicker films has indicated 1 μm scale columnar structures. Apparently the larger scale structures still contain the smaller scale morphology. It has furthermore been shown that the material behaves as if it were two phase—with high H concentration material between the columns and lower concentration in the columns.[16]

In contrast to the films containing large quantities of H, optimal films show no sign of morphology under either TEM or SEM examination. However, recent NMR results indicate clustered and unclustered (or distributed) H environments.[17] While it has been suggested that this indicates a vestige of columnar morphology, other possibilities such as small voids or even the fluctuations of a random distribution have not been eliminated.

2.3. Electronic Properties

The electronic properties of any semiconductor are dominated by the states in the band gap between the edges of the valence and conduction bands.

Because wave vector is not a meaningful parameter for electronic states in an amorphous semiconductor, it is not necessary to obtain the electronic dispersion of the band edge states as in a crystal. There is, however, no well-defined band edge in the density of states, but the distribution of states at the valence and conduction band edges extends into the gap where they become localized because of the disorder potential. Furthermore, even in the most optimal films, defect densities of $10^{15}/\mathrm{eV\,cm^3}$ exist in the gap, and these states pin the Fermi energy at near the center of the 1.8 eV gap. To explore the properties of the electronic states, the techniques of luminescence, electron spin resonance (ESR), optical absorption, deep level transient spectroscopy (DLTS), and photoconductivity have been utilized.

In this section, the relevant experimental probes of the electronic properties of undoped *a*-Si:H will be briefly reviewed. The picture of gap states properties will evolve in this presentation. It should be noted that many research programs have focused on the properties of *a*-Si:H, thus the presentation here can only touch on a few. More in-depth reviews have been presented by Street,[18] and Crandall,[19] and we recommend referring to these articles.

Consider first the optical absorption. A recent detailed work by Cody *et al.*[20] is summarized in Fig. 4. This result showed that for less optimal *a*-Si:H, the optical absorption of the band tails flattened and decreased more slowly, indicating a higher concentrations of band tail states. They also showed that the changes were attributed to disorder and not to varying amounts of H incorporated in the films. While it is difficult to absolutely define a band-gap value for *a*-Si:H, it appears that a value of 1.8 eV is representative

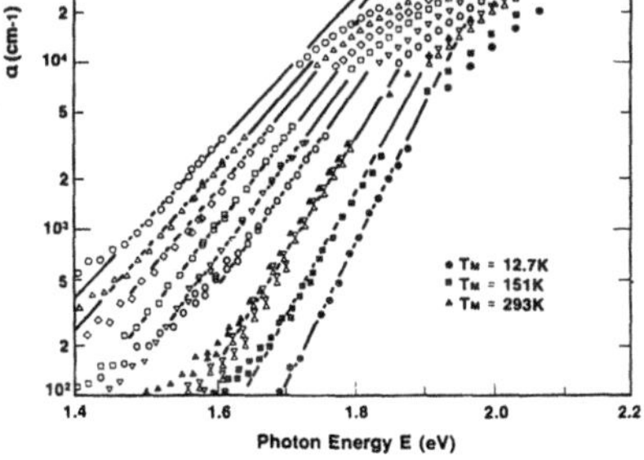

Figure 4. Optical absorption coefficient, α, as a function of photon energy. The solid symbols refer to data obtained at different measurement temperatures, T_M and the open symbols refer to isochronal anneal temperatures to evolve H (from Cody *et al.*[20])

of the regime where localizations begins and the absorption has dropped to a value of 10^{-3} cm^{-1}. Thus the absorption below this value represents localized states which will strongly affect the transport properties.

The luminescence and ESR measurements have been used to successfully study the recombination processes. Typical luminescence spectra show peaks at 0.9 eV and 1.4 eV.[18] These two peaks exhibit different temperature dependences but both are strongly quenched at room temperature. The ESR signal shows a resonance at $g = 2.0055$. For the most optimal material no signal is detectable—indicating a gap state density of $< 10^{16}$ eV^{-1} cm^{-3}. However, when illuminated with above gap radiation, the ESR (LESR) of even the most optimal samples exhibits a signal corresponding to $\sim 10^{16}$ eV^{-1} cm^{-3} states. The temperature dependence of the LESR correlates with the 0.9-eV luminescnece line and is associated with an intrinsic defect which is usually assumed to be a "dangling" Si bond. Furthermore for less optimal material, the ESR shows an anticorrelation with the 1.4-eV band. It has thus been interpreted that this band results from band-to-band luminescence from the tail states.[18]

The details of the states in the gap have been arived at by DLTS,[19,21] dispersive transport,[22] and optical absorption measured by photothermal deflection spectroscopy (PDS).[23] In addition, photoconductivity measurements have yielded results consistent with the PDS optical absorption but have been in general more difficult to interpret.[24] Without going into detail we present in Fig. 5 the gap state density of states arrived at by PDS (which is similar to that of DLTS). The features are a sharp conduction band edge with a broader valence band edge. Extending into the gap from the valence band is a band of states associated with the "dangling" Si bond defects.

The important aspects of the features of the picture presented here are that holes are rapidly trapped in the band tail states while electrons can be thermally excited out of the much shallower conduction band tails. Electron

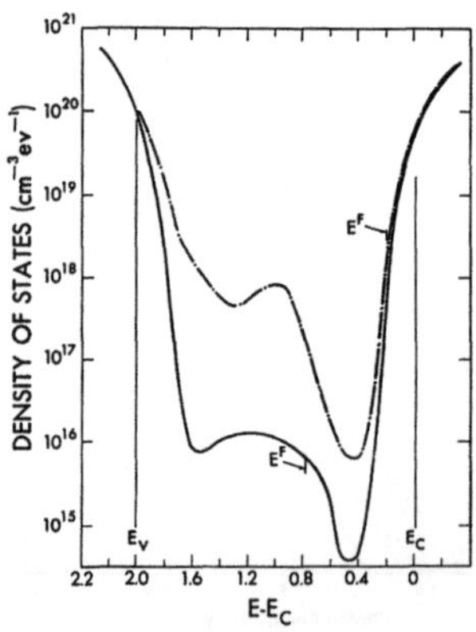

Figure 5. The density of states in the gap for undoped (solid line) and P-doped (dashed line) *a*-Si:H. The data have been determined from DLTS, Photoconductivity, and absorption measurements (from Jackson *et al.*[24])

mobilities of $\sim 2\,\text{cm}^2\,\text{V}^{-1}\,\text{s}^{-1}$ are usually obtained, but much lower hole mobilities are measured.[25] The role of the defect band is similar to deep impurity levels in crystalline semiconductors in that it causes carrier recombination.

A most significant effect in *a*-Si:H is that the Fermi energy can be shifted by incorporation of dopants.[24] This is, however, not accomplished without changes; typically the gap state densities increase by several orders of magnitude for a 0.5 eV shift in E_F. Thus in all practical configurations, doped layers must be thin ($\leq 500\,\text{Å}$) or defect recombination will dominate the transport properties.

2.4. Surfaces

There have been few studies of surface properties on this new type of semiconductor, but those that have occurred address three areas: absorbate effects,[27] transport changes due to band bending,[24] and surface or interface states. While the first two effects are most studied and relevant to many experimental device configurations, the last effect is most important in understanding Schottky barrier formation.

The presence of surface or interface states has been deduced from optical absorption measurements of very thin *a*-Si:H films.[28] For undoped samples the defect absorption did not scale with thickness. The results are summarized in Fig. 6. For the undoped samples the absorption corresponds to $\sim 10^{13}$ defects/cm and is only weakly dependent on thickness. For doped samples a stronger thickness dependence is observed (indicating bulk defects), but the extrapolation to zero thickness also exhibits a surface density of $\sim 10^{12}$ spins/cm^2. This experiment does not distinguish between top surface or substrate interface. Furthermore, the experiments were carried out on samples which had been exposed to air and thus had an oxidized surface. While more work must be carried out in this area, it is reasonable to consider that the surface exhibits defects which will affect Schottky barrier formation in ways similar to that of crystalline Si.

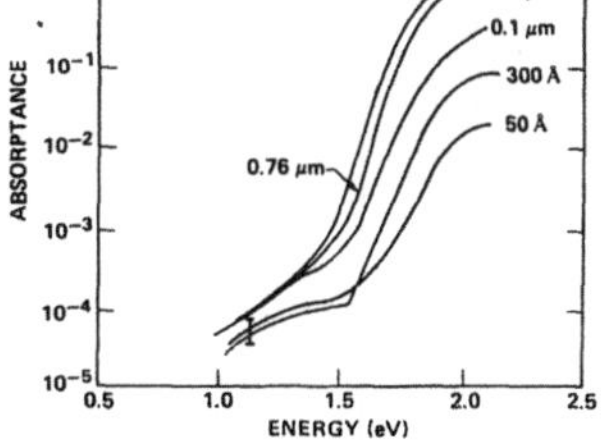

Figure 6. The absorptivity vs. photon energy for various thickness *a*-Si:H films. The region below 1.5eV which is attributed to a defect band is only weakly thickness dependent (from Jackson *et al.*[28])

3. THE SCHOTTKY BARRIER ON a-Si:H

One aspect of *a*-Si:H devices that is different from crystalline Si is that the active region is usually intrinsic or undoped. The density of gap states in all *a*-Si produced to date is significant enough to pin the Fermi energy near midgap, hence working with undoped material can be carried out in a controlled fashion. Furthermore, doping *a*-Si:H either *p* or *n* type causes a large increase in gap states densities which leads to carrier recombination. Thus the Schottky barrier of most interest is formed on the undoped material. A schematic of the band structure is shown in Fig. 7. Possible transport mechanisms are indicated by arrows in the figure. The circles in the semiconductor gap indicate defect centers near the Fermi energy. The role of these states is similar to deep centers in crystalline Si. Thus the depletion width is limited by the defect density. Another aspect of the *a*-Si:H diodes that is difficult to represent in the usual schematic diagrams is the band tail states. The states localized due to the disorder are represented by dashed lines in the band tails. The influence of the defect states on transport is discussed later.

Of significant importance is the fact that the Fermi energy is near midgap (and the gap energy is $\sim 1.8\,\mathrm{eV}$). Thus even for a case with zero built-in potential, a barrier of half the gap can be formed. Furthermore, with high work function metals, barriers greater than 1.1 eV can be achieved: this is significantly larger than similar structures formed on crystalline Si.

We now turn to considerations about the depletion region. While solutions for general gap state distributions are difficult to obtain, two cases which typify different possibilities can be solved exactly. For a sharp donor or acceptor band (as is usually the case in crystalline Si) the depletion region will exhibit a parabolic potential. Thus $V = V_d(x - w)^2$ where V_d is the built-in potential, and w is length of the depletion region. The charge density in the

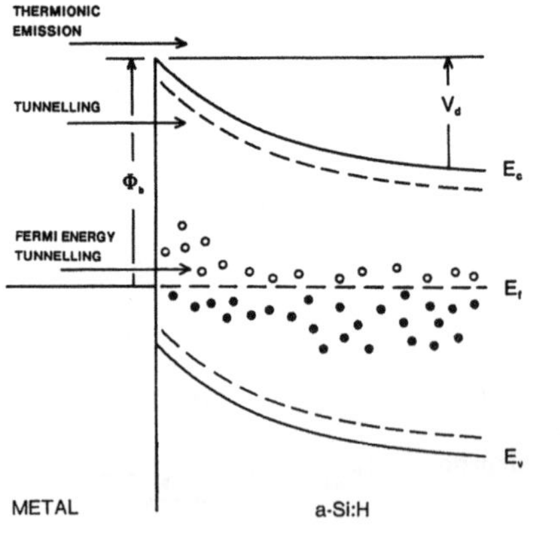

Figure 7. Schematic of the Schottky barrier on *a*-Si:H. The different possible transport mechanisms are indicated.

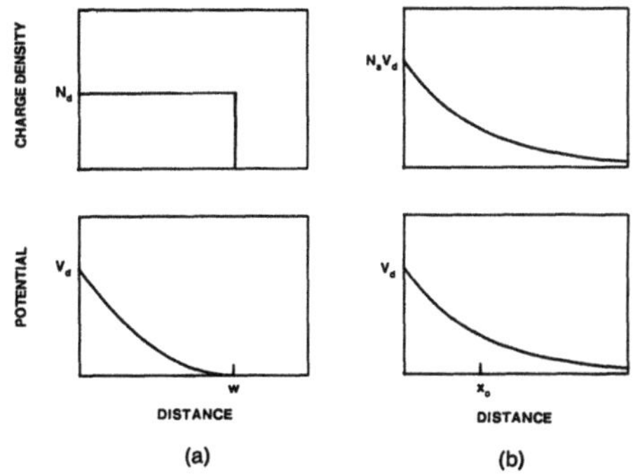

Figure 8. Schematic of charge density and potential profiles for a single-defect (donor) band (a), and a uniform distribution of defects in the gap (b).

depletion region is essentially constant and equal to the density of donor (or acceptor) centers N_d. For amorphous semiconductors a uniform distribution of states in the gap is often considered as a reasonable approximation, and this results in an exponential potential distribution $V = V_d \exp(-x/x_0)$ where x_0 is defined as the field decay length. The charge density also exhibits an exponential form with a value at the interface of $N_a V_d$ where N_a is the gap state density per eV. The potential and charge distributions for the two cases discussed here are shown in Fig. 8. The depletion region profiles have been discussed by Spear *et al.*[29]

As described in the previous section, a reasonable description of the gap states of *a*-Si:H is a defect band of $\sim$0.4 eV wide centered slightly below midgap. Thus for Schottky barriers with a built-in potential of $\sim$ 0.4 eV or less, the exponential potential is a good description of the sytem. For metals with larger work functions, larger values of the built-in potential may be obtainable. In this case the bands will exhibit a combination of exponential and parabolic behavior. Near the metal/*a*-Si:H interface, the potential will be parabolic while at the far end of the space charge region, the potential will be exponential. These results demonstrate another important difference in the properties of Schottky barriers on amorphous or crystalline Si—namely, that the characteristics of the depletion region on *a*-Si:H diodes can vary significantly for structures with different barrier heights. Furthermore, since the states in the gap are a function of deposition conditions, the Fermi energy and the depletion width are material-dependent properties.

Another aspect that is affected by the midgap Fermi energy of *a*-Si:H Schottky barriers is the minority carrier contribution. Because the Fermi energy for crystalline silicon diodes is located at the band edges, minority carrier currents are usually four orders of magnitude weaker than the majority carrier current.* The details of the minority (hole) current in *a*-Si:H diodes will

*For a description of *c*-Si diodes see Ref. 30.

be described in the next section, but the basic result is that minority carrier contributions cannot be always neglected. It is only the fact that the band gap is so much larger for *a*-Si:H that prevents the "minority" current from being nearly equal to the electron current in forward bias characteristics.

3.1. Current–Voltage Measurements

Because of the high resistance of undoped *a*-Si:H, current—voltage ($J-V$) measurements have proved to be the most reliable method of determining the barrier height of Schottky diodes. As in the case of crystalline semiconductors, the Schottky diode equation can be written as

$$J = J_0 \exp(eV/nkT)[1 - \exp(-eV/kT)] \tag{1}$$

where J_0 is the reverse bias saturation current, n is the ideality parameter, V the applied voltage, T the temperature, e the electron charge, and k the Boltzmann constant. This equation is often written as

$$J = J_0 \exp(eV/nkT) \tag{2}$$

and this expression is a good approximation for V greater than $3kT/e$. The properties of the ideality parameter can lend some insight into the current transport mechanism. For instance if the forward-bias transport is limited by "thermionic emission," then $n = 1$. If, however, the transport is diffusion limited, the carrier recombination in the depletion then can cause n to vary between 1 and 2. The details of diffusion-limited transport have been modeled by Chen and Lee.[30] It must be emphasized that all of the transport processes may be occurring in the Schottky barrier, but the current–voltage measurements will usually be dominated by only one of the mechanisms. For transport over the barrier, the diffusion and thermionic emission processes are in series, and the one that provides the highest resistance to carrier transport will dominate the $I-V$ characteristics.

Besides recombination and diffusion-limited transport, several effects can cause deviations from unity for the ideality parameter. These include interface layers, lateral inhomegeneities, and field dependence of the barrier. This last property is due to image force lowering which will cause the effective barrier to occur slightly into the semiconductor. Hence the barrier height will change with applied voltage for even the most ideal cases. Thin oxide layers between the metal and semiconductor or a defective surface layer on the *a*-Si:H can also lead to recombination at the interface. Lastly, lateral inhomogeneities cause barrier height fluctuations which can also lead to changes in the ideality values.[32]

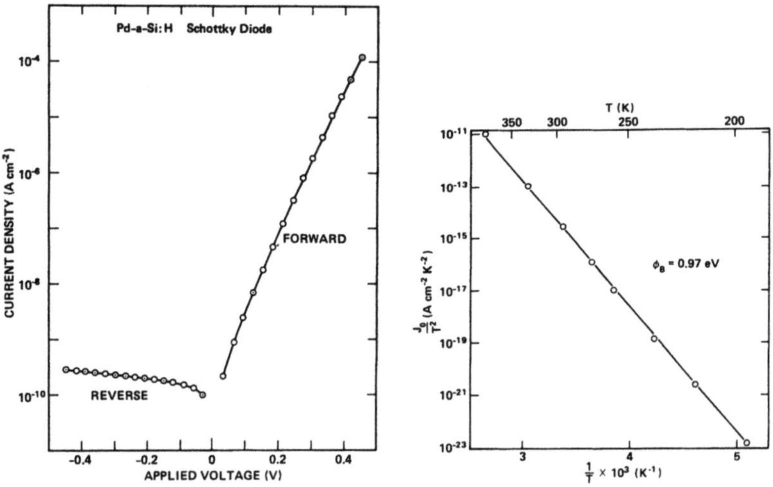

Figure 9. (a) The forward and reverse bias current–voltage characteristics for a Pd/*a*-Si:H Schottky barrier. (b) The extrapolated value of J plotted in a "thermionic emission plot" (from Thompson *et al.*[34])

Because *a*-Si:H has an electron mobility of $\sim 2\,\text{cm}^2\,\text{V}^{-1}\,\text{s}^{-1}$, it might be anticipated that the transport is diffusion limited. Several groups have demonstrated that Schottky diodes can be routinely produced with ideality parameter values of ~ 1.2.[2,33] Furthermore when steps are taken to assure an intimate, uniform interface, then ideality values of less than 1.05 have been obtained.[34] Typical J–V curves for Pd Schottky diodes exhibiting the low ideality values are shown in Fig. 9a. The interface kinetics which are responsible for the improved ideality values are discussed in the following section.

The localized band tail and gap defect states can have further profound effects on the conduction process in the Schottky barrier. A number of studies have been made on a wide range of material with different defect densities. It has been shown that field emission tunneling through the Schottky barrier from localized states can dominate transport in high defect material particularly at low temperatures.[35] Tunneling to extended electronic states can also be dominant at low temperatures in low defect density materials.[36]

If the forward bias characteristics are due to predominantly thermionic emission, then

$$J_0 = A^{**}T^2 \exp(-e\Phi_B/kT) \tag{3}$$

where Φ_B is the effective barrier height (including image force lowering) and A^{**} is Richardson's constant. Because of the disorder in *a*-Si:H which leads to localized band tail states, it is impossible to calculate A^{**}. Thus the J–V characteristics at a single temperature cannot be used. If the temperature dependence of J_0 is determined, then Φ_B can be obtained. This is accomplished by plotting $\ln(J_0/T^2)$ vs. T^{-1}. The slope will yield Φ_B (actually $-e\Phi_B/k$) while the intercept at $1/T = 0$ yields A^{**}. To apply this formalism the ideality parameter of the J–V characteristics must be less than 1.1 over the entire temperature range.

If the transport is instead diffusion limited, then the temperature dependance of J_0 is given by

$$J_0 = eN_c\mu E_\mu \exp(-e\Phi_B/kT) \tag{4}$$

where μ is the electron mobility, E_μ is the maximum field, and N_c is the effective density of states in the conduction band. Hence, the barrier height can be determined from the slope of the plot of $\ln J_0$ vs. $1/T$. The temperature dependence of J_0 can then be used to distinguish between the two transport mechanisms.

The thermionic emission characteristics of Pd Schottky diodes have been recently verified by thermionic emission plots over the temperature range of 200 to 350 K,[34] and the results are shown in Fig. 9b. At temperatures below 200 K the transport is trap limited and diffusion characteristics appear to dominate the junction characteristics.[37] In other cases Deneuville and Brodsky[38] have shown characteristics indistinguishable between thermionic emission and diffusion for Pt/a-Si:H diodes. They also found that A^{**} exhibited a value similar to that of crystalline Si. Other investigators have interpreted the temperature dependence of J_0 as indicating diffusion-limited transport, but in all these studies the value of $n \geq 1.2$.

In a study by Wronski and Carlson,[33] a series of Schottky diodes of seven different metals was examined. The barrier height was determined by both J–V and C–V measurements and plotted vs. metal work function: the

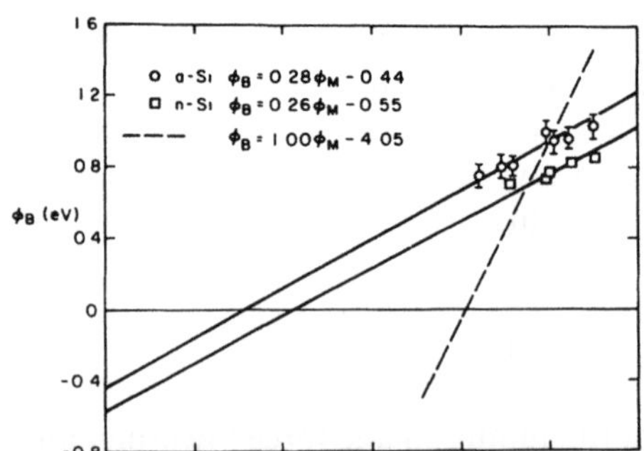

Figure 10. Plot of barrier height Φ_B for a-Si:H and single-crystal Si Schottky barriers vs. metal work function. The extrapolations suggests similar surface state densities for the diodes (from Wronski and Carlson.[33])

results are displayed in Fig. 10. While this study neglected effects due to interface kinetics, it demonstrated a similar trend with diodes produced on crystalline Si. In each case the *a*-Si:H diode exhibited a barrier height of $\sim 0.2\,\mathrm{eV}$ greater than on crystalline Si.

3.2. Capacitance Measurements

While the $J-V$ characteristics of *a*-Si:H diodes exhibited properties similar to those of crystalline Si, the capacitance properties are distinctly different. This is because of the deep localized states in the gap. To understand the frequency dependence an equivalent circuit shown in Fig. 11 has been used by several authors to analyze their data.[29,39–44] .The relationships of the circuit to the regions of the diode are also shown in Fig. 11. The two limiting case of the solution are $\omega \to 0$ and $\omega \to \infty$. For $\omega \to \infty$ there will be no response due to the states in the gap. The capacitance simply becomes the geometric, dielectric capacitance, i.e., $C = \epsilon/d$ where d is the film thickness.

The more significant region is at the $\omega = 0$ limit. At this point all the states near E_F will contribute to the change in capacitance with applied voltage. If the resistance of the space charge region dominates that of the bulk (i.e., $R_1 > R_2$), then the capacitance will be dominated by the changes in occupation of the localized states. The frequency dependence of the $C-V$ curves were first measured in detail by Snell *et al.*,[40] and similar results were obtained by several other groups. More recently Beichler *et al.*[43] measured the frequency dependence of undoped and doped samples over a frequency range of 10^{-3} to 10^4 Hz. The results are shown in Fig. 12. It was found that even at the lowest frequencies of 10^{-3} Hz, the capacitance of the undoped *a*-Si:H diodes did not saturate while at between 1 and 10 kHz the capacitance had reached the geometric limit.

While the frequency variation of the *a*-Si:H Schottky barrier is significant, it might be anticipated that a frequency can be chosen where the deep states will not respond, but where the conduction band can respond. Several authors have suggested that for frequencies in the range of 100 to 1000 Hz, $C-V$ curves are obtainable which exhibit linear regions of $1/C^2$ vs. V plots. Extrapolation of these curves yields the built-in potential or band-bending.

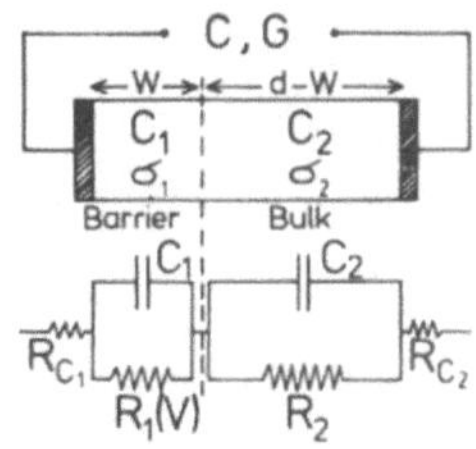

Figure 11. The equivalent circuit used to analyze the $C-V$ properties of *a*-Si:H Schottky barriers, and its relation to the specimen geometry (from Snell *et al.*[40])

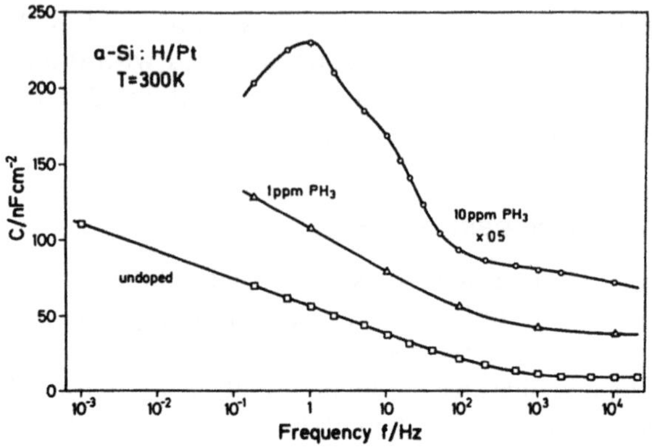

Figure 12. The frequency dependence of the capacitance for differently doped *a*-Si-H Schottky barrier diodes (from Beichler *et al.*[43])

Part of the problem becomes apparent when the detailed *C–V* curves are examined. Typical curves obtained by Snell *et al.*[40] are shown in Fig. 13. The dramatic drop in capacitance at small forward voltages makes extrapolation of $1/C^2$ plots difficult. Furthermore, since the contribution from the deep states is an unknown quantity in most cases, it is best to rely on other methods (*J–V* or internal photoemission) to support the results obtained from *C–V* measurements.

While the frequency dependence of the *C–V* properties has proved troublesome, by detailed analysis properties of the barrier and the gap state density in *a*-Si:H can be determined. It was first shown by Snell *et al.*[40] that by using density of gap states determined from other measurements, the space charge density could be determined. Further analysis by Viktorovitch *et al.*[39,42] and others showed that the density of states above E_F could actually be derived from the zero-frequency *C–V* curves. From the derived density of states, the barrier profile and the depletion width can be determined. While controversy remains on the details and limitations of the analysis, it is

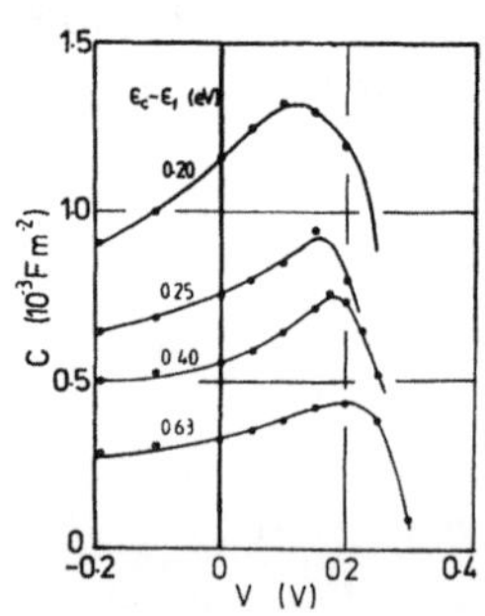

Figure 13. The barrier capacitance as a function of bias voltage measured at a frequency of 10 Hz. The energy of the Fermi level with respect to the conduction band for the differently doped samples is indicated (from Snell *et al.*[40])

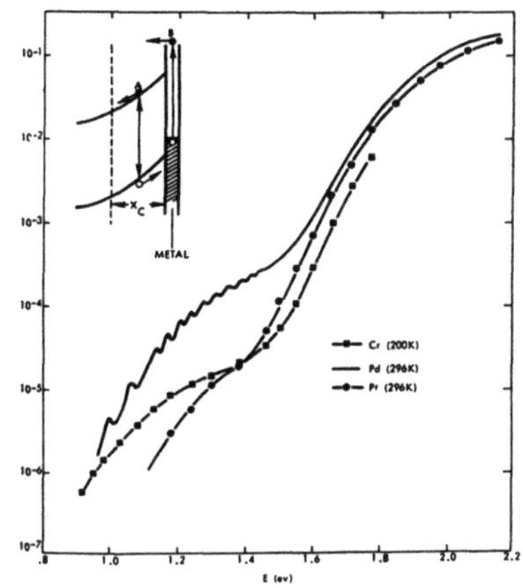

Figure 14. Photoelectric yield $Y(E)$ vs. photon energy E for different Schottky barriers on a-Si:H. The broad feature extending below 1.4 eV is due to internal photoemission (from Wronski *et al.*[45])

clear that the C–V measurements will continue to be important for characterizing the properties of both the Schottky barrier and the materials aspects of a-Si:H.

3.3. Internal Photoemission

While photoelectric measurements of the internal photoemission process have proved to be one of the most accurate methods of determination of barrier height in crystalline Si, it has been only sparingly used in a-Si:H. The first measurements were reported by Wronski *et al.*,[45] and their results are shown in Figs. 14 and 15. Following the standard theory by Fowler,[30] the internal photoemission yield (Y) is given by

$$\begin{aligned} Y(h\nu) &= A(h\nu - \Phi_B)^2 \qquad h\nu > \Phi_B \\ &= 0 \qquad h\nu < \Phi_B \end{aligned} \tag{5}$$

where A is a constant determined by the absorption of the metal and the

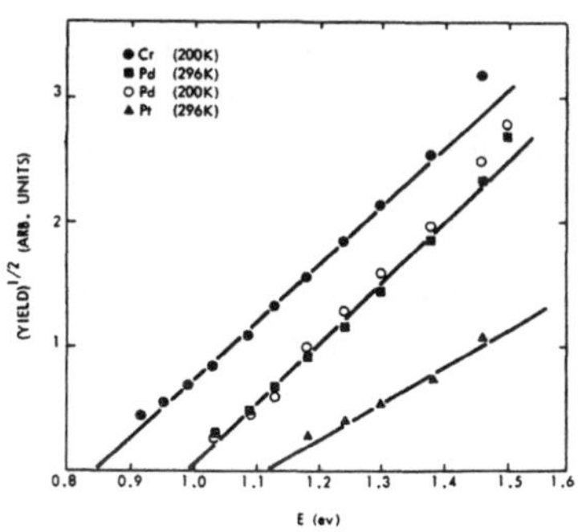

Figure 15. Plot of $[Y(E)]^{1/2}$ vs. photon energy E. The intercepts of the straight lines with the E axis indicate the potential barrier heights for the different metals. The measurement temperature is indicated (from Wronski *et al.*[45])

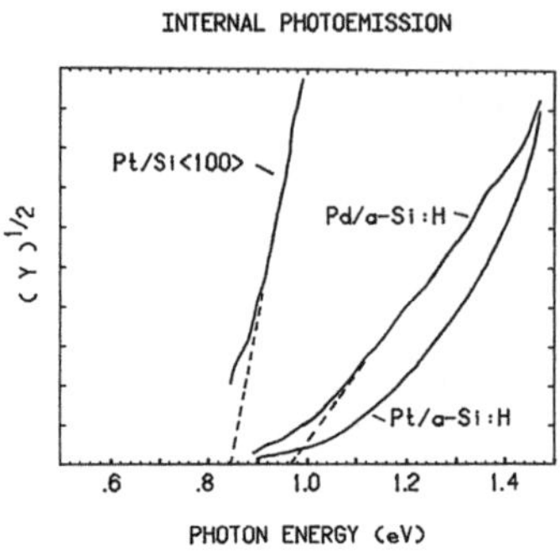

Figure 16. The internal photoemission data for Pt diodes on *a*-Si:H and ⟨100⟩ Si and Pd on *a*-Si:H. The dashed lines indicate extrapolations (where possible) to barrier height values (from Nemanich *et al.*[48]).

emission probability. The results clearly exhibited a linear behavior of $Y^{1/2}$ vs. $h\nu$ for Cr and Pd diodes.[45] The curves were extrapolated to $h\nu = 0$ and barrier heights of 0.83 and 0.98 eV were obtained for Cr and Pd. These results were consistent with values obtained from $J-V$ measurements. A similar value of Φ_B for Pd/*a*-Si:H diodes was also obtained by Yamamoto *et al.*[46] and Jackson *et al.*[47,48] In the study by Yamamoto, they noticed a better fit to $Y^{1/3}$ vs. $h\nu$. The 1/3 value is to be expected when an insulating interfacial layer is present. They found a large variation of Φ_B with temperature which also may be due to an interfacial layer. The study by Jackson *et al.* showed good agreement to $Y^{1/2}$ for Pd diodes which exhibited ideality factors less than 1.05. They found, however, that no linear dependence to the photoemission was observed for Pt/*a*-Si:H diodes. The results are shown in Fig. 16. These diodes also exhibited low ideality parameters and Pt/Si(100) and Pd/*a*-Si:H diodes fabricated the same way did show linear regions of $Y^{1/2}$ vs. $h\nu$ plots. Closer examination of the results of Wronski *et al.*,[45] indicates a similar result for their Pt/*a*-Si:H diodes. It has been suggested that tunneling or optical transitions involving localized states may be contributing to the effect.

Because comparison of $J-V$ and internal photoemission results are often made, it is worth noting that the internal photoemission results indicate the barrier at the measured temperature. In contrast, determination of Φ_B by the temperature dependence of $J-V$ yields the zero temperature value. The fact that the measured values of Φ_B are similar for both techniques is consistent with the results of Wronski *et al.*,[45] where they found no temperature dependence of the internal photoemission.

4. INTERFACE KINETICS AND ITS EFFECT ON THE SCHOTTKY BARRIER

Because *a*-Si:H is a relatively new material, there have been few studies which correlated electrical properties with structural changes occurring at the interface. The ignorance of these changes has led to reports of wide variations of electrical properties. In fact, knowledge of the varied possible reactions of metals and crystalline Si should have been caution enough to forewarn of such possibilities.

Consider first the type of interactions which occur for metals deposited on crystalline Si.[49,50] Two effects have been predominantly observed. For metals which form silicide compounds, interactions often occur which lead to silicide formation at the metal–Si interface. To form a silicide of thickness greater than 100 Å it is necessary to anneal at temperatures ranging from 200 to 600°C. For metals which do not form silicides, atomic interdiffusion often occurs.

There have been several recent studies which have focused on the reactions at the interfaces of metals on *a*-Si and *a*-Si:H. The technique which has proved most sensitive to the initial interactions has been interference enhanced Raman scattering. This technique employs a multilayer thin film configuration which yields two benefits over standard Raman scattering configurations.[51,52] These are that the signal is enhanced by more than a factor of 40, and that only vibrational excitations near the desired interface are observed. A schematic of the multilayer configuration used to investigate the Pd/*a*-Si:H interface is shown in Fig. 17. Also shown is the intensity distribution of the incident light due to the interference conditions.[52] As is evident, the light intensity is maximum at the Pd/*a*-Si:H interface, hence significant Raman scattering will be generated in this region. Furthermore, the same thin film structure causes the Raman scattering originating from this region to be emitted with constructive interference while Raman scattered light from deeper in the sample will suffer destructive interference.

The IERS technique has been applied to study the so-called "reactive" metal/*a*-Si:H interfaces of Pd and Pt.[52,48] The results are shown in Figs. 18 and 19. For Pd deposited on *a*-Si:H at room temperature, the spectra display sharp peaks at ~ 100 cm^{-1} which are ascribed to the formation of ~ 20 Å of a crystalline phase of Pd_2Si. Annealing at temperatures less than 200°C causes

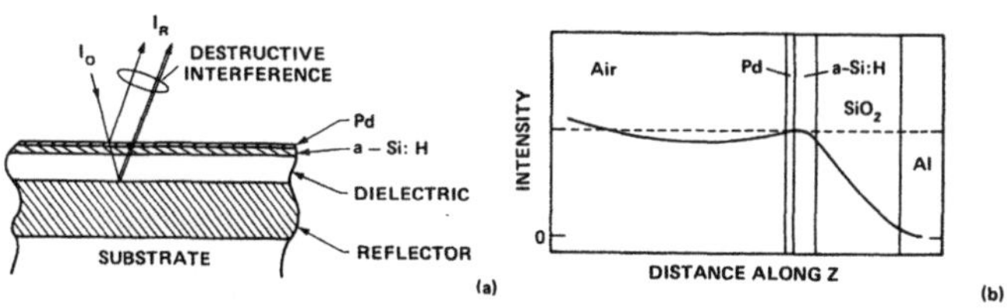

Figure 17. (a) A schematic of the four-layer sample configuration used to obtain interference enhanced Raman scattering (IERS). (b) The electric field intensity due to normally incident 514.5 nm light impinging on the multilayer sample shown in (a). The dashed line represents the lights intensity if no sample were present (from Nemanich *et al.*[52]).

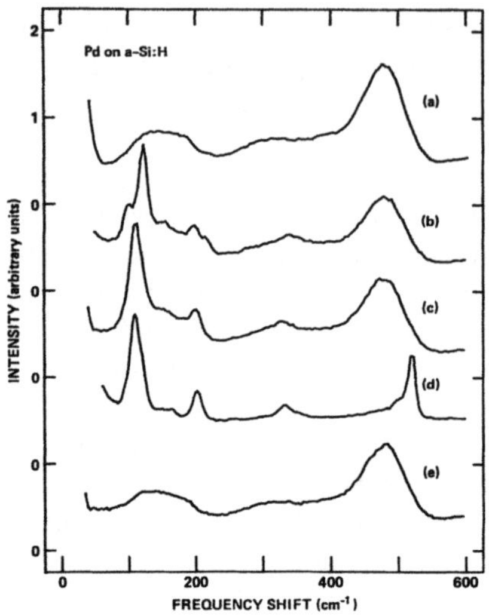

Figure 18. The Raman spectrum of *a*-Si:H, (a), compared with that of Pd on *a*-Si:H after various preparation sequences (b)–(e). All spectra were obtained using the IERS configuration. The Pd on *a*-Si:H spectra were obtained (b) from Pd deposited on freshly prepared *a*-Si:H, (c) the sample annealed to 300°C and (d) 500°C, and for (e) Pd deposited on an aged (oxidized) *a*-Si:H film (from Nemanich *et al.*[52]).

the growth of the silicide phase while annealing to $> 200°C$ causes a change in the spectral features which is attributed to a second phase of Pd_2Si. Annealing at $\sim 550°C$ causes no changes in the features due to the silicide, but the features due to the *a*-Si:H network have been replaced by a sharp line at $\sim 520\,cm^{-1}$ indicating the formation of crystalline Si. The Raman study also demonstrated the effect of a surface oxide on the *a*-Si:H. It was found that aged or oxidized surfaces completely inhibited the silicide formation.

Changes in electrical properties due to annealing of Pd/*a*-Si:H Schottky diodes have been reported by two groups.[34,53] The forward and reverse bias *J–V* characteristics reported by Thompson *et al.* are shown in Fig. 20. Also

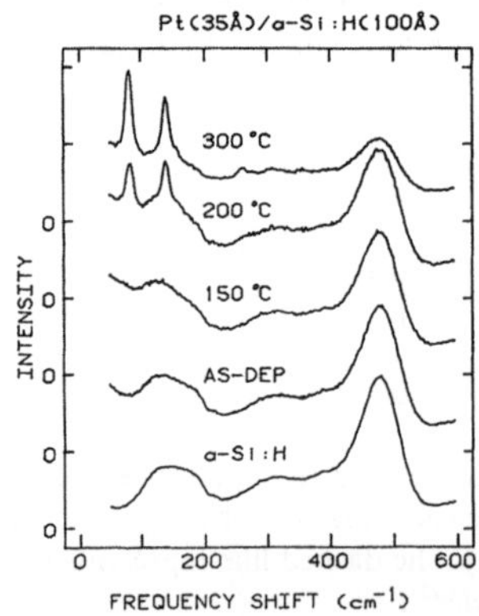

Figure 19. The Raman spectrum of *a*-Si:H compared with that of Pt deposited on *a*-Si:H after various annealing stages. The sharp features at ~ 90 and $140\,cm^{-1}$ observed after annealing to 200°C are attributed to crystalline silicide formation while the broad background in the same frequency range is attributed to a disordered Pt–Si phase (from Nemanich *et al.*[48]).

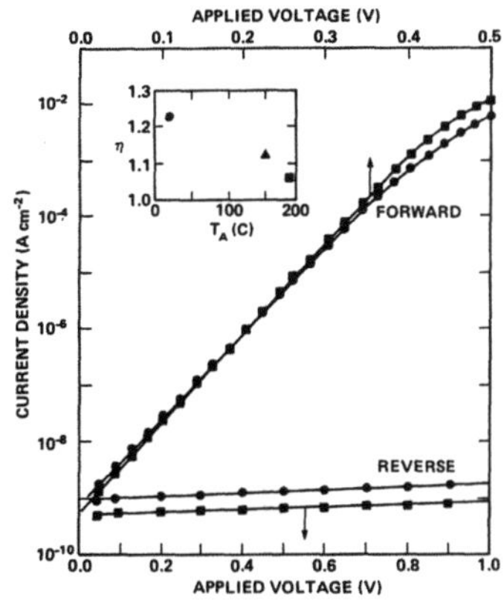

Figure 20. The current density versus voltage characteristics before (circles) and after (squares) annealing at 180°C for 15 min in vacuum. The inset shows the ideality factor n obtained at various anneal temperatures T (from Thompson *et al.*[34]).

shown is the variation of the ideality parameter with annealing temperature. This study was carried out in parallel with the Raman study, and the improved $J-V$ characteristics were directly related to the formation of the silicide. The final ideality value of ~1.05 is as good as has been achieved on crystalline Si; hence it can be concluded that there is little recombination at the interface.

The structural studies for the Pt/*a*-Si:H interface yield the unexpected result that a disordered silicide phase forms at the interface. This phase is characterized by the broad background at $< 100\,\text{cm}^{-1}$ in the Raman spectra (Fig. 19). At higher annealing temperatures, sharp features indicative of crystalline silicides are observed. The $J-V$ characteristics of the Pt/*a*-Si:H diodes are surprisingly similar to those of Pd/*a*-Si:H diodes. The as-deposited samples exhibit ideality values of ~1.2 while annealing to ~150°C causes a reduction to ~1.04.[38,48] The diodes exhibit barriers of ~1.1 eV and rectify over 10 orders of magnitude of current density.

Of materials which do not form silicides, Au and Al have been recently studied. While Raman scattering has proved useful, TEM and Auger electron spectroscopy have yielded greater insight into the atomic rearrangements. The first demonstration that TEM would prove useful was carried out for metals on *a*-Si by Herd *et al.*[54] They observed silicide formation at high temperatures for reactive metals. Additionally, for metals which do not form silicides they observed the formation of crystalline Si islands. Recently Tsai *et al.*[55–57] have coordinated TEM, Auger, IERS, and $J-V$ measurements to characterize the properties of Au and Al contacts on *a*-Si:H. While there was evidence of slight atomic interdiffusion at room temperature, upon annealing to 200 to 250°C large (~1-μm-diam.) islands of crystalline Si were observed. A micrograph of the resultant structure is shown in Fig. 21. With such drastic structural changes it would be expected that dramatic changes should occur in

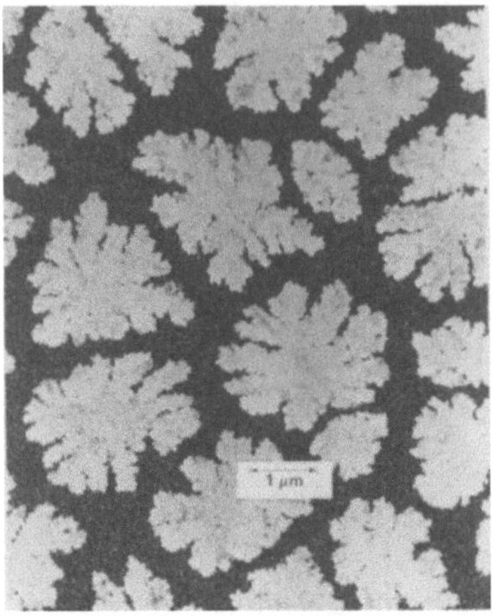

Figures 21. Bright-field TEM micrograph (9800 ×) of 8.0 nm Au on 35 nm of *a*-Si:H which shows the dendritic growth of islands after *in situ* annealing to ~ 180°C (from Tsai *et al.*[56]).

the barrier properties. The electrical properties shown in Fig. 22 indicate that this is indeed the case for Al diodes where nearly ohmic behavior is observed after annealing. However, the same is not true for Au Schottky barriers. Here the diodes actually exhibit a lower ideality behavior after annealing. Because lateral variations in work function would be manifested in a larger ideality value, it can be concluded that the current paths must not include the crystalline Si islands.

Unfortunately there is little work on the interactions of other transition metals and *a*-Si:H. But comparisons with crystalline Si results may lend insight into the problem. Most of these metals oxidize more rapidly than the near noble and noble metals. Hence greater care must be taken to assure that

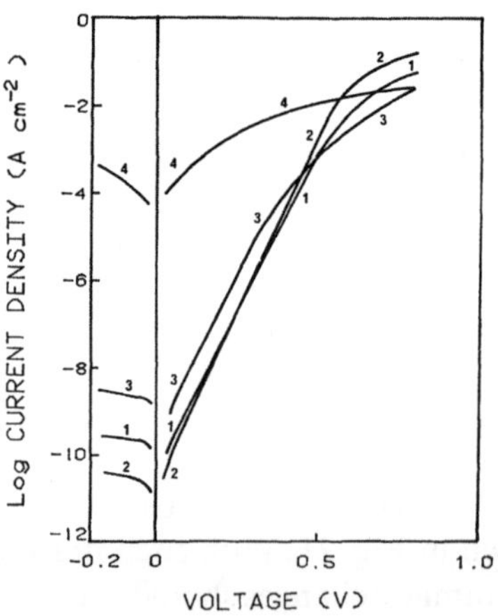

Figure 22. The current–voltage characteristics of Al and Au Schottky barriers on *a*-Si: (1) As-deposited Au diode, (2) Au diode annealed to 200°C, (3) As-deposited Al diode, and (4) Al diode annealed to 250°C (from Tsai *et al.*[57]).

an oxide-free interface is obtained. If an oxide-free interface is obtained, it is likely that the strong metal–Si bonding will preclude atomic interdiffusion for greater than ~ 10Å from the interface. Annealing at temperatures greater than 400°C should be required to form a stoichiometric silicide. At this temperature the H will evolve from the *a*-Si:H film[58] and cause deterioration of the electrical properties.

The basic result of all the studies is that atomic interdiffusion will occur for almost all metals deposited on clean surfaces. The interdiffusion occurs even at 300 K and can often result in a stable silicide compound forming at the interface. Thus it will not be correct to model the interface in terms of the metal work function alone, but the work function of the metal–Si structure at the interface may be important. The improvement of the ideality of the diodes with silicide formation can be due to at least two effects. These include removal of the top surface of the *a*-Si:H which may contain a significant number of defects and/or improving lateral uniformity. Whatever the case, it is clear that structural studies must accompany electrical measurements to understand the implications of the results.

5. APPLICATIONS

Schottky barriers have been extensively used for measurements of the basic properties of *a*-Si:H in addition to a number of device applications. Transient conductivity measurements have been used to measure drift of carriers and the influence of trapping on the electron and hole mobility. Transient techniques for the analysis of the release of carriers from deep traps have enabled the determination of the density of states in the band gap. The charge transport in Schottky barriers under illumination conditions is markedly influenced by recombination. Thus the solar cell performance is dominated by minority carrier behavior. In majority carrier devices like thin film transistors the injection properties of the source and drain contacts have a fundamental influence on the transfer and output characteristics of the devices.

5.1. Drift Mobility

The most common measurement of mobility of carriers in amorphous materials is the time-of-light technique. Hall mobility measurements cannot be used successfully in amorphous semiconductors, the Hall coefficient often giving the incorrect sign. The time-of-light measurement of electrons in *a*-Si:H is often made on Schottky barrier structures under reverse bias conditions. A light pulse producing highly absorbed photons typically from a dye laser illuminates the Schottky contact producing electrons and holes near the

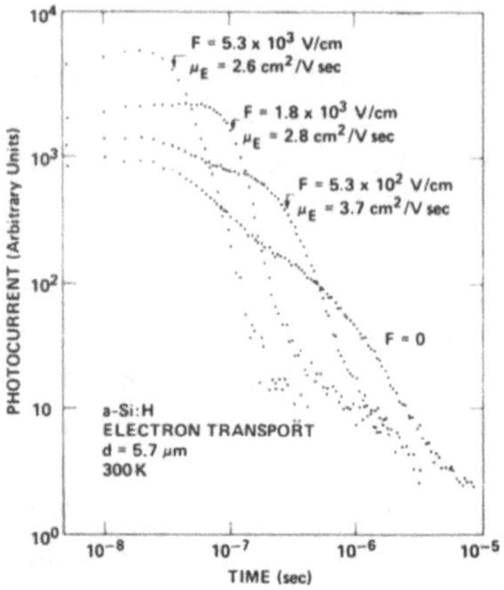

Figure 23. Electron transport measurements at different applied fields as indicated. The vertical scale is the same for each transient. The transit times are indicated by arrows and are used to obtain the drift mobility. The response at zero applied field is due to electrons drifting down the internal field (from Street[25]).

contact. The electrons drift across the *a*-Si:H and either recombine in the bulk or reach the opposite contact. The measured transit time of the charge enables the mobility to be calculated. It has been found that at room temperature the transport of holes in *a*-Si:H is dispersive resulting in a time-dependent mobility,[25] whereas the electron mobility is nondispersive.[59] It has been deduced that this behavior is due to trapping on localized tail states; the further the carrier travels in the material, the deeper it is trapped in the continuous distribution of traps below the band edge and thus the lower the effective mobility. A typical logarithmic plot indicating the transit behavior of electrons is shown in Fig. 23. The multiple trapping model for dispersive transport gives a mobility

$$\mu_d = \mu_0 t^{-(1-\alpha)} \tag{6}$$

where t is time and α is the dispersion parameter. The room-temperature electron mobility is around $2\,\mathrm{cm^2 V^{-1} s^{-1}}$ in high quality *a*-Si:H with an activation energy of 0.2 eV.[60] The hole mobility is several orders of magnitude lower with an activation energy of 0.4 eV.[60]

This time-of-flight measurement can also be used to evaluate depletion within the sample. In the normal drift mobility measurement the voltage pulse is applied before the light pulse; in order that a uniform field across the sample be maintained, the interval between the voltage and light pulse must be less than the dielectric relaxation time. Under uniform voltage conditions the mobility is obtained from the transit time τ_T:

$$\mu = d/F\tau_T \tag{7}$$

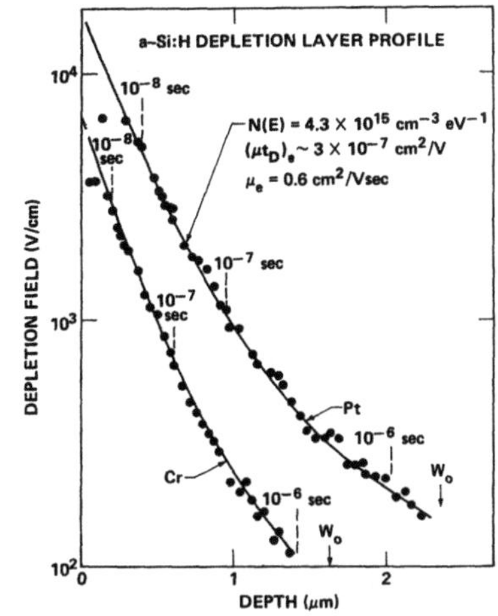

Figure 24. Examples of the internal field profile for Cr and Pt contacts to the same sample. The times marked correspond to those of the transient response. The values of W_0 are obtained from charge collection measurements. The density of states, $N(E)$, is derived from the exponential region of the field profile (from Street[25]).

where d is the thickness of the sample and F is the field. For nondispersive transport τ_T is taken as point of change of slope in the $\log J$ vs. $\log t$ plots.

If a dc field is applied to a Schottky barrier, a nonuniform field is established in the *a*-Si:H due to the depletion layer. The charge generated near the Schottky contact will drift under the action of the internal field. Thus the current through the sample is given by

$$J(t) = \eta n e_\mu F(x, t) \tag{8}$$

where η is the carrier generation efficiency. For carriers starting at $x = 0$ at time $t = 0$, x is related to t by

$$x = \int_0^t \mu F(t')dt' \tag{9}$$

$F(x)$ can therefore be obtained directly from the measurement of $J(t)$.

This technique has been used by Street[25] to measure the depletion layer profile in Schottky barriers. The results for Pt and Cr are shown in Fig. 24. Charge collection measurements ($\int_\phi = \int J\,dt$) give a direct convenient measure of the average distance moved by the carriers before deep trapping occurs and therefore provide a qualitative probe of the internal field.

5.2. Deep-Level Transient Spectroscopy

The technique of DLTS is an extremely powerful method of analyzing defects in crystalline semiconductors. Over the past few years attempts have been made to apply this technique to analyze the density of states in *a*-Si:H. Since the states are continuously distributed in the energy gap, this case presents a much more difficult situation to analyze than that of discrete traps in crystalline semiconductors. In a Schottky barrier diode, filling is accomplished either by illuminating the sample—thus generating and trapping both electrons and holes—or by pulsing the diode into forward bias, thus providing predominantly electron injection and trapping in the *a*-Si:H. Transient measurements are made of the capacitance or current in the diode as the deep traps emptied by thermal excitation to the conduction and valence bands. Crandal[19] first used these techniques in *a*-Si:H to evaluate deep traps. He performed isothermal time-resolved transients in order to analyze electron trapping; here a complete single transient is analyzed. An alternative approach is to record the transients at a fixed emission rate window, i.e., at a fixed time t_1 and t_2 following the trap filling pulse and vary the temperature. Thus the scan is recorded over a temperature range at a fixed time after the trap filling events. As mentioned above, both current transients due to trap released carriers can be measured or transient capacitance measurements can be used—the capacitance changing due to the reduction in fixed charge due to escape of the carrier from the trap. Capacitance measurements have the advantage that the sign of the trap carrier can be distinguished; however, due to the long dielectric relaxation time of undoped *a*-Si:H it is not possible to use capacitance DLTS to study undoped material. Of course it is undoped material that is normally used as the active layer in devices as it has the lowest defect density, and thus it is for this material that information on density of states is most important. Current transient spectroscopy has the disadvantage that the carrier type cannot be identified and it is sensitive to interface states at the metal–semiconductor contact. In addition transit time effects can confuse the analysis.

One important assumption made in these techniques is that the transient signal is emission rate e_n limited

$$e_n = 1/\tau_n = \sigma_n \langle V_n \rangle N_c e^{-E/kT} \tag{10}$$

where σ is the capture cross section, $\langle V_n \rangle$ is the thermal velocity for carriers, N_c is the occupation number, and E is the energy difference between the trap energy and the conduction or valence band. Any retrapping or transit time effects must not be long compared with the emission rate in order for the analysis to be correct. Crandall[19] first identified deep electron traps at 1 eV below the conduction band with capacitance DLTS. Lang *et al.*[21] have provided an extensive analysis of capacitance DLTS on doped *a*-Si:H in order to generate a model of the density of states in the band gap. Their rather complex analysis of the dynamic response of Schottky barriers involved developing expressions for the complex admittance of the device and generated differential equations following Losee's analysis which were numerically solved. From this they produced a density of states shown in Fig. 5 which were different in two fundamental points from results obtained by other techniques. Firstly, the minimum density of states was considerably lower by at least one order of magnitude. They infer that the reason for this is that the other techniques are sensitive to surface states and that the DLTS gives a more accurate determination of the bulk defect density. However, an issue with the DLTS is whether the complete trap filling occurs over the entire range of traps in the gap. It is difficult to determine from the data whether this is the case for the part of the DLTS spectrum which corresponds to the minimum in the density of states. A second feature of the data of Lang *et al.*[21] is that there is no feature corresponding to a bump or shoulder in the upper half of the gap. The existence of such a feature remains controversial. A new constant capacitance technique has recently been used to analyze defects in *a*-Si:H.[61] As the traps empty instead of recording the change in capacitance, a feedback loop is provided in the measurement circuit in order to vary the voltage on the diode to maintain a constant capacitance; thus a voltage transient is recorded. The analysis of this technique is somewhat simpler in that a solution to Poisson's equation is required only for the steady-state charge distribution.

DLTS remains a powerful technique for analysis of defects in *a*-Si:H despite some as yet unresolved issues.

5.3. Solar Cells

Schottky barriers have been widely used in solar cell structures on *a*-Si:H rendering efficiencies of 5%–6% very early in these studies.[62] In order to obtain a large open-circuit voltage, a large internal field must be created at the interface. Thus Pt/*a*-Si:H Schottky barriers were most commonly used for solar cell studies as these have the largest barrier heights. An increase in the barrier height could be obtained by oxidizing the *a*-Si:H or interposing a thin insulating layer such as TiO_x[63] or Nb_2O_5[64] between the metal and *a*-Si:H. The oxide thickness should not be greater than 20 Å otherwise it would lower the short circuit current J_{sc} and the fill factor n. The Schottky barrier solar cell has two disadvantages over the now more commonly used *p-i-n* cell. The open circuit voltage V_{oc} is limited to about 700 mV in most practical situations in an MIS cell whereas using heavily doped n and p regions and heterostructures of *a*-SiC[65] on *a*-Si:H $V_{oc} \approx 900$ mV can be achieved in *p-i-n* devices. Secondly Carlson *et al.*[66] reported instabilities in MIS Schottky barriers due to water molecules apparently changing the charge state of defects in the thin oxide. However, encapsulation can prevent this problem.

A remarkable improvement in material quality has been achieved in the last few years which has resulted in solar cell efficiencies in excess of 10% being achieved in *p-i-n* cells. The wide band gap SiC *p* layer used in the cells not only has the advantage providing a higher V_{oc} but absorbs less than a corresponding heavily *B*-doped *a*-Si:H layer which gives significant defect absorption. The aim is to provide maximum absorption in the *i* layer where the transport takes place. The minority carrier transport is the most critical parameter to obtaining a high J_{sc} and fill factor. Depletion(25) and diffusion(67) lengths in excess of 1 μm have been recently reported in *a*-Si:H. An issue that is still important in limiting solar cell performance is cross contamination of dopants in the deposition chamber due to the deposition of previous layers. Attempts have been made to avoid this problem by depositing the doped and undoped layers in a separate reaction chamber.(68)

Further improvements will be made by improving understanding of the deposition process and the incorporation of multilayer tandem cells. However, production of large-area panels for consumer applications is already under way.

5.4. Thin Film Transistors

The capability of being able to prepare uniform *a*-Si:H over large areas makes it ideally suited for many applications of thin film transistors. The devices are normally prepared with a Si_3N_4 gate electrode grown from an NH_3–silane plasma in the same reactor as the *a*-Si:H. Providing the insulator and *a*-Si:H layers are grown sequentially without breaking the vacuum, a good quality interface is formed which enables TFT with a large dynamic range to be fabricated.(69–71) On/off ratios of $> 10^6$ can be achieved for gate voltages of 15 and 0V. Arrays of devices can be produced on large-area substrates (3 × 4 in. at least) with excellent uniformity. Typical transfer

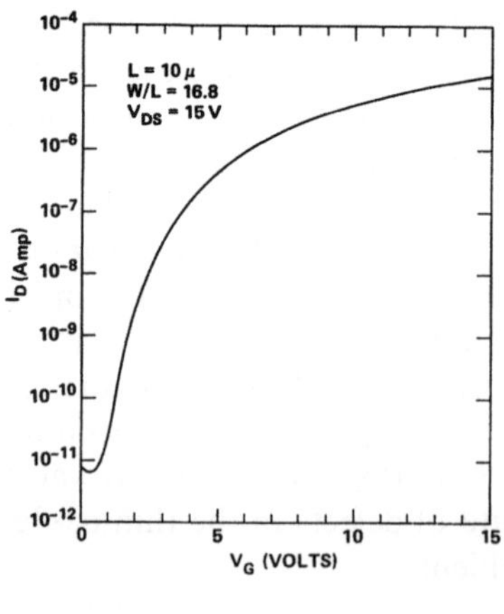

Figure 25. The transfer characteristics of the TFT (from Tuan(71)).

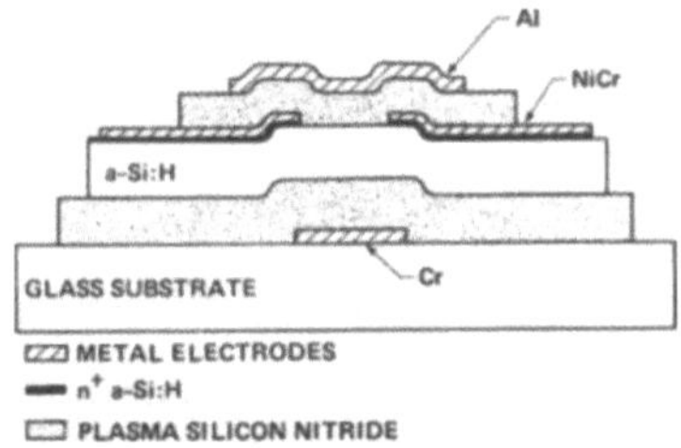

Figure 26. The schematic of a dual-gate *a*-Si:H thin film transistor (from Tuan[71]).

characteristics for the devices are shown in Fig. 25. On currents of only 10^{-6} A can be achieved unless a heavily doped n^{+} layer is provided at the source and drain contacts. Without this layer a Schottky diode is formed at these contacts which limits injection and can influence the threshold of the device. The importance and role of the contacts is illustrated in the dual-gate structure shown in Fig. 26. This looks like a symmetric structure where an accumulation layer can be formed on either the top or bottom Si_3N_4 – *a*-Si:H interface by applying a suitable voltage to the top or bottom gate. However, on examination of the low-voltage output characteristics of the device which are shown in Fig. 27, for each of the gates turned on, it is seen that the slope of the curve is much sharper for the top gate device. This is because the source and drain electrodes are deposited on the top interface and there is a series resistance existing between these contacts and the lower accumulation layer which gives this device a shallower turn on. A similar effect is obtained when Schottky barriers form the source and drain contacts due to injection problems.

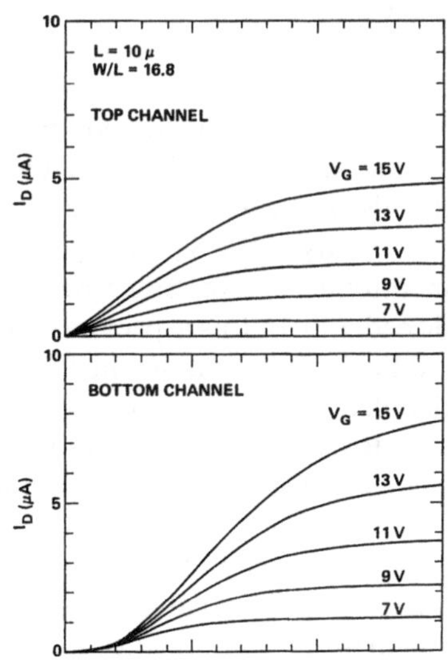

Figure 27. (a) The output characteristics of the top channel of the dual gate transistor shown in Fig. 26 with the bottom electrode grounded. (b) The output characteristics of the bottom channel with the top gate grounded (from Tuan[71]).

a-Si:H TFTs are becoming extensively used as cross point switching elements in large-area liquid crystal displays and other large-area applications.

6. CONCLUDING REMARKS

Because *a*-Si:H is a unique semiconductor material, understanding of the physical properties has proceeded very rapidly. This has occurred because of the interesting scientific problems and because of technological applications. Our perception of the state of study of Schottky barriers is that it is in the infancy of understanding. But because of the same considerations which have driven the rapid progress in the research of *a*-Si:H, it should be anticipated that similar progress will now occur in the *a*-Si:H Schottky barrier. The technological importance of these structures will almost certainly lead to empirical studies, and in fact, many researches to date fall in this category. To gain full understanding and to achieve the full potential of the material, careful UHV experiments will almost certainly be required. This field will benefit strongly by comparison with crystalline Si Schottky barriers.

There are several problems which are obviously important. The first of these is whether the forward bias current is described by the thermionic emission or diffusion model for all metals. While both models have been used to describe results, the differences may be due to surface preparation or that for different barrier heights, different transport mechanisms apply. While defect properties have been considered in describing the C–V measurements, a detailed model of the effects of localized band tail states has not been presented.

Close comparisons with crystalline silicon should also prove important in understanding the Schottky barrier formation—after all, the two semiconductors (crystalline Si and amorphous Si) are indeed alike on the short-range atomic scale. And of course it will be increasingly necessary to understand the details of the atomic interactions at the metal–semiconductor interface.

Study of the Schottky barrier on doped *a*-Si:H will impact all the aspects mentioned above. Since research and device applications have focused on doped *a*-Si:H, this aspect has taken on increased importance. Simple empirical studies will be important for obtaining some of the basic properties before detailed studies can focus on critical aspects.

Because of the basic understanding of Schottky barriers already obtained, this structure will become more common as a research configuration. Transient photoconductivity and DLTS are areas already benefiting from the current understanding. While *p-i-n* structures seem to have displaced Schottky barriers as likely candidates for solar cells, the photoresponse may still prove useful in some applications. A new and possibly most important application of *a*-Si:H is in large-area thin film electronics. Schottky barrier configurations will almost certainly be used in some applications, but even more important may be understanding the "ohmic" contact to *a*-Si:H.

In this manuscript, we have attempted to relate recent studies to the problems of the Schottky barrier. The work to date has yielded a simple understanding of the Schottky barrier on *a*-Si:H and also served to point out critical aspects. We see the research in the area of Schottky barriers on *a*-Si:H as an expanding area in the near future. The work will undoubtedly be tied to research and technological requirements and advances.

REFERENCES

1. R.C. Chittick, J.H. Alexander, and H.F. Sterling, *J. Electrochem. Soc.* **116**, 77 (1969).
2. C.R. Wronski, D.E. Carlson, and R.E. Daniel, *Appl. Phys. Lett.* **29**, 602 (1976).
3. R.J. Loveland, W.E. Spear, and A. Al-Sharbaty, *J. Non-Cryst. Solids* **13**, 55 (1973).
4. G.A.N. Connell and J.R. Pawlik, *Phys. Rev. B* **13**, 787 (1976).
5. J.C. Knights, *Jpn. J. Appl. Phys.* **18**, suppl. 18–1, 101 (1979).
6. M.P. Rosenblum, M.J. Thompson, and R.A. Street, in *Tetrahedrally Bonded Amorphous Semiconductors*, (R.A. Street, D.K. Biegelsen, and J.C. Knights, eds.), American Institute of Physics, New York (1981), p. 42.
7. B.A. Scott, J.A. Reimer, R.M. Plecenik, E.E. Simonyi, and W. Reuter, *Appl. Phys. Lett.* **40**, 973 (1982).
8. M.H. Brodsky, M. Cardona, and J.J. Cuomo, *Phys. Rev. B* **16**, 3556 (1977).
9. J.C. Knights, G. Lucovsky, and R.J. Nemanich, *Phil. Mag. B* **37**, 467 (1978).
10. G. Lucovsky, R.J. Nemanich, and J.C. Knights, *Phys. Rev. B* **19**, 2064 (1979).
11. M. Cardona, *Phys. Status Solidi B* **118**, 463 (1983).
12. S.C. Moss and D. Adler, *Comments Solid State Phys.* **5**, 47 (1973).
13. N. Maley, L.J. Pilione, S.T. Kshirsagar, and J.S. Lannin, *Physica* **118B**, 880 (1983).
14. R.A. Street, J.C. Knights, and D.K. Biegelson, *Phys. Rev. B* **18**, 1880 (1978).
15. J.C. Knights and R.A. Lujan, *Appl. Phys. Lett.* **35**, 244 (1979).
16. R.J. Nemanich, D.K. Biegelson, and M.P. Rosenblum, *J. Phys. Soc. Jpn.* **49A**, 1189 (1980).
17. J.A. Reimer, R.W. Vaughan, and J.C. Knights, *Phys. Rev. B* **24**, 3360 (1981).
18. R.A. Street, *Adv. Phys.* **30**, 593 (1981).
19. R.S. Crandall, *J. Phys. (Paris)* **42**, Suppl. 10, C4-413 (1981).
20. G.D. Cody, T. Tiedje, B. Abeles, B. Brooks, and Y. Goldstein, *Phys. Rev. Lett.* **47**, 1480 (1981).
21. D.V. Lang, J.D. Cohen, and J.P. Harbison, *Phys. Rev. B* **25**, 5285 (1982).
22. T. Tiedje and A. Rose, *Solid State Commun.* **37**, 49 (1981).
23. W.B. Jackson, *Solid State Commun.* **44**, 477 (1982).
24. W.B. Jackson, R.J. Nemanich, and N.M. Amer, *Phys. Rev. B* **27**, 4861 (1983).
25. R.A. Street, *Phys. Rev. B* **27**, 4924 (1983).
26. W.E. Spear and P.G. LeComber, *Solid State Commun.* **17**, 1193 (1975).
27. M. Tanelean, *Phil. Mag. B* **45**, 435 (1982).
28. W.B. Jackson, D.K. Biegelsen, R.J. Nemanich, and J.C. Knights, *Appl. Phys. Lett.* **42**, 105 (1983).
29. W.E. Spear, P.G. LeComber, and A.J. Snell, *Phil. Mag. B* **38**, 303 (1978).
30. E.H. Rhoderick, *Metal–Semiconductor Contacts*, Clarendon Press, Oxford (1978).
31. I. Chen and S. Lee, *J. Appl. Phys.* **53**, 1045 (1982).
32. J.L. Freeouf, T.N. Jackson, S.E. Laux, and J.M. Woodall, *J. Vac. Sci. Technol.* **21**, 570 (1982).
33. C.R. Wronski and D.E. Carlson, *Solid State Commun.* **23**, 421 (1977).
34. M.J. Thompson, N.M. Johnson, R.J. Nemanich, and C.C. Tsai, *Appl. Phys. Lett.* **39**, 274 (1981).

35. M.J. Thompson, M.M. Alkaisi, and J. Allison, *IEE Proc.* **127**, 213 (1980).
36. M.M. Alkaisi and M.J. Thompson, *Solar Cells* **1**, 91 (1979/1980).
37. A. Madan, W. Czubatyj, J. Yang, M.S. Shur, and M.P. Shaw, *Appl. Phys. Lett.* **40**, 234 (1982).
38. A. Deneuville and M.H. Brodsky, *J. Appl. Phys.* **50**, 1414 (1979).
39. P. Viktorovitch, D. Jousse, A. Chenevas–Paule, and L. Vieux–Rochas, *Rev. Phys. Appl.* **14**, 201 (1979).
40. A.J. Snell, K.D. Mackenzie, P.G. LeComber, and W.E. Spear, *Phil Mag. B* **40**, 1 (1979).
41. A.J. Snell, K.D. Mackenzie, P.G. LeComber, and W.E. Spear, *J. Non-Cryst. Solids* **35** & **36**, 593 (1980).
42. P. Viktorovitch and D. Jousse, *J. Non-Cryst. Solids* **35** & **36**, 569 (1980).
43. J. Beichler, W. Fuhs, H. Mell, and H.M. Welsch, *J. Non-Cryst. Solids* **35** & **36**, 587 (1980).
44. H. Fernandez–Canque, J. Allison, and M.J. Thompson, *J. Appl. Phys.* (in press).
45. C.R. Wronski, B. Abeles, G.D. Cody, and T. Tiedje, *Appl. Phys. Lett.* **37**, 96 (1980).
46. T. Yamamoto, Y. Mishima, M. Hirose, and Y. Osaka, *Jpn. J. Appl. Phys.* **20**, Supp. 20–2, 185 (1981).
47. W.B. Jackson, R.J. Nemanich, and M.J. Thompson (to be published).
48. R.J. Nemanich, M.J. Thompson, W.B. Jackson, C.C. Tsai, and B.L. Stafford, *J. Vac. Sci. Technol.* **B1**, 519 (1983).
49. K.N. Tu and J.W. Mayer, in *Thin Films Interdiffusion and Reactions* (J.M. Poate, K.N. Tu, and J.W. Mayer, eds.), John Wiley and Sons (1978), p. 359.
50. G. Ottaviani, *J. Vac. Sci. Technol.* **16**, 1112 (1979).
51. G.A.N. Connell, R.J. Nemanich, and C.C. Tsai, *Appl. Phys. Lett.* **36**, 31 (1980).
52. R.J. Nemanich, C.C. Tsai, M.J. Thompson, and T.W. Sigmon, *J. Vac. Sci. Technol.* **19**, 685 (1981).
53. S.M. Pietruszko, K.L. Narasimhan, and S. Guha, *J. Vac. Sci. Technol.* **20**, 801 (1982).
54. S.R. Herd, P. Chaudhari, and M.H. Brodsky, *J. Non-Cryst. Solids* **7**, 309 (1972).
55. C.C. Tsai, M.J. Thompson, and R.J. Nemanich, *J. Phys. (Paris)* **42**, Suppl. 10, C4-1077 (1981).
56. C.C. Tsai, R.J. Nemanich, and M.J. Thompson, *J. Vac. Sci. Technol.* **21**, 632 (1982).
57. C.C. Tsai, R.J. Nemanich, M.J. Thompson, and B.L. Stafford, *Physica* **117B** & **118B**, 953 (1983).
58. C.C. Tsai and R.J. Nemanich, *J. Non-Cryst. Solids* **35–36**, 1203 (1980).
59. T. Tiedje, J.M. Cebulka, D.L. Morel, and B. Abeles, *Phys. Rev. Lett.* **46**, 1425 (1981).
60. T. Tiedje, A. Rose, and J.M. Cebulka, *AIP Conf. Proc.*, Vol. 73, American Institute of Physics, New York (1981), p. 197.
61. N.M. Johnson, *Appl. Phys. Lett.* **42**, 981 (1983).
62. D.E. Carlson and C.R. Wronski, *Appl. Phys. Lett.* **28**, 671 (1976).
63. J.I.B. Wilson, J. McGill, and S. Kinmond, *Nature* **272**, 152 (1978).
64. A. Madan, J. McGill, W. Czubatyj, J. Yang, and S.R. Ovshinsky, *Appl. Phys. Lett.* **37**, 826 (1980).
65. A. Matsuda, S. Yamasaki, K. Nakagawa, H. Okushi, K. Tanaka, S. Ilzima, M. Masunura, and H. Yamamoto, *Jpn. J. Appl. Phys. Lett.* **19**, L305 (1980).
66. D.E. Carlson and C.W. Magee, in proc. *2nd E.C. Photovoltaic Solar Energy Conference* (R. Van Overstraeten and W. Palz, eds.). Reidel, Boston (1979), p. 312.
67. D.E. Carlson (private communication).
68. Y. Kuwano and M. Ohnishi, *J. Phys. (Paris)* **42**, Suppl. 10, C4-1155 (1981).
69. A.J. Snell, K.D. Mackenzie, W.E. Spear, and P.G. LeComber, *Appl. Phys.* **24**, 357 (1981).
70. M.J. Powell, B.C. Easton, and O.F. Hill, *Appl. Phys. Lett.* **38**, 794 (1981).
71. H.C. Tuan, M.J. Thompson, N.M. Johnson, and R.A. Lujan, *IEEE Elect. Dev. Lett.* **EDL-3**, 357 (1982).

Index

MIX
Papier aus verantwortungsvollen Quellen
Paper from responsible sources
FSC® C105338

If you have any concerns about our products,
you can contact us on
ProductSafety@springernature.com

In case Publisher is established outside the EU,
the EU authorized representative is:
Springer Nature Customer Service Center GmbH
Europaplatz 3, 69115 Heidelberg, Germany

Printed by Libri Plureos GmbH
in Hamburg, Germany